Cosmochemistry
The Melting Pot of the Elements

The evolution of the chemical composition of the Universe is a story lasting billions of years, which started just seconds after the Big Bang. Understanding why celestial objects have the chemical composition that we observe is an amazing but very difficult task. It requires the knowledge of which elements were generated at the birth of the Universe, how galaxies and stars form and evolve, how many stars are produced, how and where the elements are synthesised, and a host of astrophysical processes and observations of varying relevance-truly a titanic endeavour that we have just begun to tackle during the last decades.

This book contains the lectures delivered at the XIII Canary Islands Winter School of Astrophysics dedicated to Cosmochemistry. Written by seven prestigious astrophysics researchers, it covers cosmological and stellar nucleosynthesis, abundance determinations in stars and ionised nebulae, chemical composition of nearby and distant galaxies, and models of chemical evolution of galaxies and intracluster medium. The main scientific originality of this school has been to gather and review in a single event the tremendous improvements in the field of Cosmochemistry in the last decade, especially since the advent of space observatories and very large ground-based telescopes.

CAMBRIDGE CONTEMPORARY ASTROPHYSICS

Series editors
José Franco, Steven M. Kahn, Andrew R. King and Barry F. Madore

Titles available in this series

Gravitational Dynamics, *edited by O. Lahav, E. Terlevich
and R. J. Terlevich* (ISBN 0 521 56327 5)

High-sensitivity Radio Astronomy, *edited by N. Jackson and R. J. Davis*
(ISBN 0 521 57350 5)

Relativistic Astrophysics, *edited by B. J. T. Jones and D. Marković*
(ISBN 0 521 62113 5)

Advances in Stellar Evolution, *edited by R. T. Rood and A. Renzini*
(ISBN 0 521 59184 8)

Relativistic Gravitation and Gravitational Radiation,
edited by J.-A. Marck and J.-P. Lasota
(ISBN 0 521 59065 5)

Instrumentation for Large Telescopes,
edited by J. M. Rodríguez Espinosa, A. Herrero and F. Sánchez
(ISBN 0 521 58291 1)

Stellar Astrophysics for the Local Group,
edited by A. Aparicio, A. Herrero and F. Sánchez
(ISBN 0 521 63255 2)

Nuclear and Particle Astrophysics, *edited by J. G. Hirsch and D. Page*
(ISBN 0 521 63010 X)

Theory of Black Hole Accretion Discs,
edited by M. A. Abramowicz, G. Björnsson and J. E. Pringle
(ISBN 0 521 62362 6)

Interstellar Turbulence,
edited by J. Franco and A.Carramiñana
(ISBN 0 521 65131 X)

Globular Clusters,
edited by C. Martínez Roger, I. Pérez Fournón and F. Sánchez
(ISBN 0 521 77058 0)

The Formation of Galactic Bulges,
edited by C. M. Carollo, H. C. Ferguson and R. F. G. Wyse
(ISBN 0 521 66334 2)

Very Low-Mass Stars and Brown Dwarfs,
edited by R. Rebolo and M. R. Zapatero-Osorio
(ISBN 0 521 66335 0)

Cosmochemistry
The Melting Pot of the Elements

XIII Canary Islands Winter School of Astrophysics
Puerto de la Cruz, Tenerife, Spain
November 19-30, 2001

Edited by: C. Esteban, R. J. García López,
A. Herrero and F. Sánchez

Instituto de Astrofísica de Canarias

CAMBRIDGE
UNIVERSITY PRESS

CAMBRIDGE UNIVERSITY PRESS
Cambridge, New York, Melbourne, Madrid, Cape Town, Singapore,
São Paulo, Delhi, Dubai, Tokyo, Mexico City

Cambridge University Press
The Edinburgh Building, Cambridge CB2 8RU, UK

Published in the United States of America by Cambridge University Press, New York

www.cambridge.org
Information on this title: www.cambridge.org/9780521169592

First published 2004
First paperback edition 2010

A catalogue record for this publication is available from the British Library

ISBN 978-0-521-82768-3 Hardback
ISBN 978-0-521-16959-2 Paperback

Contents

Primordial Alchemy: From The Big Bang To The Present Universe

G. Steigman

Stellar Nucleosynthesis

N. Langer

Observational Aspects Of Stellar Nucleosynthesis

D. L. Lambert

Abundance Determinations In HII Regions And Planetary Nebulae

G. Stasińska

Element Abundances In Nearby Galaxies

D. R. Garnett

Chemical Evolution Of Galaxies And Intracluster Medium

F. Matteucci

Element Abundances Through The Cosmic Ages

M. Pettini

Participants

Barmina, Roberto	Sternwarte-Ludwig Maximilian Univ. (Germany)
Benítez, Georgina	UNAM (México)
Bensby, Thomas	Lund Observatory (Sweden)
Bonastre Mulero, Sonia	Universidad de Barcelona (Spain)
Boschetti, Carla Stefania	Universitá di Padova (Italy)
Cabezón Gómez, Rubén Martín	Universidad Politécnica de Cataluña (Spain)
Calura, Francesco	Universitá di Trieste (Italy)
Carigi, Leticia	UNAM (México)
Castellanos, Marcelo	Universidad Autónoma de Madrid (Spain)
Christlieb, Norbert	University of Hamburg (Germany)
Ciprini, Stefano	Universitá di Perugia (Italy)
Cirigliano, Daniela	Inst. d'Astrophysique Spatiale (France)
Collet, Remo	University of Uppsala (Sweden)
Domínguez Cerdeña, Carolina	Universidad de La Laguna (Spain)
Egorova, Irina	Odessa National University (Ukraine)
Eliche Mora, Mª Carmen	Instituto de Astrofísica de Canarias (Spain)
Esteban, César	Instituto de Astrofísica de Canarias (Spain)
Ferrarotti, Andrea Silvina	Institut für Theoretische Astrophysik (Germany)
García Gil, Alejandro	Instituto de Astrofísica de Canarias (Spain)
García López, Ramón	Instituto de Astrofísica de Canarias (Spain)
García Pérez, Ana Elia	University of Uppsala (Sweden)
Gavilán Bouzas, Marta	Universidad Autónoma de Madrid (Spain)
Gil-Merino, Rodrigo	University of Postdam (Germany)
González Hernández, Jonay	Instituto de Astrofísica de Canarias (Spain)
Haberzettl, Lutz	Ruhr - Universität Bochum (Germany)
Herrero, Artemio	Instituto de Astrofísica de Canarias (Spain)
Humhprey, Andrew	University of Hertfordshire (UK)
Izzard, Rob	University of Cambridge (UK)
Jamet, Luc	Observatoire de Meudon (France)
Jungwiert, Bruno	Astronomical Institute (Czech Republic)
Kandevosian, Lilit	Institute Haibusak (Armenia)
Karlsson, Torgny	University of Uppsala (Sweden)
Korn, Andreas	Institut für Astronomie und Astrophysik (Germany)
López Sánchez, Ángel R.	Instituto de Astrofísica de Canarias (Spain)
López Martín, Luis	Observatoire de Paris (France)
Lucatello, Sara	Universitá di Padova (Italy)
de Lucia, Gabriella	Max Planck Institut für Astrophysik (Germany)
Lyubchyk, Yuri	Main Astronomical Observatory (Ukraine)
Madrid, Juan Pablo	Space Telescope Science Institute (USA)
Maier, Christian	Max Planck Institut für Astronomie (Germany)
Marquez, Isabel	Inst. Astrofísica de Andalucía (Spain)
Masegosa, Josefa M.	Instituto de Astrofísica de Andalucía (Spain)
Mizuno-Wiedner, Michelle	University of Uppsala (Sweden)
Niemczura, Ewa	Wroclaw University (Poland)
Pérez Montero, Enrique	Universidad Autónoma de Madrid (Spain)
Petrovic, Jelena	Sterrenkundig Instituut Utrecht (The Netherlands)
Przybilla, Norbert	Universität Sternwarte München (Germany)
Rae, Jamie	University of Leeds (UK)

Participants

Recio Blanco, Alejandra	Universitá di Padova (Italy)
Relaño, Mónica	Instituto de Astrofísica de Canarias (Spain)
Reyniers, Maarten	Instituut voor Sterrenkunde (Belgium)
Rocha-Pinto, Helio Jaques	Universidade de São Paulo (Brazil)
Rozas Amador, José M^a	Universidad de Barcelona (Spain)
Salvato, Mara	IAP Astrophysikalisches Institut (Germany)
Sánchez Blázquez, Patricia	Universidad Complutense de Madrid (Spain)
Satorre, Miguel A.	Universidad Politécnica de Valencia (Spain)
Scannapieco, Cecilia	Inst. de Astronomía y Física del Espacio (Argentine)
Simón Díaz, Sergio	Instituto de Astrofísica de Canarias (Spain)
Sosa Brito, Rafael M.	Instituto de Astrofísica de Canarias (Spain)
Sung-Chul, Yoon	Sterrenkundig Instituut Utrecht (The Netherlands)
Tantalo, Rosaria	Universitá di Padova (Italy)
Tanvuia, Laura	University of Vienna (Austria)
Trancho Lemes, Gelys	Gemini Observatory (Chile)
Trundle, Carrie	Queen University Belfast (Ireland)
Urbaneja Pérez, Miguel A.	Instituto de Astrofísica de Canarias (Spain)
Yong, David	University of Texas at Austin (USA)

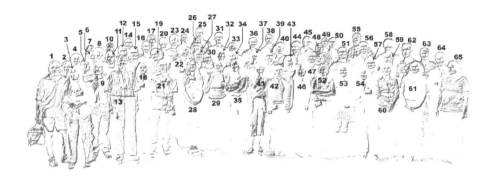

1	Max Pettini	23	Andrew Humhprey	45	Norbert Christlieb
2	Andrea Ferrarotti	24	Jonay González Hernández	46	Laura Tanvuia
3	Francesca Matteucci	25	Alejandro García Gil	47	Maarten Reyniers
4	Michelle Mizuno-Wiedner	26	Lutz Haberzettl	48	Helio Jaques Rocha-Pinto
5	Ana Elia García Pérez	27	Grazyna Stasinska	49	Ewa Niemczura
6	Marta Gavilán Bouzas	28	Leticia Carigi	50	Yuri Lyubchyk
7	Cecilia Scannapieco	29	María Carmen Eliche Moral	51	Norbert Przybilla
8	Rafael Sosa Brito	30	Luis López Martín	52	Jelena Petrovic
9	Jamie Rae	31	Miguel Urbaneja Pérez	53	Georgina Benítez
10	Sally Wright	32	Carolina Domínguez Cerdeña	54	Sonia Bonastre Mulero
11	Remo Collet	33	Alejandra Recio Blanco	55	David Lambert
12	Luc Jamet	34	Sara Lucatello	56	Juan Pablo Madrid
13	Gary Steigman	35	Gelys Trancho Lemes	57	Rubén Martín Cabezón Gómez
14	Enrique Pérez Montero	36	Christian Maier	58	Rodrigo Gil-Merino
15	Torgny Karlsson	37	Rosaria Tantalo	59	José María Rozas Amador
16	Marcelo Castellanos	38	Francesco Calura	60	Carrie Trundle
17	Rob Izzard	39	Roberto Barmina	61	Nieves Villoslada
18	Patricia Sánchez Blázquez	40	Yoon Sung-Chul	62	Thomas Bensby
19	Andreas Korn	41	Carla Stefania Boschetti	63	César Esteban
20	Sergio Simón Díaz	42	Gabriella De Lucia	64	Ramón García López
21	David Yong	43	Irina Egorova	65	Lourdes González
22	Stefano Ciprini	44	Bruno Jungwiert		

Preface

The distribution of elements in the cosmos is the result of many different physical processes in the history of the Universe, from Big Bang to present times. Its study provides us with a powerful tool for understanding the physical conditions of the primordial cosmos, the physics of nucleosynthesis processes that occur in different objects and places, and the formation and evolution of stars and galaxies. Cosmochemistry is a fundamental topic for many different branches of Astrophysics as Cosmology, Stellar Structure and Evolution, Interstellar Medium, and Galaxy Formation and Evolution.

The advances made in the last decade of the XXth century in the study of the chemical evolution of the Universe have been really spectacular. On one hand, they have been brought by the availability of large-aperture ground-based telescopes and space borne telescopes (working in both the visible and other regions of the electromagnetic spectrum), and on the other hand by advances in theory and numerical modelling techniques in many fields of astrophysics such as stellar evolution stellar atmospheres, the physics of ionised plasmas and atomic and molecular physics.

According to the predictions of the most commonly accepted cosmological models, most of the light elements, especially deuterium and helium, were produced during the first minutes after the Big Bang. Comparison between observed and predicted light-element abundances is one of the classical fundamental tests of cosmological models. Stellar evolutionary models have advanced considerably in recent years. It is particularly noteworthy that it is only since the 1990s that we began to have available quantitative determinations of the production of chemical elements in stars of any initial mass and various initial metallicities. The appropriate treatment of convection and rotation are recent important advances with serious implications on the expected stellar nucleosynthesis. The determination of stellar abundances has improved spectacularly due to the availability of spectra from new very large telescopes and the basic improvements in the physical theory of stellar atmospheres. With the new generation of telescopes we have the possibility to observe distant Galactic subgiants, subdwarfs in the Magellanic Clouds, and even supergiants in galaxies as far as those in the Virgo Cluster. The chemical composition of the interstellar and intergalactic gas is far better known due to the gigantic amount of data from ground-based large telescopes, space observatories and new detectors. Deep optical and infrared emission line spectra permit to derive chemical abundances for distant and chemically pristine extragalactic HII regions, extragalactic planetary nebulae, as well as for the hot intergalactic or intracluster gas from the recent space X-ray observations. Of special importance is the determination of the chemical composition of clouds and emission-line galaxies at the high redshift universe (Lyman alpha forest, damped Lyman alpha systems, Lyman break galaxies). These data for distant objects give us the possibility to study the galaxy formation and evolution and the process of chemical enrichment in the early epochs of the Universe. The models of chemical evolution of the solar neighbourhood, the Milky Way and other galaxies have also improved in order to explain the new observational data. The origin and time evolution of radial abundance gradients and gradients of abundance ratios in spiral galaxies is a key problem that many theoreticians have tried to explain with different ingredients and invoking different physical mechanisms. The role of gas flows in the chemical evolution of galaxies is also a major ingredient for the models and has been intensely explored in the last years.

Most of the scientific advances commented above have taken place in closed areas traditionally independent from each other. We consider that the main scientific originality of the XIII Canary Islands Winter School of Astrophysics, organised by the Instituto de

Astrofísica de Canarias (IAC), is to gather and review the tremendous observational and theoretical improvements experienced in the field of Cosmochemistry in the last decade. To accomplish this successfully we have edited this book that contains the lectures delivered at the Winter School by seven leading scientists working actively on each of the major aspects covered. We hope this book will be a useful source of information for a wide range of young -and not so young- astronomers and astrophysicists.

César Esteban, Ramón García López, Artemio Herrero & Francisco Sánchez

Instituto de Astrofísica de Canarias

September 2003

Acknowledgements

The editors want to express their warmest gratitude to the lecturers for their efforts in preparing their classes and the chapters of this book. We also wish to thank our efficient secretaries: Nieves Villoslada and Lourdes González, and many staff members of the IAC: Jesús Burgos, Begoña López Betancor, Carmen del Puerto, Ramón Castro, Terry Mahoney, Jesús Jiménez, Miguel Briganti and the technicians of Servicios Informáticos Comunes of the IAC. We acknowledge Gabriel Gómez for preparing the final version of this book for Cambridge University Press.

Finally, we also wish to thank the following institutions for their generous and indispensable support: The European Commission, the Spanish Ministry of Science and Technology, Iberia, the local governments (cabildos) of the islands of Tenerife and La Palma, and the Puerto de la Cruz Town Council.

Primordial Alchemy:
From The Big Bang To The Present Universe

By GARY STEIGMAN

Departments of Physics and Astronomy,
The Ohio State University, Columbus, OH 43210, USA

Of the light nuclides observed in the universe today, D, ^{3}He, ^{4}He, and ^{7}Li are relics from its early evolution. The primordial abundances of these relics, produced via Big Bang Nucleosynthesis (BBN) during the first half hour of the evolution of the universe provide a unique window on Physics and Cosmology at redshifts $\sim 10^{10}$. Comparing the BBN-predicted abundances with those inferred from observational data tests the consistency of the standard cosmological model over ten orders of magnitude in redshift, constrains the baryon and other particle content of the universe, and probes both Physics and Cosmology beyond the current standard models. These lectures are intended to introduce students, both of theory and observation, to those aspects of the evolution of the universe relevant to the production and evolution of the light nuclides from the Big Bang to the present. The current observational data is reviewed and compared with the BBN predictions and the implications for cosmology (*e.g.*, universal baryon density) and particle physics (*e.g.*, relativistic energy density) are discussed. While this comparison reveals the stunning success of the standard model(s), there are currently some challenges which leave open the door for more theoretical and observational work with potential implications for astronomy, cosmology, and particle physics.

1. Introduction

The present universe is expanding and is filled with radiation (the 2.7 K Cosmic Microwave Background – CMB) as well as "ordinary" matter (baryons), "dark" matter and, "dark energy". Extrapolating back to the past, the early universe was hot and dense, with the overall energy density dominated by relativistic particles ("radiation dominated"). During its early evolution the universe hurtled through an all too brief epoch when it served as a primordial nuclear reactor, leading to the synthesis of the lightest nuclides: D, ^{3}He, ^{4}He, and ^{7}Li. These relics from the distant past provide a unique window on the early universe, probing our standard models of cosmology and particle physics. By comparing the predicted primordial abundances with those inferred from observational data we may test the standard models and, perhaps, uncover clues to modifications or extensions beyond them.

These notes summarize the lectures delivered at the XIII Canary Islands Winter School of Astrophysics: "Cosmochemistry: The Melting Pot of Elements". The goal of the lectures was to provide both theorists and observers with an overview of the evolution of the universe from its earliest epochs to the present, concentrating on the production, evolution, and observations of the light nuclides. Standard Big Bang Nucleosynthesis (SBBN) depends on only one free parameter, the universal density of baryons; fixing the primordial abundances fixes the baryon density at the time of BBN. But, since baryons are conserved (at least for these epochs), fixing the baryon density at a redshift $\sim 10^{10}$, fixes the present-universe baryon density. Comparing this prediction with other, independent probes of the baryon density in the present and recent universe offers the opportunity to test the consistency of our standard, hot big bang cosmological model.

Since these lectures are intended for a student audience, they begin with an overview

1

of the physics of the early evolution of the Universe in the form of a "quick and dirty" mini-course on Cosmology (§2). The experts may wish to skip this material. With the necessary background in place, the second lecture (§3) discusses the physics of primordial nucleosynthesis and outlines the abundances predicted by the standard model. The third lecture considers the evolution of the abundances of the relic nuclides from BBN to the present and reviews the observational status of the primordial abundances (§4). As is to be expected in such a vibrant and active field of research, this latter is a moving target; the results presented here represent the status in November 2001. Armed with the predictions and the observations, the fourth lecture (§5) is devoted to the confrontation between them. As is by now well known, this confrontation is a stunning success for SBBN. However, given the precision of the predictions and of the observational data, it is inappropriate to ignore some of the potential discrepancies. In the end, these may be traceable to overly optimistic error budgets, to unidentified systematic errors in the abundance determinations, to incomplete knowledge of the evolution from the big bang to the present or, to new physics beyond the standard models. In the last lecture I present a selected overview of BBN in some non-standard models of Cosmology and Particle Physics (§6). Although I have attempted to provide a *representative* set of references, I am aware they are incomplete and I apologize in advance for any omissions.

2. The Early Evolution of the Universe

Observations of the present universe establish that, on sufficiently large scales, galaxies and clusters of galaxies are distributed homogeneously and they are expanding isotropically. On the assumption that this is true for the large scale universe throughout its evolution (at least back to redshifts $\sim 10^{10}$, when the universe was a few hundred milliseconds old), the relation between space-time points may be described uniquely by the Robertson – Walker metric

$$ds^2 = c^2 dt^2 - a^2(t)\left(\frac{dr^2}{1 - \kappa r^2} + r^2 d\Omega^2\right), \qquad (2.1)$$

where r is a *comoving* radial coordinate and θ and ϕ are *comoving* spherical coordinates related by

$$d\Omega^2 \equiv d\theta^2 + sin^2\theta d\phi^2. \qquad (2.2)$$

A useful alternative to the comoving radial coordinate r is Θ, defined by

$$d\Theta \equiv \frac{dr}{(1 - \kappa r^2)^{1/2}}. \qquad (2.3)$$

The 3-space curvature is described by κ, the curvature constant. For closed (bounded), or "spherical" universes, $\kappa > 0$; for open (unbounded), or "hyperbolic" models, $\kappa < 0$; when $\kappa = 0$, the universe is spatially flat or "Euclidean". It is the "scale factor", $a = a(t)$, which describes how physical distances between comoving locations change with time. As the universe expands, a increases while, for comoving observers, r, θ, and ϕ remain fixed. The growth of the separation between comoving observers is solely due to the growth of a. Note that **neither** a **nor** κ is observable since a rescaling of κ can always be compensated by a rescaling of a.

Photons and other massless particles travel on geodesics: $ds = 0$; for them (see eq. 2.1) $d\Theta = \pm \, cdt/a(t)$. To illustrate the significance of this result consider a photon travelling from emission at time t_e to observation at a later time t_o. In the course of its journey through the universe the photon traverses a comoving radial distance $\Delta\Theta$, where

$$\Delta\Theta = \int_{t_e}^{t_o} \frac{cdt}{a(t)} \, . \tag{2.4}$$

Some special choices of t_e or t_o are of particular interest. For $t_e \to 0$, $\Delta\Theta \equiv \Theta_H(t_o)$ is the comoving radial distance to the "Particle Horizon" at time t_0. It is the comoving distance a photon could have travelled (in the absence of scattering or absorption) from the beginning of the expansion of the universe until the time t_o. The "Event Horizon", $\Theta_E(t_e)$, corresponds to the limit $t_o \to \infty$ (provided that Θ_E is finite!). It is the comoving radial distance a photon will travel for the entire future evolution of the universe, after it is emitted at time t_e.

2.1. Redshift

Light emitted from a comoving galaxy located at Θ_g at time t_e will reach an observer situated at $\Theta_o \equiv 0$ at a later time t_o, where

$$\Theta_g(t_o, t_e) = \int_{t_e}^{t_o} \frac{cdt}{a(t)} \, . \tag{2.5}$$

Equation 2.5 provides the relation among Θ_g, t_o, and t_e. For a comoving galaxy, Θ_g is unchanged so that differentiating eq. 2.5 leads to

$$\frac{dt_o}{dt_e} = \frac{a_o}{a_e} = \frac{\nu_e}{\nu_o} = \frac{\lambda_o}{\lambda_e} \, . \tag{2.6}$$

This result relates the evolution of the universe (a_o/a_e) as the photon travels from emission to observation, to the change in its frequency (ν) or wavelength (λ). As the universe expands (or contracts!), wavelengths expand (contract) and frequencies decrease (increase). The redshift of a spectral line is defined by relating the wavelength at emission (the "lab" or "rest-frame" wavelength λ_e) to the wavelength observed at a later time t_o, λ_o.

$$z \equiv \frac{\lambda_o - \lambda_e}{\lambda_e} \implies 1 + z = \frac{a_o}{a_e} = \frac{\nu_e}{\nu_o} \, . \tag{2.7}$$

Since the energies of photons are directly proportional to their frequencies, as the universe expands photon energies redshift to smaller values: $E_\gamma = h\nu \implies E_\gamma \propto (1+z)^{-1}$. For **all** particles, massless or not, de Broglie told us that wavelength and momentum are inversely related, so that: $p \propto \lambda^{-1} \implies p \propto (1+z)^{-1}$. All momenta redshift; for non-relativistic particles (*e.g.*, galaxies) this implies that their "peculiar" velocities redshift: $v = p/M \propto (1+z)^{-1}$.

2.2. Dynamics

Everything discussed so far has been "geometrical", relying only on the form of the Robertson-Walker metric. To make further progress in understanding the evolution of the universe, it is necessary to determine the time dependence of the scale factor $a(t)$. Although the scale factor is not an observable, the expansion rate, the Hubble parameter, $H = H(t)$, is.

$$H(t) \equiv \frac{1}{a}\left(\frac{da}{dt}\right) \, . \tag{2.8}$$

The present value of the Hubble parameter, often referred to as the Hubble "constant", is $H_0 \equiv H(t_0) \equiv 100 \, h$ km s^{-1}Mpc^{-1} (throughout, unless explicitly stated otherwise, the subscript "0" indicates the present time). The inverse of the Hubble parameter provides an expansion timescale, $H_0^{-1} = 9.78 \, h^{-1}$ Gyr. For the HST Key Project (Freedman *et al.* 2001) value of $H_0 = 72$ km s^{-1}Mpc^{-1} ($h = 0.72$), $H_0^{-1} = 13.6$ Gyr.

The time-evolution of H describes the evolution of the universe. Employing the Robertson-Walker metric in the Einstein equations of General Relativity (relating matter/energy content to geometry) leads to the Friedmann equation

$$H^2 = \frac{8\pi}{3}G\rho - \frac{\kappa c^2}{a^2}. \tag{2.9}$$

It is convenient to introduce a *dimensionless* density parameter, Ω, defined by

$$\Omega \equiv \frac{8\pi G\rho}{3H^2}. \tag{2.10}$$

We may rearrange eq. 2.9 to highlight the relation between matter content and geometry

$$\kappa c^2 = (aH)^2(\Omega - 1). \tag{2.11}$$

Although, in general, a, H, and Ω are all time-dependent, eq. 2.11 reveals that if ever $\Omega < 1$, then it will always be < 1 **and** in this case the universe is open ($\kappa < 0$). Similarly, if ever $\Omega > 1$, then it will always be > 1 **and** in this case the universe is closed ($\kappa > 0$). For the special case of $\Omega = 1$, where the density is equal to the "critical density" $\rho_{\mathrm{crit}} \equiv 3H^2/8\pi G$, Ω is always unity and the universe is flat (Euclidean 3-space sections; $\kappa = 0$).

The Friedmann equation (eq. 2.9) relates the time-dependence of the scale factor to that of the density. The Einstein equations yield a second relation among these which may be thought of as the surrogate for energy conservation in an expanding universe.

$$\frac{d\rho}{\rho} + 3(1 + \frac{p}{\rho})\frac{da}{a} = 0. \tag{2.12}$$

For "matter" (non-relativistic matter; often called "dust"), $p \ll \rho$, so that $\rho/\rho_0 = (a_0/a)^3$. In contrast, for "radiation" (relativistic particles) $p = \rho/3$, so that $\rho/\rho_0 = (a_0/a)^4$. Another interesting case is that of the energy density and pressure associated with the vacuum (the quantum mechanical vacuum is not empty!). In this case $p = -\rho$, so that $\rho = \rho_0$. This provides a term in the Friedmann equation entirely equivalent to Einstein's "cosmological constant" Λ. More generally, for $p = w\rho$, $\rho/\rho_0 = (a_0/a)^{3(1+w)}$.

Allowing for these three contributions to the total energy density, eq. 2.9 may be rewritten in a convenient dimensionless form

$$(\frac{H}{H_0})^2 = \Omega_{\mathrm{M}}(\frac{a_0}{a})^3 + \Omega_{\mathrm{R}}(\frac{a_0}{a})^4 + \Omega_{\Lambda} + (1 - \Omega)(\frac{a_0}{a})^2, \tag{2.13}$$

where $\Omega \equiv \Omega_{\mathrm{M}} + \Omega_{\mathrm{R}} + \Omega_{\Lambda}$.

Since our universe is expanding, for the early universe ($t \ll t_0$) $a \ll a_0$, so that it is the "radiation" term in eq. 2.13 which dominates; the early universe is radiation-dominated (RD). In this case $a \propto t^{1/2}$ and $\rho \propto t^{-2}$, so that the age of the universe or, equivalently, its expansion rate is fixed by the radiation density. For thermal radiation, the energy density is only a function of the temperature ($\rho_{\mathrm{R}} \propto T^4$).

2.2.1. *Counting Relativistic Degrees of Freedom*

It is convenient to write the total (radiation) energy density in terms of that in the CMB photons

$$\rho_{\mathrm{R}} \equiv (\frac{g_{eff}}{2})\rho_\gamma, \tag{2.14}$$

where g_{eff} counts the "effective" relativistic degrees of freedom. Once g_{eff} is known or specified, the time – temperature relation is determined. If the temperature is measured in energy units (kT), then

$$t(\text{sec}) = (\frac{2.4}{g_{eff}^{1/2}})T_{\text{MeV}}^{-2} \,.$$ (2.15)

If more relativistic particles are present, g_{eff} increases and the universe would expand faster so that, at **fixed** T, the universe would be younger. Since the synthesis of the elements in the expanding universe involves a competition between reaction rates and the universal expansion rate, g_{eff} will play a key role in determining the BBN-predicted primordial abundances.

- *Photons*

Photons are vector bosons. Since they are massless, they have only two degress of freedom: $g_{eff} = 2$. At temperature T their number density is $n_\gamma = 411(T/2.726K)^3$ cm$^{-3} = 10^{31.5}T_{\text{MeV}}^3$ cm^{-3}, while their contribution to the total radiation energy density is $\rho_\gamma = 0.261(T/2.726K)^4$ eV cm^{-3}. Taking the ratio of the energy density to the number density leads to the average energy per photon $\langle E_\gamma \rangle = \rho_\gamma/n_\gamma = 2.70 \ kT$. All other relativistic **bosons** may be simply related to photons by

$$\frac{n_B}{n_\gamma} = \frac{g_B}{2}(\frac{T_B}{T_\gamma})^3\,, \qquad \frac{\rho_B}{\rho_\gamma} = \frac{g_B}{2}(\frac{T_B}{T_\gamma})^4\,, \qquad \langle E_B \rangle = 2.70 \ kT_B \,.$$ (2.16)

The g_B are the boson degrees of freedom (1 for a scalar, 2 for a vector, etc.). In general, some bosons may have decoupled from the radiation background and, therefore, they will not necessarily have the same temperature as do the photons ($T_B \neq T_\gamma$).

- Relativistic Fermions

Accounting for the difference between the Fermi-Dirac and Bose-Einstein distributions, relativistic fermions may also be related to photons

$$\frac{n_F}{n_\gamma} = \frac{3}{4}\frac{g_F}{2}(\frac{T_F}{T_\gamma})^3\,, \qquad \frac{\rho_F}{\rho_\gamma} = \frac{7}{8}\frac{g_F}{2}(\frac{T_F}{T_\gamma})^4\,, \qquad \langle E_F \rangle = 3.15 \ kT_F \,.$$ (2.17)

g_F counts the fermion degrees of freedom. For example, for electrons (spin up, spin down, electron, positron) $g_F = 4$, while for neutrinos (lefthanded neutrino, righthanded antineutrino) $g_F = 2$.

Accounting for all of the particles present at a given epoch in the early (RD) evolution of the universe,

$$g_{eff} = \Sigma_B \ g_B(\frac{T_B}{T_\gamma})^4 + \frac{7}{8} \ \Sigma_F \ g_F(\frac{T_F}{T_\gamma})^4 \,.$$ (2.18)

For example, for the standard model particles at temperatures $T_\gamma \approx$ few MeV there are photons, electron-positron pairs, and three "flavors" of lefthanded neutrinos (along with their righthanded antiparticles). At this stage all these particles are in equilibrium so that $T_\gamma = T_e = T_\nu$ where $\nu \equiv \nu_e, \nu_\mu, \nu_\tau$. As a result

$$g_{eff} = 2 + \frac{7}{8}(4 + 3 \times 2) = \frac{43}{4}\,,$$ (2.19)

leading to a time – temperature relation: $t = 0.74 \ T_{\text{Mev}}^{-2}$ sec.

As the universe expands and cools below the electron rest mass energy, the $e^\pm$ pairs annihilate, heating the CMB photons, but **not** the neutrinos which have already decoupled. The decoupled neutrinos continue to cool by the expansion of the universe ($T_\nu \propto a^{-1}$), as do the photons which now have a higher temperature $T_\gamma = (11/4)^{1/3}T_\nu$ ($n_\gamma/n_\nu = 11/3$). During these epochs

$$g_{eff} = 2 + \frac{7}{8} \times 3 \times 2(\frac{4}{11})^{4/3} = 3.36\,,$$ (2.20)

leading to a modified time – temperature relation: $t = 1.3 \ T_{\text{Mev}}^{-2}$ sec.

2.2.2. *"Extra" Relativistic Energy*

Suppose there is some new physics beyond the standard model of particle physics which leads to "extra" relativistic energy so that $\rho_R \rightarrow \rho'_R \equiv \rho_R + \rho_X$; hereafter, for convenience of notation, the subscript R will be dropped. It is useful, and conventional, to account for this extra energy in terms of the equivalent number of extra neutrinos: $\Delta N_\nu \equiv \rho_X/\rho_\nu$ (Steigman, Schramm, & Gunn 1977 (SSG); see also Hoyle & Tayler 1964, Peebles 1966, Shvartsman 1969). In the presence of this extra energy, prior to $e^\pm$ annihilation

$$\frac{\rho'}{\rho_\gamma} = \frac{43}{8}\left(1 + \frac{7\Delta N_\nu}{43}\right) = 5.375\,(1 + 0.1628\,\Delta N_\nu). \tag{2.21}$$

In this case the early universe would expand faster than in the standard model. The pre-$e^\pm$ annihilation speedup in the expansion rate is

$$S_{pre} \equiv \frac{t}{t'} = \left(\frac{\rho'}{\rho}\right)^{1/2} = (1 + 0.1628\,\Delta N_\nu)^{1/2}. \tag{2.22}$$

After $e^\pm$ annihilation there are similar, but quantitatively different changes

$$\frac{\rho'}{\rho_\gamma} = 1.681\,(1 + 0.1351\,\Delta N_\nu), \qquad S_{post} = (1 + 0.1351\,\Delta N_\nu)^{1/2}. \tag{2.23}$$

Armed with an understanding of the evolution of the early universe and its particle content, we may now proceed to the main subject of these lectures, primordial nucleosynthesis.

3. Big Bang Nucleosynthesis and the Primordial Abundances

Since the early universe is hot and dense, interactions among the various particles present are rapid and equilibrium among them is established quickly. But, as the universe expands and cools, there are departures from equilibrium; these are at the core of the most interesting themes of our story.

3.1. *An Early Universe Chronology*

At temperatures above a few MeV, when the universe is tens of milliseconds old, interactions among photons, neutrinos, electrons, and positrons establish and maintain equilibrium ($T_\gamma = T_\nu = T_e$). When the temperature drops below a few MeV the weakly interacting neutrinos decouple, continuing to cool and dilute along with the expansion of the universe ($T_\nu \propto a^{-1}$, $n_\nu \propto T_\nu^3$, and $\rho_\nu \propto T_\nu^4$).

3.1.1. *Neutron – Proton Interconversion*

Up to now we haven't considered the baryon (nucleon) content of the universe. At these early times there are neutrons and protons present whose relative abundance is determined by the usual weak interactions.

$$p + e^- \Longleftrightarrow n + \nu_e, \quad n + e^+ \Longleftrightarrow p + \bar{\nu}_e, \quad n \Longleftrightarrow p + e^- + \bar{\nu}_e. \tag{3.24}$$

As time goes by and the universe cools, the lighter protons are favored over the heavier neutrons and the neutron-to-proton ratio decreases, initially as $n/p \propto \exp(-\Delta m/T)$, where $\Delta m = 1.29$ MeV is the neutron-proton mass difference. As the temperature drops below roughly 0.8 MeV, when the universe is roughly one second old, the rate of the two-body collisions in eq. 3.24 becomes slow compared to the universal expansion rate and deviations from equilibrium occur. This is often referred to as "freeze-out", but it should be noted that the n/p ratio continues to decrease as the universe expands, albeit at a slower rate than if the ratio tracked the exponential. Later, when the universe is several hundred seconds old, a time comparable to the neutron lifetime ($\tau_n = 885.7 \pm 0.8$ sec.), the n/p ratio resumes falling exponentially: $n/p \propto \exp(-t/\tau_n)$. Notice that the n/p ratio at BBN depends on the competition between the weak interaction rates and the early universe expansion rate so that any deviations from the standard model (*e.g.*, $\rho \to \rho + \rho_X$) will change the relative numbers of neutrons and protons available for building more complex nuclides.

3.1.2. *Building The Elements*

At the same time that neutrons and protons are interconverting, they are also colliding among themselves to create deuterons: $n + p \Longleftrightarrow D + \gamma$. However, at early times when the density and average energy of the CMB photons is very high, the newly-formed deuterons find themselves bathed in a background of high energy gamma rays capable of photodissociating them. As we shall soon see, there are more than a billion photons for every nucleon in the universe so that before a neutron or a proton can be added to D to begin building the heavier nuclides, the D is photodissociated. This bottleneck to BBN beginning in earnest persists until the temperature drops sufficiently so that there are too few photons energetic enough to photodissociate the deuterons before they can capture nucleons to launch BBN. This occurs after $e^{\pm}$ annihilation, when the universe is a few minutes old and the temperature has dropped below 80 keV (0.08 MeV).

Once BBN begins in earnest, neutrons and protons quickly combine to form D, ^{3}H, ^{3}He, and ^{4}He. Here, there is another, different kind of bottleneck. There is a gap at mass-5; there is no stable mass-5 nuclide. To jump the gap requires ^{4}He reactions with D or ^{3}H or ^{3}He, all of which are positively charged. The coulomb repulsion among these colliding nuclei suppresses the reaction rate ensuring that virtually all of the neutrons available for BBN are incorporated in ^{4}He (the most tightly bound of the light nuclides), and also that the abundances of the heavier nuclides are severely depressed below that of ^{4}He (and even of D and ^{3}He). Recall that ^{3}H is unstable and will decay to ^{3}He. The few reactions which manage to bridge the mass-5 gap mainly lead to mass-7 (^{7}Li, or ^{7}Be which later, when the universe has cooled further, will capture an electron and decay to ^{7}Li); the abundance of ^{6}Li is below that of the more tightly bound ^{7}Li by one to two orders of magnitude. There is another gap at mass-8. This absence of any stable mass-8 nuclides ensures there will be no astrophysically interesting production of heavier nuclides.

The primordial nuclear reactor is short-lived, quickly encountering an energy crisis. Because of the falling temperatures and the coulomb barriers, nuclear reactions cease rather abruptly when the temperature drops below roughly 30 keV, when the universe is about 20 minutes old. As a result there is "nuclear freeze-out" since no already existing nuclides are destroyed (except for those that are unstable and decay) and no new nuclides are created. In ~ 1000 seconds BBN has run its course.

3.2. *The SBBN-Predicted Abundances*

The primordial abundances of D, ^{3}He, and ^{7}Li(^{7}Be) are rate limited, depending sensitively on the competition between the nuclear reactions rates and the universal expansion rate. As a result, these nuclides are potential baryometers since their abundances are sensitive to the universal density of nucleons. As the universe expands, the nucleon density decreases so it is useful to compare the nucleon density to that of the CMB photons $\eta \equiv n_N/n_\gamma$. Since this ratio will turn out to be very small, it is convenient to introduce

$$\eta_{10} \equiv 10^{10}(n_N/n_\gamma) = 274\Omega_B h^2 \,. \tag{3.25}$$

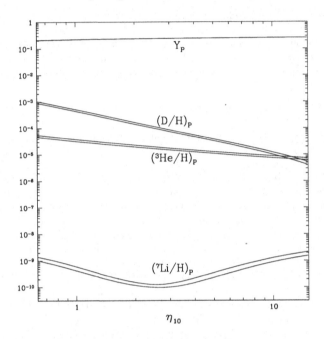

FIGURE 1. The SBBN-predicted primordial abundances of D, ^{3}He, ^{7}Li (by number with respect to hydrogen), and the ^{4}He mass fraction Y as a function of the nucleon abundance η_{10}. The widths of the bands reflect the theoretical uncertainties.

As the universe evolves (post-$e^{\pm}$ annihilation) this ratio is accurately preserved so that $\eta_{BBN} = \eta_0$. Testing this relation over ten orders of magnitude in redshift, over a range of some ten billion years, can provide a confirmation of or a challenge to the standard model.

In contrast to the other light nuclides, the primordial abundance of ^{4}He (mass fraction Y) is relatively insensitive to the baryon density, but since virtually all neutrons available at BBN are incorporated in ^{4}He, it does depend on the competition between the weak interaction rate (largely fixed by the accurately measured neutron lifetime) and the universal expansion rate (which depends on g_{eff}). The higher the nucleon density, the earlier can the D-bottleneck be breached. At early times there are more neutrons and, therefore, more ^{4}He will be synthesized. This latter effect is responsible for the very slow (logarithmic) increase in Y with η. Given the standard model relation between time and temperature and the nuclear and weak cross sections and decay rates measured in the laboratory, the evolution of the light nuclide abundances may be calculated and the frozen-out relic abundances predicted as a function of the one free parameter, the nucleon density or η. These are shown in Fig. 1.

Not shown on Fig. 1 are the relic abundances of ^{6}Li, ^{9}Be, ^{10}B, and ^{11}B, all of which, over the same range in η, lie offscale, in the range $10^{-20} - 10^{-13}$.

The reader may notice the abundances appear in Fig. 1 as bands. These represent the theoretical uncertainties in the predicted abundances. For D/H and ^{3}He/H they are at the $\sim 8\%$ level, while they are much larger, $\sim 12\%$, for ^{7}Li. The reader may not notice that a band is also shown for ^{4}He, since the uncertainty in Y is only at the

$\sim 0.2\%$ level ($\sigma_Y \approx 0.0005$). The results shown here are from the BBN code developed and refined over the years by my colleagues at The Ohio State University. They are in excellent agreement with the published results of the Chicago group (Burles, Nollett & Turner 2001) who, in a reanalysis of the relevant published cross sections have reduced the theoretical errors by roughly a factor of three for D and ^{3}He and a factor of two for ^{7}Li. The uncertainty in Y is largely due to the (very small) uncertainty in the neutron lifetime.

The trends shown in Fig. 1 are easy to understand based on our previous discussion. D and ^{3}He are burned to ^{4}He. The higher the nucleon density, the faster this occurs, leaving behind fewer nuclei of D or ^{3}He. The very slight increase of Y with η is largely due to BBN starting earlier, at higher nucleon density (more complete burning of D, ^{3}H, and ^{3}He to ^{4}He) and neutron-to-proton ratio (more neutrons, more ^{4}He). The behavior of ^{7}Li is more interesting. At relatively low values of $\eta \lesssim 3$, mass-7 is largely synthesized as ^{7}Li (by ^{3}H$(\alpha,\gamma)^7$Li reactions) which is easily destroyed in collisons with protons. So, as η increases at low values, ^{7}Li/H decreases. However, at relatively high values of $\eta \gtrsim 3$, mass-7 is largely synthesized as ^{7}Be (via ^{3}He$(\alpha,\gamma)^7$Be reactions) which is more tightly bound than ^{7}Li and, therefore, harder to destroy. As η increases at high values, the abundance of ^{7}Be increases. Later in the evolution of the universe, when it is cooler and neutral atoms begin to form, ^{7}Be will capture an electron and β-decay to ^{7}Li.

3.3. *Variations On A Theme: Non-Standard BBN*

Before moving on, let's take a diversion to which we'll return again in §6. Suppose the standard model is modified through the addition of extra relativistic particles ($\Delta N_\nu > 0$; SSG). Equivalently (ignoring some small differences), it could be that the gravitational constant in the early universe differs from its present value ($G \rightarrow G' \neq G$). Depending on whether $G' > G$ or $G' < G$, the early universe expansion rate can be speeded up or slowed down compared to the standard rate. For concreteness, let's assume that $S > 1$. Now, there will be less time to destroy D and ^{3}He, so their relic abundances will increase relative to the SBBN prediction. There is less time for neutrons to transform to protons. With more neutrons available, more ^{4}He will be synthesized. The changes in ^{7}Li are more complex. At low η there is less time to destroy ^{7}Li, so the relic ^{7}Li abundance increases. At high η there is less time to produce ^{7}Be, so the relic ^{7}Li (mass-7) abundance decreases.

Since the ^{4}He mass fraction is relatively insensitive to the baryon density, it provides an excellent probe of any changes in the expansion rate. The faster the universe expands, the less time for neutrons to convert to protons, the more ^{4}He will be synthesized. The increase in Y for "modest" changes in S is roughly $\Delta Y \approx 0.16(S-1) \approx 0.013\Delta N_\nu$. In Fig. 2 are shown the BBN-predicted Y versus the BBN-predicted Deuterium abundance (relative to Hydrogen) for three choices of N_ν ($N_\nu \equiv 3 + \Delta N_\nu$).

4. Observational Status of the Relic Abundances

Armed with the SBBN-predicted primordial abundances, as well as with those in a variation on the standard model, we now turn to the observational data. The four light nuclides of interest, D, ^{3}He, ^{4}He, and ^{7}Li follow different evolutionary paths in the post-BBN universe. In addition, the observations leading to their abundance determinations are different for all four. Neutral D is observed in absorption in the UV; singly-ionized ^{3}He is observed in emission in galactic H II regions; both singly- and doubly-ionized ^{4}He are observed in emission via their recombinations in extragalactic H II regions; ^{7}Li is observed in absorption in the atmospheres of very metal-poor halo stars. The different histories and observational strategies provides some insurance that systematic errors affecting the

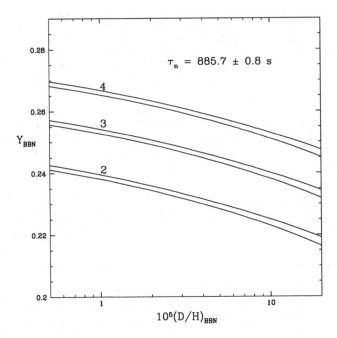

FIGURE 2. The BBN-predicted primordial ^{4}He mass fraction Y as a function of the BBN-predicted primordial Deuterium abundance (by number relative to Hydrogen) for three choices of N_ν. The width of the bands represents the theoretical uncertainty, largely due to that of the neutron lifetime τ_n.

inferred primordial abundances of any one of the light nuclides will not influence the inferred abundances of the others.

4.1. *Deuterium*

The post-BBN evolution of D is simple. As gas is incorporated into stars the very loosely bound deuteron is burned to ^{3}He (and beyond). Any D which passes through a star is destroyed. Furthermore, there are no astrophysical sites where D can be produced in an abundance anywhere near that which is observed (Epstein, Lattimer, & Schramm 1976). As a result, as the universe evolves and gas is cycled through generations of stars, Deuterium is only destroyed. Therefore, observations of the deuterium abundance anywhere, anytime, provide lower bounds on its primordial abundance. Furthermore, if D can be observed in "young" systems, in the sense of very little stellar processing, the observed abundance should be very close to the primordial value. Thus, while there are extensive data on deuterium in the solar system and the local interstellar medium (ISM) of the Galaxy, it is the handful of observations of deuterium absorption in high-redshift (hi-z), low-metallicity (low-Z), QSO absorption-line systems (QSOALS) which are, potentially the most valuable. In Fig. 3 the extant data (circa November 2001) are shown for D/H as a function of redshift from the work of Burles & Tytler (1998a,b), O'Meara *et al.* (2001), D'Odorico *et al.* (2001), and Pettini & Bowen (2001). Also shown for comparison are the local ISM D/H (Linsky & Wood 2000) and that for the presolar

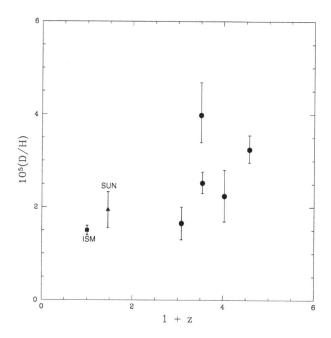

FIGURE 3. The deuterium abundance, D/H, versus redshift, z, from observations of QSOALS (filled circles). Also shown for comparison are the D-abundances for the local ISM (filled square) and the solar system ("Sun"; filled triangle).

nebula as inferred from solar system data (Geiss & Gloeckler 1998, Gloeckler & Geiss 2000).

On the basis of our discussion of the post-BBN evolution of D/H, it would be expected that there should be a "Deuterium Plateau" at high redshift. If, indeed, one is present, the dispersion in the limited set of current data hide it. Alternatively, to explore the possibility that the D-abundances may be correlated with the metallicity of the QSOALS, we may plot the observed D/H versus the metallicity, as measured by [Si/H], for these absorbers. This is shown in Fig. 4 where there is some evidence for an (unexpected!) increase in D/H with decreasing [Si/H]; once again, the dispersion in D/H hides any plateau.

Aside from observational errors, there are several sources of systematic error which may account for the observed dispersion. For example, the Lyα absorption of H I in these systems is saturated, potentially hiding complex velocity structure. Usually, but not always, this velocity structure can be revealed in the higher lines of the Lyman series and, expecially, in the narrower metal-absorption lines. Recall, also, that the lines in the Lyman series of D I are identical to those of H I, only shifted by ≈ 81 km/s. Given the highly saturated H I Lyα, it may be difficult to identify which, and how much, of the H I corresponds to an absorption feature identified as D I Lyα. Furthermore, are such features really D I or, an interloping, low column density H I-absorber? After all, there are many more low-, rather than high-column density H I systems. Statistically, the highest column density absorbers may be more immune to these systematic errors. Therefore, in Fig. 5 are shown the very same D/H data, now plotted against the neutral

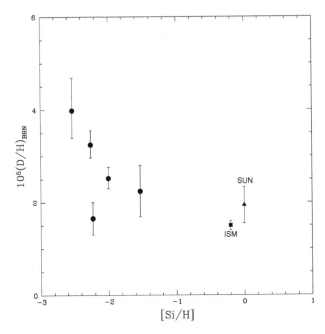

FIGURE 4. The deuterium abundance, D/H, versus metallicity ([Si/H]) for the same QSOALS as in Fig. 3 (filled circles). Also shown for comparison are the D-abundances for the local ISM (filled square) and the solar system ("Sun"; filled triangle).

hydrogen column density. The three highest column density absorbers (Damped Lyα Absorbers: DLAs) fail to reveal the D-plateau, although it may be the case that *some* of the D associated with the two lower column density systems may be attributable to an interloper, which would reduce the D/H inferred for them.

Actually, the situation is even more confused. The highest column density absorber, from D'Odorico *et al.* (2001), was reobserved by Levshakov *et al.* (2002) and revealed to have a more complex velocity structure. As a result, the D/H has been revised from 2.24×10^{-5} to 3.2×10^{-5} to 3.75×10^{-5}. To this theorist, at least, this evolution suggests that the complex velocity structure in this absorber renders it suspect for determining primordial D/H. The sharp-eyed reader may notice that if this D/H determination if removed from Fig. 5, there is a hint of an *anticorrelation* between D/H and N_H among the remaining data points, suggesting that interlopers may be contributing to (but not necessarily dominating) the inferred D I column density.

However, the next highest column density H I-absorber (Pettini & Bowen 2001), has the lowest D/H ratio, at a value indistinguishable from the ISM and solar system abundances. Why such a high-z, low-Z system should have destroyed so much of its primordial D so early in the evolution of the universe, apparently without producing very many heavy elements, is a mystery. If, for no really justifiable reason, this system is arbitrarily set aside, only the three "UCSD" systems of Burles & Tytler (1998a,b) and O'Meara *et al.* (2001) remain. The weighted mean for these three absorbers is D/H = 3.0×10^{-5}. O'Meara *et al.* note the larger than expected dispersion, even for this subset of D-abundances, and they suggest increasing the formal error in the mean, leading to:

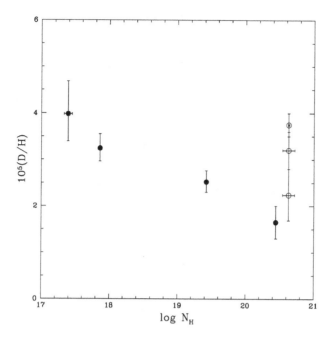

FIGURE 5. The same QSOALS D/H data as in Fig. 3 & Fig. 4 versus the H I column density (log scale). The open circle symbols are for the original (D'Odorico *et al.* 2001) and the revised (Levshakov *et al.* 2002) D/H values for Q0347-3819.

$(D/H)_P = 3.0 \pm 0.4 \times 10^{-5}$. I will be even more cautious; when the SBBN predictions are compared with the primordial abundances inferred from the data, I will adopt: $(D/H)_P = 3.0^{+1.0}_{-0.5} \times 10^{-5}$. Since the primordial D abundance is sensitive to the baryon abundance $(D/H \propto \eta^{-1.6})$, even these perhaps overly generous errors will still result in SBBN-derived baryon abundances which are accurate to 10 – 20%.

4.2. *Helium-3*

The post-BBN evolution of ^{3}He is much more complex than that of D. Indeed, when D is incorporated into a star it is rapidly burned to ^{3}He, increasing the ^{3}He abundance. The more tightly bound ^{3}He, with a larger coulomb barrier, is more robust than D to nuclear burning. Nonetheless, in the hotter interiors of most stars ^{3}He is burned to ^{4}He and beyond. However, in the cooler, outer layers of most stars, and throughout most of the volume of the cooler, lower mass stars, ^{3}He is preserved (Iben 1967, Rood, Steigman & Tinsley 1976; Iben & Truran 1978, Dearborn, Schramm & Steigman 1986; Dearborn, Steigman & Tosi 1996). As a result, prestellar ^{3}He is enhanced by the burning of prestellar D, and some, but not all, of this ^{3}He survives further stellar processing. However, there's more to the story. As stars burn hydrogen to helium and beyond, some of their newly synthesized ^{3}He will avoid further processing so that the cooler, lower mass stars should be significant post-BBN sources of ^{3}He.

Aside from studies of meteorites and in samples of the lunar soil (Reeves *et al.* 1973, Geiss & Gloeckler 1998, Gloeckler & Geiss 2000), ^{3}He is only observed via its hyperfine line (of singly-ionized ^{3}He) in interstellar H II regions in the Galaxy. It is, therefore,

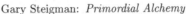

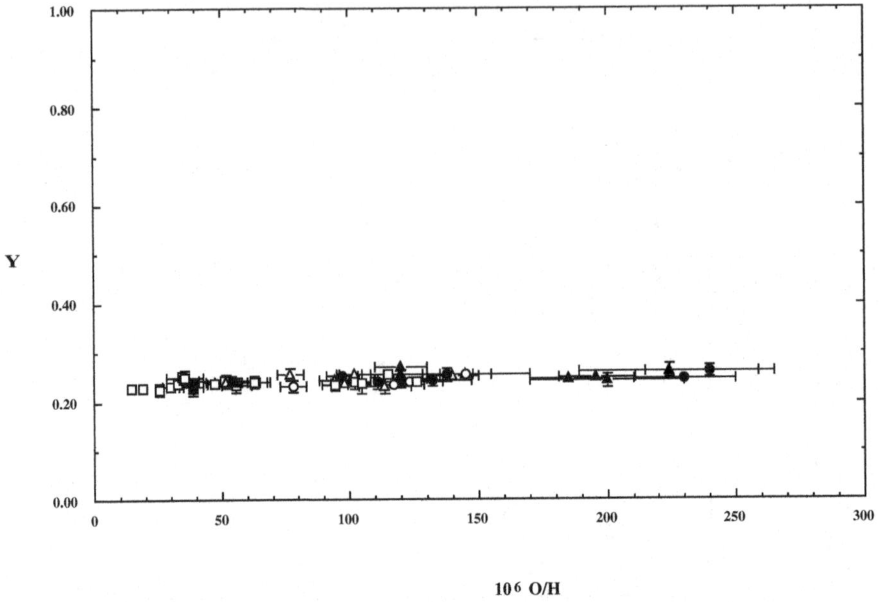

10^6 O/H

FIGURE 6. The ^{4}He mass fraction, Y, inferred from observations of low-metallicity, extragalactic H II regions versus the oxygen abundance in those regions.

unavoidable that models of stellar yields and Galactic chemical evolution are required in order to go from here and now (ISM) to there and then (BBN). It has been clear since the early work of Rood, Steigman and Tinsley (1976) that according to such models, ^{3}He should have increased from the big bang and, indeed, since the formation of the solar system (see, *e.g.*, Dearborn, Steigman & Tosi 1996 and further references therein). For an element whose abundance increases with stellar processing, there should also be a clear gradient in abundance with galactocentric distance. Neither of these expectations is borne out by the data (Rood, Bania & Wilson 1992; Balser *et al.* 1994, 1997, 1999; Bania, Rood & Balser 2002) which shows no increase from the time of the formation of the solar system, nor any gradient within the Galaxy. The most likely explanation is that before the low mass stars can return their newly processed ^{3}He to the interstellar medium, it is mixed to the hotter interior and destroyed (Charbonnel 1995, Hogan 1995). Whatever the explanation, the data suggest that for ^{3}He there is a delicate balance between production and destruction. As a result, the model-dependent uncertainties in extrapolating from the present data to the primordial abundances are large, limiting the value of ^{3}He as a barometer. For this reason I will not dwell further on ^{3}He in these lectures; for further discussion and references, the interested reader is referred to the excellent review by Tosi (2000).

4.3. *Helium-4*

Helium-4 is the second most abundant nuclide in the universe, after hydrogen. In the post-BBN epochs the net effect of gas cycling though generations of stars is to burn hydrogen to helium, increasing the ^{4}He abundance. As with deuterium, a ^{4}He "plateau" is expected at low metallicity. Although ^{4}He is observed in the Sun and in Galactic H II regions, the most relevant data for inferring its primordial abundance (the plateau value)

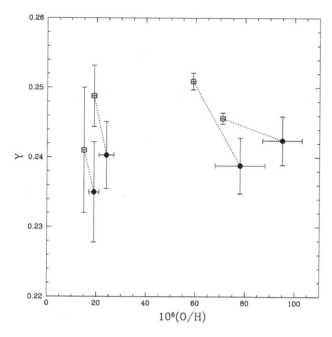

FIGURE 7. The Peimbert, Peimbert, & Luridiana (2002) reanalysis of the Helium-4 abundance data for 4 of the IT H II regions. The open circles are the IT abundances, while the filled circles are from PPL.

is from observations of the helium and hydrogen recombination lines in low-metallicity, extragalactic H II regions. The present inventory of such observations is approaching of order 100. It is, therefore, not surprising that even with modest observational errors for any individual H II region, the statistical uncertainty in the inferred primordial abundance may be quite small. Especially in this situation, care must be taken with hitherto ignored or unaccounted for corrections and systematic errors.

In Fig. 6 is shown a compilation of the data used by Olive & Steigman (1995) and Olive, Skillman, & Steigman (1997), along with the independent data from Izotov, Thuan, & Lipovetsky (1997) and Izotov & Thuan (1998). To track the evolution of the ^{4}He mass fraction, Y is plotted versus the H II region oxygen abundance. These H II regions are all metal-poor, ranging from $\sim 1/2$ down to $\sim 1/30$ of solar (for a solar oxygen abundance of O/H $= 5 \times 10^{-4}$; Allende Prieto, Lambert, & Asplund 2001). A key feature of Fig. 6 is that independent of whether there is a statistically significant non-zero slope to the Y vs. O/H relation, there is a ^{4}He plateau! Since Y is increasing with metallicity, the relic abundance can either be bounded from above by the lowest metallicity regions, or the Y vs. O/H relation determined observationally may be extrapolated to zero metallicity (a not very large extrapolation, $\Delta Y \approx -0.001$).

The good news is that the data reveal a well-defined primordial abundance for ^{4}He. The bad news is that the scale of Fig. 6 hides the very small statistical errors, along with a dichotomy between the OS/OSS and ITL/IT primordial helium abundance determinations ($Y_P(OS) = 0.234 \pm 0.003$ versus $Y_P(IT) = 0.244 \pm 0.002$). Furthermore, even if one adopts the IT/ITL data, there are corrections which should be applied which change the inferred primordial ^{4}He abundance by more than their quoted statistical errors (see, *e.g.*, Steigman, Viegas & Gruenwald 1997; Viegas, Gruenwald & Steigman 2000; Sauer & Jedamzik 2002, Gruenwald, Steigman & Viegas 2002 (GSV); Peimbert, Peimbert & Luridiana 2002). In recent high quality observations of a relatively metal-rich SMC H II region, Peimbert, Peimbert & Ruiz (2000; PPR) derive $Y_{SMC} = 0.2405 \pm 0.0018$. This is already *lower* than the IT-inferred *primordial* ^{4}He abundance. Further, when PPR extrapolate this abundance to zero-metallicity, they derive $Y_P(PPR) = 0.2345 \pm 0.0026$, lending some indirect support for the lower OS/OSS value.

Recently, Peimbert, Peimbert, & Luridiana (2002; PPL) have reanalyzed the data from four of the IT H II regions. When correcting for the H II region temperatures and the temperature fluctuations, PPL derive systematically lower helium abundances as shown in Fig. 7. PPL also combine their redetermined abundances for these four H II regions with the recent accurate determination of Y in the more metal-rich SMC H II region (PPR). These five data points are consistent with zero slope in the Y vs. O/H relation, leading to a primordial abundance $Y_P = 0.240 \pm 0.001$. However, this very limited data set is also consistent with $\Delta Y \approx 40(O/H)$. In this case, the extrapolation to zero metallicity, starting at the higher SMC metallicity, leads to the considerably smaller estimate of $Y_P \approx 0.237$.

It seems clear that until new data address the unresolved systematic errors afflicting the derivation of the primordial helium abundance, the true errors must be much larger than the statistical uncertainties. For the comparisons between the predictions of SBBN and the observational data to be made in the next section, I will adopt the Olive, Steigman & Walker (2000; OSW) compromise: $Y_P = 0.238 \pm 0.005$; the inflated errors are an attempt to account for the poorly-constrained systematic uncertaintiess.

4.4. *Lithium-7*

Lithium-7 is fragile, burning in stars at a relatively low temperature. As a result, the majority of interstellar ^{7}Li cycled through stars is destroyed. For the same reason, it is difficult for stars to create new ^{7}Li and return it to the ISM before it is destroyed by nuclear burning. As the data in Fig. 8 reveal, only relatively late in the evolution of the Galaxy, when the metallicity approaches solar, does the lithium abundance increase noticeably. However, the intermediate-mass nuclides ^{6}Li, ^{7}Li, ^{9}Be, ^{10}B, and ^{11}B can be synthesized via Cosmic Ray Nucleosynthesis (CRN), either by alpha-alpha fusion reactions, or by spallation reactions (nuclear breakup) between protons and alpha particles on the one hand and CNO nuclei on the other. In the early Galaxy, when the metallicity is low, the post-BBN production of lithium is expected to be subdominant to the pregalactic, BBN abundance. This is confirmed in Fig. 8 by the "Spite Plateau" (Spite & Spite 1982), the absence of a significant slope in the Li/H versus [Fe/H] relation at low metallicity. This plateau is a clear signal of the primordial origin of the low-metallicity lithium abundance. Notice, also, the enormous *spread* among the lithium abundances at higher metallicity. This range in Li/H results from the destruction/dilution of lithium on the surfaces of the observed stars, implying that it is the *upper envelope* of the Li/H versus [Fe/H] relation which preserves the history of the Galactic lithium evolution. Note, also, that at low metallicity this dispersion is much narrower, suggesting that the corrections for depletion/dilution are much smaller for the Pop II stars.

As with the other relic nuclides, the dominant uncertainties in estimating the primordial abundance of ^{7}Li are not statistical, they are systematic. Lithium is observed in the atmospheres of cool stars (see Lambert (2001) in these lectures). It is the metal-poor, Pop II halo stars that are of direct relevance for the BBN abundance of ^{7}Li. Uncertainties in the lithium equivalent width measurements, in the temperature scales for

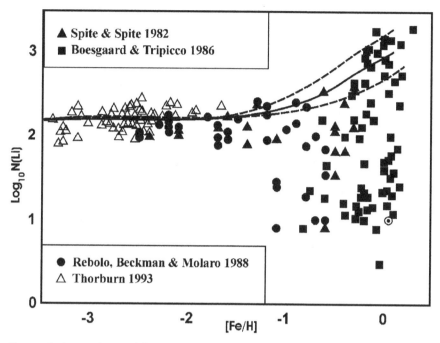

FIGURE 8. A compilation of the lithium abundance data from stellar observations as a function of metallicity. $N(Li) \equiv 10^{12}(Li/H)$ and [Fe/H] is the usual metallicity relative to solar. Note the "Spite Plateau" in Li/H for [Fe/H] $\lesssim -2$.

these Pop II stars, and in their model atmospheres, dominate the overall error budget. For example, Ryan *et al.* (2000), using the Ryan, Norris & Beers (1999) data, infer $[Li]_P \equiv 12+\log(Li/H) = 2.1$, while Bonifacio & Molaro (1997) and Bonifacio, Molaro & Pasquini (1997) derive $[Li]_P = 2.2$, and Thorburn (1994) finds $[Li]_P = 2.3$. From recent observations of stars in a metal-poor globular cluster, Bonifacio *et al.* (2002) derive $[Li]_P = 2.34 \pm 0.056$. But, there's more.

The very metal-poor halo stars used to define the lithium plateau are very old. They have had the most time to disturb the prestellar lithium which may survive in their cooler, outer layers. Mixing of these outer layers with the hotter interior where lithium has been destroyed will dilute the surface abundance. Pinsonneault *et al.* (1999, 2002) have shown that rotational mixing may decrease the surface abundance of lithium in these Pop II stars by $0.1 - 0.3$ dex while maintaining a rather narrow *dispersion* among their abundances (see also, Chaboyer *et al.* 1992; Theado & Vauclair 2001, Salaris & Weiss 2002).

In Pinsonneault *et al.* (2002) we adopted for our baseline (Spite Plateau) estimate $[Li] = 2.2 \pm 0.1$; for an overall depletion factor we chose 0.2 ± 0.1 dex. Combining these *linearly*, we derived an estimate of the primordial lithium abundance of $[Li]_P = 2.4 \pm 0.2$. I will use this in the comparison between theory and observation to be addressed next.

5. Confrontation Of Theoretical Predictions With Observational Data

As the discussion in the previous section should have made clear, the attempts to use a variety of observational data to infer the BBN abundances of the light nuclides is fraught with evolutionary uncertainties and dominated by systematic errors. It may be folly to represent such data by a "best" value along with normally distributed errors. Nonetheless, in the absence of a better alternative, this is what will be done in the following.

• Deuterium

From their data along the lines-of-sight to three QSOALS, O'Meara *et al.* (2001) recommend $(D/H)_P = 3.0 \pm 0.4 \times 10^{-5}$. While I agree this is likely a good estimate for the *central* value, the spread among the extant data (see Fig. 3 – Fig. 5) favors a larger uncertainty. Since D is only destroyed in the post-BBN universe, the solar system and ISM abundances set a floor to the primordial value. Keeping this in mind, I will adopt asymmetric errors ($\sim 1\sigma$): $(D/H)_P = 3.0^{+1.0}_{-0.5} \times 10^{-5}$.

• Helium-4

In our discussion of ^{4}He as derived from hydrogen and helium recombination lines in low-metallicity, extragalactic H II regions it was noted that the inferred primordial mass fraction varied from $Y_P = 0.234 \pm 0.003$ (OS/OSS), to $Y_P = 0.238 \pm 0.003$ (PPL and GSV), to $Y_P = 0.244 \pm 0.002$ (IT/ITL). Following the recommendation of OSW, here I will choose as a compromise $Y_P = 0.238 \pm 0.005$.

• Lithium-7

Here, too, the spread in the level of the "Spite Plateau" dominates the formal errors in the means among the different data sets. To this must be added the uncertainties due to temperature scale and model atmospheres, as well as some allowance for dilution or depletion over the long lifetimes of the metal-poor halo stars. Attempting to accomodate all these sources of systematic uncertainty, I adopt the Pinsonneault *et al.* (2002) choice of $[Li]_P = 2.4 \pm 0.2$.

As discussed earlier, the stellar and Galactic chemical evolution uncertainties afflicting ^{3}He are so large as to render the use of ^{3}He to probe or test BBN problematic; therefore, I will ignore ^{3}He in the subsequent discussion. There are a variety of equally valid approaches to using D, ^{4}He, and ^{7}Li to test and constrain the standard models of cosmology and particle physics (SBBN). In the approach adopted here deuterium will be used to constrain the baryon density (η or, equivalently, $\Omega_B h^2$). Within SBBN, this leads to predictions of Y_P and $[Li]_P$. Indeed, once the primordial deuterium abundance is chosen, η may be eliminated and both Y_P and $[Li]_P$ predicted directly, thereby testing the consistency of SBBN.

5.1. *Deuterium – The Baryometer Of Choice*

Recall that D is produced (in an astrophysically interesting abundance) **ONLY** during BBN. The predicted primordial abundance is sensitive to the baryon density (D/H $\propto \eta^{-1.6}$). Furthermore, during post-BBN evolution, as gas is cycled through stars, deuterium is **ONLY** destroyed, so that for the "true" abundance of D anywhere, at any time, $(D/H)_P \geq (D/H)_{TRUE}$. That's the good news. The bad news is that the spectra of H I and D I are identical, except for the wavelength/velocity shift in their spectral lines. As a result, the true D-abundance may differ from that inferred from the observations if *some* of the presumed D I is actually an H I interloper masquerading as D I: $(D/H)_{TRUE} \leq (D/H)_{OBS}$. Because of these opposing effects, the connection between $(D/H)_{OBS}$ and $(D/H)_P$ is not predetermined; the data themselves which must tell us

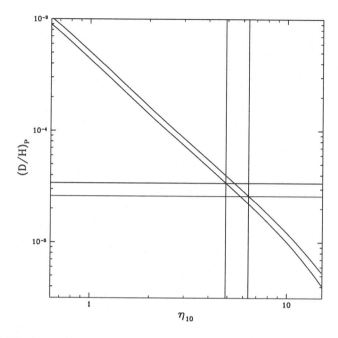

FIGURE 9. The diagonal band is the SBBN-predicted deuterium abundance (by number relative to hydrogen) as a function of the nucleon-to-photon ratio η_{10} (the width of the band accounts for the theoretical uncertainties in the SBBN prediction). The horizontal band is the $\pm 1\sigma$ range in the adopted primordial deuterium abundance. The vertical band is, approximately, the corresponding SBBN-predicted η range.

how to relate the two. With this caveat in mind, it will be assumed that the value of $(D/H)_P$ identified above is a fair estimate of the primordial D abundance. Using it, the SBBN-predicted baryon abundance may be determined. The result of this comparison is shown in Fig. 9 where, approximately, the overlap between the SBBN-predicted band and that from the data fix the allowed range of η.

5.2. *SBBN Baryon Density – The Baryon Density At 20 Minutes*

The universal abundance of baryons which follows from SBBN and our adopted primordial D-abundance is: $\eta_{10} = 5.6^{+0.6}_{-1.2}$ ($\Omega_B h^2 = 0.020^{+0.002}_{-0.004}$). For the HST Key Project recommended value for H_0 ($h = 0.72 \pm 0.08$; Freedman *et al.* 2001), the fraction of the present universe critical density contributed by baryons is small, $\Omega_B \approx 0.04$. In Fig. 10 is shown a comparison among the various determinations of the present mass/energy density (as a fraction of the critical density), baryonic as well as non-baryonic. It is clear from Fig. 10 that the present universe ($z \lesssim 1$) baryon density inferred from SBBN far exceeds that inferred from emission/absorption observations (Persic & Salucci 1992, Fukugita, Hogan & Peebles 1998). The gap between the upper bound to luminous baryons and the BBN band is the "dark baryon problem": at present, most of the baryons in the universe are dark. Evidence that although dark, the baryons are, indeed, present comes from the absorption observed in the Lyα forest at redshifts $z \approx 2-3$ (see, *e.g.*, Weinberg *et al.* 1997). The gap between the BBN band and the band labelled by Ω_M is the "dark

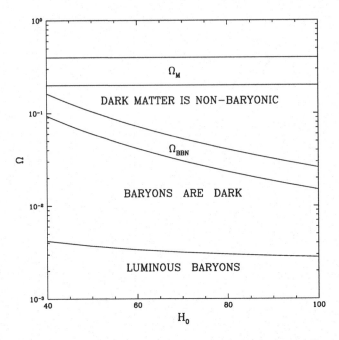

FIGURE 10. The various contributions to the present universal mass/energy density, as a fraction of the critical density (Ω), as a function of the Hubble parameter (H_0). The curve labelled Luminous Baryons is an estimate of the upper bound to those baryons seen at present ($z \lesssim 1$) either in emission or absorption (see the text). The band labelled BBN represents the D-predicted SBBN baryon density. The band labelled by "M" ($\Omega_M = 0.3 \pm 0.1$) is an estimate of the current mass density in nonrelativistic particles ("Dark Matter").

matter problem": the mass density inferred from the structure and movements of the galaxies and galaxy clusters far exceeds the SBBN baryon contribution. Most of the mass in the universe must be nonbaryonic. Finally, the gap from the top of the Ω_M band to $\Omega = 1$ is the "dark energy problem".

5.3. *CMB Baryon Density – The Baryon Density At A Few Hundred Thousand Years*

As discussed in the first lecture, the early universe is hot and dominated by relativistic particles ("radiation"). As the universe expands and cools, nonrelativistic particles ("matter") come to dominate after a few hundred thousand years, and any preexisting density perturbations can begin to grow under the influence of gravity. On length scales determined by the density of baryons, oscillations ("sound waves") in the baryon-photon fluid develop. At a redshift of $z \sim 1100$ the electron-proton plasma combines ("recombination) to form neutral hydrogen which is transparent to the CMB photons. Free to travel throughout the post-recombination universe, these CMB photons preserve the record of the baryon-photon oscillations as small temperature fluctuations in the CMB spectrum. Utilizing recent CMB observations (Lee *et al.* 2001; Netterfield *et al.* 2002; Halverson *et al.* 2002), many groups have inferred the intermediate age universe baryon density. The work of our group at OSU (Kneller *et al.* 2001) is consistent with more

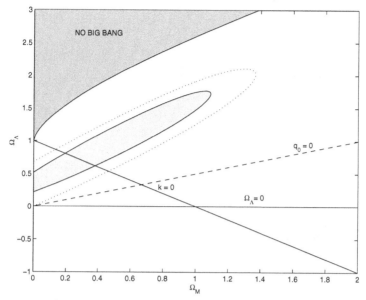

FIGURE 11. The 68% (solid) and 95% (dotted) contours in the Ω_Λ – Ω_M plane consistent with the SNIa data (see the text). Geometrically flat models lie along the line labelled k = 0.

detailed analyses and is the one I adopt for the purpose of comparison with the SBBN result: $\eta_{10} = 6.0 \pm 0.6$; $\Omega_B h^2 = 0.022 \pm 0.002$.

5.4. *The Baryon Density At 10 Gyr*

Although the majority of baryons in the recent/present universe are dark, it is still possible to constrain the baryon density indirectly using observational data (see, *e.g.*, Steigman, Hata & Felten 1999, Steigman, Walker & Zentner 2000; Steigman 2001). The magnitude-redshift relation determined by observations of type Ia supernovae (SNIa) constrain the relation between the present matter density (Ω_M) and that in a cosmological constant (Ω_Λ). The allowed region in the Ω_Λ – Ω_M plane derived from the observations of Perlmutter *et al.* (1997), Schmidt *et al.* (1998), and Perlmutter *et al.* (1999) are shown in Fig. 11.

If, in addition, it is *assumed* that the universe is flat ($\kappa = 0$; an assumption supported by the CMB data), a reasonably accurate determination of Ω_M results: $\Omega_M(\text{SNIa; Flat}) = 0.28^{+0.08}_{-0.07}$ (Steigman, Walker & Zentner 2000; Steigman 2001). But, how to go from the matter density to the baryon density? For this we utilize rich clusters of galaxies, the largest collapsed objects, which provide an ideal probe of the baryon *fraction* in the present universe f_B. X-ray observations of the hot gas in clusters, when corrected for the baryons in stars (albeit not for any dark cluster baryons), can be used to estimate f_B. Using the Grego *et al.* (2001) observations of the Sunyaev-Zeldovich effect in clusters, Steigman, Kneller & Zentner (2002) estimate f_B and derive a present-universe ($t_0 \approx 10$ Gyr; $z \lesssim 1$) baryon density: $\eta_{10} = 5.1^{+1.8}_{-1.4}$ ($\Omega_B h^2 = 0.019^{+0.007}_{-0.005}$).

5.5. *Baryon Density Concordance*

In Fig. 12 are shown the likelihood distributions for the three baryon density determinations discussed above. It is clear that these disparate determinations, relying on

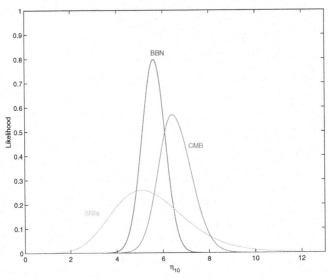

FIGURE 12. The likelihood distributions, normalized to equal areas under the curves, for the baryon-to-photon ratios (η_{10}) derived from BBN ($\sim$ 20 minutes), from the CMB ($\sim$ few hundred thousand years), and for the present universe ($t_0 \sim$ 10 Gyr; $z \lesssim 1$).

completely different physics and from widely separated epochs in the evolution of the universe are in excellent agreement, providing strong support for the standard, hot big bang cosmological model and for the standard model of particle physics. Although it has been emphasized many times in these lectures that the errors are likely dominated by evolutionary and systematic uncertainties and, therefore, are almost certainly not normally distributed, it is hard to avoid the temptation to combine these three independent estimates. Succumbing to temptation: $\eta_{10} = 5.8^{+0.4}_{-0.6}$ ($\Omega_B h^2 = 0.021^{+0.0015}_{-0.0020}$).

5.6. *Testing The Consistency Of SBBN*

As impressive as is the agreement among the three independent estimates of the universal baryon density, we should not be lured into complacency. The apparent success of SBBN should impel us to test the standard model even further. How else to expose possible systematic errors which have heretofore been hidden from view or, to find the path beyond the standard models of cosmology and particle physics? To this end, in Fig. 13 are compared the SBBN ^{4}He (Y) and D (D/H) abundance predictions along with the estimates for Y and D/H adopted here. The agreement is not very good. Indeed, while for $\eta_{10} = 5.8^{+0.4}_{-0.6}$, (D/H)$_{\rm SBBN} = 2.8^{+0.4}_{-0.7} \times 10^{-5}$, in excellent agreement with the O'Meara *et al.* (2001) estimate, the corresponding ^{4}He abundance is predicted to be $Y_{\rm SBBN} = 0.248 \pm 0.001$, which is 2σ above our OSW-adopted primordial abundance. Indeed, the SBBN-predicted ^{4}He abundance based directly on deuterium is also 2σ above the IT/ITL estimate. Here is a potential challenge to the internal consistency of SBBN. Given that systematic errors dominate, it is difficult to decide how seriously to take this challenge. In fact, if ^{4}He and D, in concert with SBBN, are each employed as baryometers, their likelihood distributions for η are consistent at the 7% level.

An *apparent success* of (or, a *potential challenge* to) SBBN emerges from a comparison between D and ^{7}Li. In Fig. 14 is shown the SBBN-predicted relation between

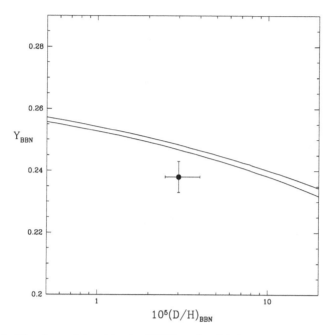

FIGURE 13. The diagonal band is the SBBN-predicted helium-4 mass fraction versus the SBBN-predicted deuterium abundance (by number relative to hydrogen). The width of the band accounts for the theoretical uncertainties in the SBBN predictions. Also shown by the filled circle and error bars are the primordial ^{4}He and D abundance estimates adopted here.

primordial D and primordial ^{7}Li along with the relic abundance estimates adopted here (Pinsonneault *et al.* 2002). Also shown for comparison is the Ryan *et al.* (2000) primordial lithium abundance estimate. The higher, depletion/dilution-corrected lithium abundance of PSWN is in excellent agreement with the SBBN-D abundance, while the lower, RBOFN value poses a challenge to SBBN.

6. BBN In Non-Standard Models

As just discussed in §5.6, there is some tension between the SBBN-predicted abundances of D and ^{4}He and their primordial abundances inferred from current observational data (see Fig. 13). Another way to see the challenge is to superpose the data on the BBN predictions from Fig. 2, where the Y_P versus D/H relations are shown for several values of N_ν (SSG). This is done in Fig. 15 where it is clear that the data prefer **nonstandard** BBN, with N_ν closer to 2 than to the standard model value of 3.

It is easy to understand this result on the basis of the earlier discussion (see §3.3). The adopted abundance of D serves, mainly, to fix the baryon density which, in turn, determines the SBBN-predicted ^{4}He abundance. The corresponding predicted value of Y_P is too large when compared to the data. A universe which expands *more slowly* ($S < 1$; $N_\nu < 3$) will permit more neutrons to transmute into protons before BBN commences, resulting in a smaller ^{4}He mass fraction. However, there are two problems (at least!) with this "solution". The main issue is that there **are** three "flavors" of

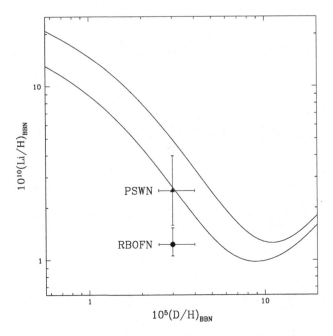

FIGURE 14. The band is the SBBN-predicted lithium abundance (by number relative to hydrogen) versus the SBBN-predicted deuterium abundance. The width of the band accounts for the theoretical uncertainties in the SBBN predictions. Also shown is the Pinsonneault *et al.* (2002; PSWN) primordial lithium abundance estimate adopted here (filled triangle) along with the Ryan *et al.* (2000; RBOFN) estimate (filled circle).

light neutrinos, so that $N_\nu \geq 3$ ($\Delta N_\nu \geq 0$). The second, probably less serious problem is that a slower expansion permits an increase in the ^{7}Be production, resulting in an increase in the predicted relic abundance of lithium. For $(D/H)_P = 3.0 \times 10^{-5}$ and $Y_P = 0.238$, the best fit values of η and ΔN_ν are: $\eta_{10} = 5.3$ ($\Omega_B h^2 = 0.019$) and $N_\nu = 2.3$ ($\Delta N_\nu = -0.7$). For this combination the BBN-predicted lithium abundance is $[Li]_P = 2.53$ (($Li/H)_P = 3.4 \times 10^{-10}$), somewhat higher than, but still in agreement with the PSWN estimate of $[Li]_P = 2.4 \pm 0.2$, but much higher than the RBOFN value of $[Li]_P = 2.1 \pm 0.1$. Although the tension between the observed and SBBN-predicted lithium abundances may not represent a serious challenge (at present), the suggestion that $\Delta N_\nu < 0$ must be addressed. One possibility is that the slower expansion of the radiation-dominated early universe could result from a non-minimally coupled scalar field ("extended quintessence") whose effect is to change the effective gravitational constant ($G \rightarrow G' < G$; see §3.3). For a discussion of such models and for further references see, *e.g.*, Chen, Scherrer & Steigman (2001).

6.1. *Degenerate BBN*

There is another alternative to SBBN which, although currently less favored, does have a venerable history: BBN in the presence of a background of **degenerate** neutrinos. First, a brief diversion to provide some perspective. In the very early universe there were a large number of particle-antiparticle pairs of all kinds. As the baryon-antibaryon pairs

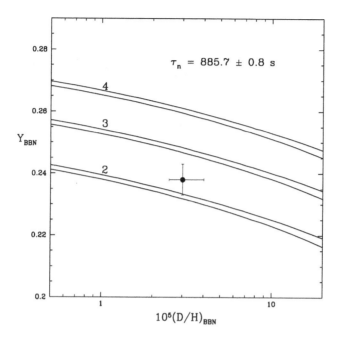

FIGURE 15. The BBN-predicted primordial ^{4}He mass fraction Y as a function of the BBN-predicted primordial Deuterium abundance (by number relative to Hydrogen) for three choices of N_ν. Also shown by the filled circle and error bars are the primordial abundances adopted here (§5).

or, their quark-antiquark precursors, annihilated, only the baryon *excess* survived. This baryon number excess, proportional to η, is very small ($\eta \lesssim 10^{-9}$). It is reasonable, but by no means compulsory, to assume that the lepton number *asymmetry* (between leptons and antileptons) is also very small. Charge neutrality of the universe ensures that the electron asymmetry is of the same order as the baryon asymmetry. But, what of the asymmetry among the several neutrino flavors?

Since the relic neutrino background has never been observed directly, not much can be said about its asymmetry. However, if there is an excess in the number of neutrinos compared to antineutrinos (or, vice-versa), "neutrino degeneracy", the total energy density in neutrinos (plus antineutrinos) is increased. As a result, during the early, radiation-dominated evolution of the universe, $\rho \to \rho' > \rho$, and the universal expansion rate increases ($S > 1$). Constraints on how large S can be do lead to some weak bounds on neutrino degeneracy (see, *e.g.*, Kang & Steigman 1992 and references therein). This effect occurs for degeneracy in all neutrino flavors (ν_e, ν_μ, and ν_τ). For **fixed** baryon density, $S > 1$ leads to an increase in D/H (less time to destroy D), more ^{4}He (less time to transform neutrons into protons), and a decrease in lithium (at high η there is less time to produce ^{7}Be). Recall that for $S = 1$ (SBBN), an increase in η results in less D (more rapid destruction), which can compensate for $S > 1$. Similarly, an increase in baryon density will increase the lithium yield (more rapid production of ^{7}Be), also tending to compensate for $S > 1$. But, at higher η, more ^{4}He is produced, further exacerbating the effect of a more rapidly expanding universe.

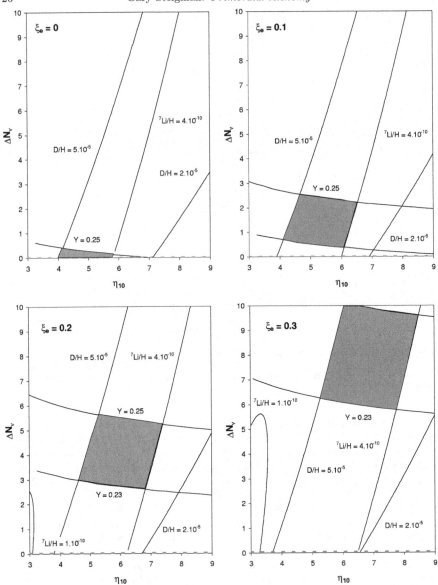

FIGURE 16. Isoabundance contours for D, ^{4}He, and ^{7}Li in the $\Delta N_\nu - \eta_{10}$ plane for four choices of the ν_e degeneracy parameter ξ_e. The shaded areas identify the ranges of parameters consistent with the abundances indicated (see the text).

However, electron-type neutrinos play a unique role in BBN, mediating the neutron-proton transformations via the weak interactions (see eq. 3.24). Suppose, for example, there are more ν_e than $\bar\nu_e$. If μ_e is the ν_e chemical potential, then $\xi_e \equiv \mu_e/kT$ is the "neutrino degeneracy parameter"; in this case, $\xi_e > 0$. The excess of ν_e will drive down the neutron-to-proton ratio, leading to a *reduction* in the primordial ^{4}He mass fraction. Thus, a combination of **three** adjustable parameters, η, ΔN_ν, and ξ_e may be varied to "tune" the primordial abundances of D, ^{4}He, and ^{7}Li. In Kneller *et al.* (2001; KSSW), we chose a range of primordial abundances similar to those adopted here ($2 \leq 10^5(D/H)_P \leq 5$; $0.23 \leq Y_P \leq 0.25$; $1 \leq 10^{10}(Li/H)_P \leq 4$) and explored the consistent ranges of η, $\xi_e \geq 0$, and $\Delta N_\nu \geq 0$. Our results are shown in Fig. 16.

It is clear from Fig. 16 that for a large range in η, a combination of ΔN_ν and ξ_e **can** be found so that the BBN-predicted abundances will lie within our adopted primordial abundance ranges. However, there are constraints on η and ΔN_ν from the CMB temperature fluctuation spectrum (see KSSW for details and further references). Although the CMB temperature fluctuation spectrum is insensitive to ξ_e, it will be modified by any changes in the universal expansion rate. While SBBN ($\Delta N_\nu = 0$) is consistent with the combined constraints from BBN and the CMB (see §5.5) for $\eta_{10} \approx 5.8$ ($\Omega_B h^2 \approx 0.021$), values of ΔN_ν as large as $\Delta N_\nu \lesssim 6$ are also allowed (KSSW).

7. Summary

As observations reveal, the present universe is filled with radiation and is expanding. According to the standard, hot big bang cosmological model the early universe was hot and dense and, during the first few minutes in its evolution, was a primordial nuclear reactor, synthesizing in significant abundances the light nuclides D, ^{3}He, ^{4}He, and ^{7}Li. These relics from the Big Bang open a window on the early evolution of the universe and provide probes of the standard models of cosmology and of particle physics. Since the BBN-predicted abundances depend on the competition between the early universe expansion rate and the weak- and nuclear-interaction rates, they can be used to test the standard models as well as to constrain the universal abundances of baryons and neutrinos. This enterprise engages astronomers, astrophysicists, cosmologists, and particle physicists alike. A wealth of new observational data has reinvigorated this subject and stimulated much recent activity. Much has been learned, revealing many new avenues to be explored (a key message for the students at this school – and for young researchers everywhere). The current high level of activity ensures that many of the detailed, quantitative results presented in these lectures will need to be revised in the light of new data, new analyses of extant data, and new theoretical ideas. Nonetheless, the underlying physics and the approaches to confronting the theoretical predictions with the observational data presented in these lectures should provide a firm foundation for future progress.

Within the context of the standard models of cosmology and of particle physics (SBBN) the relic abundances of the light nuclides depend on only one free parameter, the baryon-to-photon ratio (or, equivalently, the present-universe baryon density parameter). With one adjustable parameter and three relic abundances (four if ^{3}He is included), SBBN is an overdetermined theory, potentially falsifiable. The current status of the comparison between predictions and observations reviewed here illuminates the brilliant success of the standard models. Among the relic light nuclides, deuterium is the baryometer of choice. For $N_\nu = 3$, the SBBN-predicted deuterium abundance agrees with the primordial-D abundance derived from the current observational data for $\eta_{10} = 5.6^{+0.6}_{-1.2}$ ($\Omega_B = 0.020^{+0.002}_{-0.004}$). This baryon abundance, from the first 20 minutes of the evolution of the universe, is in excellent agreement with independent determinations from the CMB ($\sim$ few hundred thousand years) and in the present universe ($\sim$ 10 Gyr).

It is premature, however, to draw the conclusion that the present status of the comparison between theory and data closes the door on further interesting theoretical and/or observational work. As discussed in these lectures, there is some tension between the SBBN-predicted abundances and the relic abundances derived from the observational data. For the deuterium-inferred SBBN baryon density, the expected relic abundances of ^{4}He and ^{7}Li are somewhat higher than those derived from current data. The "problems" may lie with the data (large enough data sets? underestimated errors?) or, with the path from the data to the relic abundances (systematic errors? evolutionary corrections?). For example, has an overlooked correction to the H II region-derived ^{4}He abundances resulted in a value of Y_P which is systematically too small (*e.g.*, underlying stellar absorption)? Are there systematic errors in the absolute level of the lithium abundance on the Spite Plateau or, has the correction for depletion/dilution been underestimated? In these lectures the possibility that the fault may lie with the cosmology was also explored. In one simple extension of SBBN, the early universe expansion rate is allowed to differ from that in the standard model. It was noted that to reconcile D, and ^{4}He would require a *slower* than standard expansion rate, difficult to reconcile with simple particle physics extensions beyond the standard model. Furthermore, if this should be the resolution of the tension between D and ^{4}He, it would exacerbate that between the predicted and observed lithium abundances. The three abundances could be reconciled in a further extension involving neutrino degeneracy (an asymmetry between electron neutrinos and their antiparticles). But, three adjustable parameters to account for three relic abundances is far from satisfying. Clearly, this active and exciting area of current research still has some surprises in store, waiting to be discovered by astronomers, astrophysicists, cosmologists and particle physicists. The message to the students at this school – and those everywhere – is that much interesting observational and theoretical work remains to be done. I therefore conclude these lectures with a personal list of questions I would like to see addressed.

• Where (at what value of D/H) is the primordial deuterium plateau, and what is(are) the reason(s) for the currently observed spread among the high-z, low-Z QSOALS D-abundances?

• Are there stellar observations which could offer complementary insights to those from H II regions on the question of the primordial ^{4}He abundance, perhaps revealing unidentified or unquantified systematic errors in the latter approach? Is Y_P closer to 0.24 or 0.25?

• What is the level of the Spite Plateau lithium abundance? Which observations can pin down the systematic corrections due to model stellar atmospheres and temperature scales and which may reveal evidence for, and quantify, early-Galaxy production as well as stellar depletion/destruction?

• If further observational and associated theoretical work should confirm the current tension among the SBBN-predicted and observed primordial abundances of D, ^{4}He, ^{7}Li, what physics beyond the standard models of cosmology and particle physics has the potential to resolve the apparent conflicts? Are those models which modify the early, radiation-dominated universe expansion rate consistent with observations of the CMB temperature fluctuation spectrum? If neutrino degeneracy is invoked, is it consistent with the neutrino properties (masses and mixing angles) inferred from laboratory experiments as well as the solar and cosmic ray neutrino oscallation data?

To paraphrase Spock, work long and prosper!

It is with heartfelt sincerity that I thank the organizers (and their helpful, friendly, and efficient staff) for all their assistance and hospitality. Their thoughtful planning and cheerful attention to detail ensured the success of this school. It is with much fondness that I recall the many fruitful interactions with the students and with my fellow lecturers and I thank them too. I would be remiss should I fail to acknowledge the collaborations with my OSU colleagues, students, and postdocs whose work has contributed to the material presented in these lectures. The DOE is gratefully acknowledged for support under grant DE-FG02-91ER-40690.

REFERENCES

ALLENDE PRIETO, C., LAMBERT, D. L., & ASPLUND, M., 2001, ApJ, 556, L63

BALSER, D., BANIA, T., BROCKWAY, C., ROOD, R. T., & WILSON, T., 1994, ApJ, 430, 667

BALSER, D., BANIA, T., ROOD, R. T., & WILSON, T., 1997, ApJ, 483, 320

BALSER, D., BANIA, T., ROOD, R. T., & WILSON, T., 1999, ApJ, 510, 759

BANIA, T., ROOD, R. T., & BALSER, D., 2002, Nature, 415, 54

BOESGAARD, A. M., TRIPICCO, M. J., 1986, ApJ, 303, 724

BONIFACIO, P. & MOLARO, P., 1997, MNRAS, 285, 847

BONIFACIO, P., MOLARO, P., & PASQUINI, L., 1997, MNRAS, 292, L1

BONIFACIO, P., *et al.*, 2002, A&A, 390, 91

BURLES, S. NOLLETT, K. M., & TURNER, M. S., 2001, Phys. Rev. D63, 063512

BURLES, S. & TYTLER, D., 1998a, ApJ, 499, 699

BURLES, S. & TYTLER, D., 1998b, ApJ, 507, 732

CHABOYER, B. C., *et al.*, 1992, ApJ, 388, 372

CHARBONNEL, C., 1995, ApJ, 453, L41

CHEN, X., SCHERRER, R. J., & STEIGMAN, G., 2001, Phys. Rev. D63, 123504

DEARBORN, D. S. P., SCHRAMM, D. N., & STEIGMAN, G., 1986, ApJ, 302, 35

DEARBORN, D. S. P., STEIGMAN, G., & TOSI, M., 1996, ApJ, 465, 887

D'ODORICO, S., DESSAUGES-ZAVADSKY, M., & MOLARO, P., 2001, A&A, 368, L21

EPSTEIN, R., LATTIMER, J., & SCHRAMM, D. N., 1976, Nature, 263, 198

FREEDMAN, W., *et al.*, 2001, ApJ, 553, 47 (HST Key Project)

FUKUGITA, M., HOGAN, C. J., & PEEBLES, P. J. E., 1998, ApJ, 503, 518

GEISS, J., & GLOECKLER, G., 1998, Space Sci. Rev., 84, 239

GLOECKLER, G., & GEISS, J., 2000, Proceedings of IAU Symposium 198, The Light Elements and Their Evolution (L. da Silva, M. Spite, and J. R. Medeiros eds.; ASP Conference Series) p. 224

GREGO, L., *et al.*, 2001, ApJ, 552, 2

GRUENWALD, R., STEIGMAN, G., & VIEGAS, S. M., 2002, ApJ, 567, 931 (GSV)

HALVERSON, N. W., *et al.*, 2002, ApJ, 568, 38

HOGAN, C. J., 1995, ApJ, 441, L17

HOYLE, F. & TAYLER, R. J., 1964, Nature, 203, 1108

IBEN, I., 1967, ApJ, 147, 650

IBEN, I. & TRURAN, J. W., 1978, ApJ, 220, 980

IZOTOV, Y. I. & THUAN, T. X., 1998, ApJ, 500, 188 (IT)

IZOTOV, Y. I., THUAN, T. X., & LIPOVETSKY, V. A., 1997, ApJS, 108, 1 (ITL)

KANG, H. S. & STEIGMAN, G., 1992, Nucl. Phys. B372, 494

KNELLER, J. P., SCHERRER, R. J., STEIGMAN, G., & WALKER, T. P., 2001, Phys. Rev. D64, 123506 (KSSW)

LEE, A. T., *et al.*, 2001, ApJ, 561, L1

LEVSHAKOV, S. A., DESSAUGES-ZAVADSKY, M., D'ODORICO, S., & MOLARO, P., 2002, ApJ, 565, 696; see also the preprint(s): astro-ph/0105529 (v1 & v2)

LINSKY, J. L. & WOOD, B. E., 2000, Proceedings of IAU Symposium 198, The Light Elements and Their Evolution (L. da Silva, M. Spite, and J. R. Medeiros eds.; ASP Conference Series) p. 141

NETTERFIELD, C. B., *et al.*, 2002, ApJ, 571, 604

OLIVE, K. A., SKILLMAN, E., & STEIGMAN, G., 1997, ApJ, 483, 788 (OSS)

OLIVE, K. A. & STEIGMAN, G., 1995, ApJS, 97, 49 (OS)

OLIVE, K. A., STEIGMAN, G., & WALKER, T. P., 2000, Phys. Rep., 333, 389 (OSW)

O'MEARA, J. M., TYTLER, D., KIRKMAN, D., SUZUKI, N., PROCHASKA, J. X., LUBIN, D., & WOLFE, A. M., 2001, ApJ, 552, 718

PEEBLES, P. J. E., 1966, ApJ, 146, 542

PEIMBERT, A., PEIMBERT, M., & LURIDIANA, V., 2002, ApJ, 565, 668 (PPL)

PEIMBERT, M., PEIMBERT, A., & RUIZ, M. T., 2000, ApJ, 541, 688 (PPR)

PERLMUTTER, S., *et al.*, 1997, ApJ, 483, 565

PERLMUTTER, S., *et al.*, 1999, ApJ, 517, 565

PERSIC, M. & SALUCCI, P., 1992, MNRAS, 258, 14P

PETTINI, M. & BOWEN, D. V., 2001, ApJ, 560, 41

PINSONNEAULT, M. H., WALKER, T. P., STEIGMAN, G., & NARAYANAN, V. K., 1999, ApJ, 527, 180

PINSONNEAULT, M. H., STEIGMAN, G., WALKER, T. P., & NARAYANAN, V. K., 2002, ApJ, 574, 398 (PSWN)

REBOLO, R., MOLARO, P. & BECKMAN, J. E.,, 1988, A&A, 192, 192

REEVES, H., AUDOUZE, J., FOWLER, W. A., & SCHRAMM, D. N., 1973, ApJ, 179, 979

ROOD, R. T., BANIA, T. M., & WILSON, T. L., 1992, Nature, 355, 618

ROOD, R. T., STEIGMAN, G., & TINSLEY, B. M., 1976, ApJ, 207, L57

RYAN, S. G., BEERS, T. C., OLIVE, K. A., FIELDS, B. D., & NORRIS, J. E., 2000, ApJ, 530, L57 (RBOFN)

RYAN, S. G., NORRIS, J. E., & BEERS, T. C., 1999, ApJ, 523, 654

SALARIS, M. & WEISS, A., 2002, A&A, 388, 492

SAUER, D. & JEDAMZIK, K., 2002, A&A, 381, 361

SCHMIDT, B. P., *et al.*, 1998, ApJ, 507, 46

SHVARTSMAN, V. F., 1969, JETP Lett., 9, 184

SPITE, M. & SPITE, F., 1982, Nature, 297, 483

STEIGMAN, G., 2001, To appear in the Proceedings of the STScI Symposium, "The Dark Universe: Matter, Energy, and Gravity" (April 2 – 5, 2001), ed. M. Livio; astro-ph/0107222

STEIGMAN, G., SCHRAMM, D. N., & GUNN, J. E., 1977, Phys. Lett. B66, 202 (SSG)

STEIGMAN, G., HATA, N., & FELTEN, J. E., 1999, ApJ, 510, 564

STEIGMAN, G., KNELLER, J. P., & ZENTNER, A., 2002, Revista Mexicana de Astronomia y Astrofisica, 12, 265

STEIGMAN, G., VIEGAS, S. M., & GRUENWALD, R., 1997, ApJ, 490, 187

STEIGMAN, G., WALKER, T. P., & ZENTNER, A., 2000, preprint, astro-ph/0012149

THORBURN, J. A., 1994, ApJ, 421, 318

THEADO, S. & VAUCLAIR, S., 2001, A&A, 375, 70

TOSI, M., 2000, Proceedings of IAU Symposium 198, The Light Elements and Their Evolution (L. da Silva, M. Spite, and J. R. Medeiros eds.; ASP Conference Series) p. 525

VIEGAS, S. M., GRUENWALD, R., & STEIGMAN, G., 2000, ApJ, 531, 813

WEINBERG, D. H., MIRALDA-ESCUDÉ, J., HERNQUIST, L., & KATZ, N., 1997, ApJ, 490, 564

Stellar Nucleosynthesis

By NORBERT LANGER

Astronomical Institut, Utrecht University, The Netherlands

After recalling general knowledge about nuclear reactions and stellar evolution, we highlight aspects of stellar nucleosynthesis and the underlying physics of stellar evolution where progress has been achieved during the last years. In §2, we discuss the bulk nucleosynthesis in massive stars, especially of oxygen which is the most prominent massive star tracer, before we outline effects of rotation in those stars. §3 describes some recent developments in the field of s-process nucleosynthesis, §4 deals with the relevance of close binary systems for nucleosynthesis, and §5 is concerned with the most massive stars.

1. Introduction

We know 290 stable isotopes. With the exception of the nine lightest ones, they are all synthesised in the deep interior of stars. In order to study the evolutionary history of the abundance of all these nuclei, it is most efficient to group them such that the formation of the isotopes in each group can be understood through the same process. Following the legendary approach of Burbidge *et al.* (1957), one can break down the nucleosynthesis into half a dozen processes, which can be split further considering more details, but which leave only very few nuclei unexplained. While in what follows we will connect nucleosynthesis processes with evolutionary stages of stars, it is worth pointing out that Burbidge *et al.* were able to draw many of their conclusions just from the solar system abundance distribution (cf. Anders & Grevesse 1989; and Table 1) and from nuclear physics considerations.

Before we start approaching the nucleosynthesis problem from the astrophysical side, let us briefly sketch the basic picture as we have it today. The nucleosynthesis during the Big Bang was constrained to the production of the four stable isotopes of hydrogen and helium, and to ^{7}Li (see Steigman, this volume). Since then, the abundances of these isotopes have not changed much, except for ^{7}Li which apparently was enriched by about one order of magnitude due to stellar production (Casuso & Beckman 2000).

The remaining isotopes lighter than carbon, i.e., ^{6}Li, ^{9}Be, and ^{10,11}B, are practically neither produced during the Big Bang nor in stars; they are so fragile that they are even quickly destroyed in the stellar interior (with the exception of ^{11}B, some of which may be produced in supernovae by neutrino disintegration of ^{12}C; cf. Woosley & Weaver 1995). These isotopes would barely exist (actually, they do barely exist, as their abundance is five orders of magnitude below that of e.g. the CNO nuclei) were they not produced by cosmic ray spallation of CNO nuclei (Reeves *et al.* 1970, Meneguzzi *et al.* 1971). For the current discussion on the primary versus secondary nature of these light elements see Fields & Olive (1999).

The isotopes in the range of carbon to iron group nuclei are synthesised through charged particle reactions in stars, with some slight modifications due to neutron captures. Low and intermediate mass stars are producing the major part of ^{12}C, ^{13}C, and ^{14}N — and perhaps of ^{19}F, Mowlavi *et al.* (1998) — (cf. van den Hoek & Groenewegen 1997, Marigo 2001), with all the other isotopes made in massive stars, during hydrostatic evolutionary phases (roughly up to calcium) or explosive burning during the supernova explosion (Woosley & Weaver 1995, Thielemann *et al.* 1996, Rauscher *et al.* 2002).

The isotopes beyond the iron group have too high charge numbers to be formed through

TABLE 1. The 25 most abundant nuclei, and the processes which formed them. Here, NSE refers to *nuclear statistical equilibrium* (cf. Clayton, 1968) which occurs in explosive burning at high temperature, producing iron group nuclei.

Rank	Z	Element	A	mass fraction $\cdot 10^6$	Origin
1	1	H	1	705700	Big Bang
2	2	He	4	275200	Big Bang, H-burn.
3	8	O	16	9592	He-burning
4	6	C	12	3032	He-burning
5	10	Ne	20	1548	C-burning
6	26	Fe	56	1169	NSE
7	7	N	14	1105	CNO-cycle
8	14	Si	28	653	O-burning
9	12	Mg	24	513	C-burning
10	16	S	32	396	O-burning
11	10	Ne	22	207	He-burning
12	12	Mg	26	79	s-process
13	18	Ar	36	77	O-, Si-burning
14	26	Fe	54	72	NSE, Si-burning
15	12	Mg	25	69	C-burning
16	20	Ca	40	60	Si-burning
17	13	Al	27	58	C-burning
18	28	Ni	58	49	NSE, Si-burning
19	6	C	13	37	CNO-cycle
20	2	He	3	35	Big Bang, pp-chain
21	14	Si	29	34	C-, Ne-burning
22	11	Na	23	33	C-burning
23	26	Fe	57	28	NSE, s-process
24	14	Si	30	23	s-process
25	1	H	2	23	Big Bang

charged particle reactions: the Coulomb barrier (cf. §1.1) requires so high fusion temperatures that the inevitably present photons would become energetic enough to quickly destroy all nuclei into protons and neutrons. These isotopes fall into three categories, the s-, r-, and p-process nuclei. The p-process nuclei are rare (10 ... 100 times rarer than s- or r-nuclei of similar mass) proton-rich nuclei which can *not* be formed from iron group nuclei through neutron captures. The s- and r-nuclei can be formed by a "slow" and a "rapid" neutron capture process, respectively, as postulated by Burbidge *et al.* (1957). While the s-process has two distinct components, the lighter s-nuclei in the range $56 < A \leq 90$ (the so called weak component) from massive stars, and the heavier s-nuclei ($A > 90$; the so called main component) formed in low and intermediate mass stars, the astrophysical sites and number of components in the r- and p-process are still debated and will not be discussed here (cf. Cowan *et al.* 1985, Rayet *et al.* 1995, Hoffman *et al.* 1996, Qian 2000, Wasserburg & Qian 2000, Goriely *et al.* 2002).

In the following, after recalling the most basic knowledge about nuclear reactions and stellar evolution, we highlight aspects of stellar nucleosynthesis and the underlying physics of stellar evolution where progress has been achieved during the last years. In §2, we discuss the bulk nucleosynthesis in massive stars, especially of oxygen which is *the* massive star tracer, before we outline effects of rotation in those stars. §3 describes recent developments in the field of s-process nucleosynthesis, §4 deals with the relevance of close binary systems for nucleosynthesis, and §5 concerns the most massive stars.

1.1. *Nuclear Reactions*

Most stars are powered by thermonuclear reactions, a consequence of which is stellar nucleosynthesis. Let us briefly recall a few basic facts.

An atomic nucleus with mass number A and charge number Z has a binding energy $E_B(A_Z)$ which is defined as follows.

$$E_B(A_Z) := [(A - Z)m_n + Zm_p - M(A_Z)] \cdot c^2 \qquad (1.1)$$

Here, m_n and m_p are the neutron and proton masses, respectively. I.e., $E_B(A_Z)$ evaluates the difference in mass between the nucleus and the sum of the masses of the isolated nuclei which make it up. This mass difference multiplied with the square of the speed of light c gives the energy which is released by putting the isolated nuclei together in one nucleus, according to Einstein's $E = mc^2$. The quantity

$$\Delta M = (A - Z)m_n + Zm_p - M(A_Z) \qquad (1.2)$$

is called the mass defect. Note that neither the mass nor the binding energy of the electrons is considered here.

For a thermonuclear reaction involving four kinds of particles, a target nucleus 1, a projectile 2, a synthesised nucleus 4, and an ejected light particle 3, written as $1(2,3)4$ (or $1 + 2 \rightarrow 3 + 4$), the Q-value is the amount of energy released per reaction, defined as $Q = (M_1 + M_2 - M_3 - M_4)c^2$. For $Q > 0$ the reaction is called exothermic, for $Q < 0$ endothermic.

A nuclear reaction *rate* r_{xy} between two kinds of particles is defined as

$$r_{xy} := N_x N_y v \sigma(v) \qquad (1.3)$$

where N_x and N_y are the number densities of both kinds of particles, v is their relative velocity (particles meet each other more often when the velocity is larger), and $\sigma(v)$ is the cross section of the reaction, which can be a complicated function of the relative velocity or particle energy.

As the velocity distribution for the particles in a star during a nuclear fusion stage is Maxwellian, the evaluation of the reaction rate needs to take this into account:

$$r_{xy} = N_x N_y < \sigma v >, \qquad (1.4)$$

with

$$< \sigma v > := \int_0^\infty \Phi(v) v \sigma(v) dv \qquad (1.5)$$

and the Maxwell-Boltzmann distribution

$$\Phi(v) = 4\pi v^2 \left(\frac{m}{2\pi kT}\right)^{3/2} \exp\left(-\frac{mv^2}{2kT}\right). \qquad (1.6)$$

Here, $m = \frac{m_x m_y}{m_x + m_y}$ is the so called reduced particle mass. With $E = \frac{1}{2}mv^2$ it follows

$$< \sigma v > = \left(\frac{8}{\pi m}\right)^{\frac{1}{2}} \frac{1}{(kT)^{\frac{3}{2}}} \int_0^\infty \sigma(E) E \exp\left(-\frac{E}{kT}\right) dE. \qquad (1.7)$$

It is now a rather simple exercise to estimate the threshold temperature required to obtain a certain charged particle reaction within classical physics: the projectile energy needs to be larger than the repulsive Coulomb potential between the two nuclei $E_C(r) = \frac{Z_x Z_y e^2}{r}$. It turns out that in order to have the proton+proton reaction occurring at a

TABLE 2. Exponent of the temperature dependence of the Maxwellian averaged nuclear cross section (middle column) for four different nuclear reactions, evaluated at a temperature of $1.5 \cdot 10^7$ K. The last column gives the Coulomb threshold energy.

$T = 1.5 \cdot 10^7 K$	$d \log < \sigma v > /dT$	E_c
$p + p$	3.9	0.55 MeV
$p + {}^{14}N$	20	2.27 MeV
$\alpha + {}^{12}C$	42	3.43 MeV
${}^{16}O + {}^{16}O$	182	14.07 MeV

rate comparable to that inside the Sun requires a temperature of $T \simeq 6 \cdot 10^9$ K. How can the Sun fuse hydrogen at only $1.5 \cdot 10^7$ K?

It needs quantum mechanics, and in particular the *tunnel effect* as shown by Gamow around 1928, to understand the thermonuclear energy production in the Sun and in stars. Particles with an energy below the critical Coulomb energy may still penetrate the Coulomb "wall" with a finite tunnelling probability $P = \exp(-2\pi\eta)$, with $\eta = \frac{Z_x Z_y e^2}{\hbar v}$.

As the tunnelling probability is an increasing function of the particle energy, and the Maxwell-Boltzmann distribution function is a decreasing function of particle energy, the convolution of both functions has a maximum, the so called Gamow-peak. The Gamow-peak occurs at an energy E_0 *above* the thermal energy kT, at

$$E_0 = \left(\frac{bkT}{2} \right)^{2/3} = 1.22 \left(Z_x Z_y m T_6^2 \right)^{1/3} \text{KeV} , \tag{1.8}$$

where $T_6 = T/10^6$ K, and $b := (2m)^{1/2} \pi e^2 Z_x Z_y / \hbar$.

A consequence of the steepness of both functions defining the Gamow-peak is an extreme sensitivity of the Maxwellian-averaged cross section to temperature:

$$\boxed{< \sigma\nu > \sim T^{\frac{E_0}{kT} - \frac{2}{3}}} . \tag{1.9}$$

Table 2 gives numerical examples which show that the temperature sensitivity increases drastically for temperatures which imply larger ratios of thermal energy versus Coulomb threshold energy.

A consequence of Eq.(1.9) is that a fixed temperature can be assigned to each thermonuclear fusion stage, independent of the stellar properties (cf. Table 3). For example, all helium burning stars perform helium burning at $T \simeq 2 \cdot 10^8$ K. Were the temperature slightly lower, the burning would stop. Were it slightly larger, would the star explode.

1.2. *Stellar Evolution*

The energy loss of most stars is compensated by nuclear fusion, but their evolution is determined by the virial theorem. Consider a gaseous sphere of mass M in one dimensional spherical coordinates, with the spatial coordinate r being zero at the center of our sphere and having the value R at its edge. The Lagrangian mass coordinate is defined as $M_r := \int_0^r 4\pi r'^2 \rho dr'$. We also define a volume element as $dV = \frac{1}{\rho} dM_r$. With density ρ, pressure P, and internal energy density u, we have for the potential and thermal energy of our sphere:

$$E_{\text{pot}} = - \int_0^M \frac{GM_r}{r} dM_r \tag{1.10}$$

and

$$E_{\text{th}} = \int_0^M u \, dM_r = \int_0^M c \frac{P}{\rho} dM_r \,, \tag{1.11}$$

with $c = \frac{3}{2}$ for an ideal gas. For hydrostatic equilibrium

$$\frac{dP}{dM_r} = -\frac{GM_r}{4\pi r^4} \,, \tag{1.12}$$

multiplication of both sides with $4\pi r^3$ and integration over the whole mass of the sphere yields for the left hand side

$$\int_0^M 4\pi r^3 \frac{dP}{dM_r} dM_r = \left[4\pi r^3 p \right]_0^M - \int_0^M 12\pi r^2 \frac{\partial r}{\partial M_r} P \, dM_r$$

$$= -\int_0^M \frac{3P}{\rho} dM_r \quad = -\frac{3}{c} \int_0^M u \, dM_r = -\frac{3}{c} E_{\text{th}} \,, \tag{1.13}$$

while the right hand side becomes

$$-\int_0^M 4\pi r^3 \frac{GM_r}{4\pi r^2} dM_r = -\int_0^M \frac{GM_r}{r} dM_r = E_{\text{pot}} \,. \tag{1.14}$$

Thus, we obtained the virial theorem: $-E_{\text{pot}} = \frac{3}{c} E_{\text{th}}$, where $c = u\rho/P$ is a constant of order 1.

It is a consequence of the virial theorem that a contracting star obtains a higher pressure, and a higher internal energy. We shall see that the generalisation that it also obtains a higher temperature is *not* true. Instead, both situations can occur, a heating or a cooling of the star, depending on the equation of state which holds in the stellar interior.

Let us assume that all physical quantities in our star can be well characterised by their spatial mean values (indicated by bars over the variables), i.e., that our star is not structured by nuclear shell sources. For such more complicated objects, the following considerations hold nevertheless for their cores. We approximate the potential energy of our star as

$$E_{\text{pot}} = -\int_0^M \frac{GM_r}{r} dM_r \sim \frac{G}{\bar{r}} M^2 \,, \tag{1.15}$$

and the thermal energy as

$$E_{\text{th}} = \int_0^M c \frac{P}{\rho} dM_r = \int_0^M cP \cdot dV \sim \bar{P} \cdot V \sim \bar{P} \, \bar{r}^3 \,. \tag{1.16}$$

Then it follows from the virial theorem that

$$\frac{M^2}{\bar{r}} \sim \bar{P} \, \bar{r}^3 \tag{1.17}$$

or

$$\bar{P} \sim M^2 \bar{r}^{-4} \sim M^2 \left(\sqrt[3]{\frac{M}{\bar{\rho}}} \right)^{-4} = \bar{\rho}^{4/3} M^{2/3} \,, \tag{1.18}$$

i.e.,

$$\boxed{\bar{P} \sim \bar{\rho}^{\frac{4}{3}} M^{2/3}} \,. \tag{1.19}$$

The proportionality expressed as Eq. (1.19) is very meaningful. It predicts that the evolutionary tracks of stars in the $\log P - \log \rho$-diagram are extremely simple: they are straight lines with a slope of $4/3$, as long as their mass is constant. It should be stressed

TABLE 3. The main thermonuclear burning stages of massive stars.

	main fuel	main ashes	time scale yr	T 10^6 K	mass limit $M_\odot$	remarks
H-burning	^{1}H	^{4}He	10^7 yr	30	0.08	
He-burning	^{4}He	^{12}C, ^{16}O	10^6 yr	200	0.4	
C-burning	^{12}C	^{16}O, ^{20}Ne	500	700	1.0	ν-cooling
Ne-burning	^{20}Ne	^{16}O, ^{24}Mg	1	1500	1.3	ν-cooling
O-burning	^{16}O	^{28}Si, ^{32}S	0.3	2000	1.35	ν-cooling
Si-burning	^{28}Si	^{52}Cr, ^{56}Fe	0.01	3500	1.35	NSE possible
NSE	any	iron group	10^{-7}	5000	–	explosive

that this result is independent of the equation of state which happens to be valid in the stellar interior.

By referring to an equation of state, it is possible to conclude on the temperature evolution of stars. It is only necessary to consider the slopes of isothermals for the various equations of state in the $\log P - \log \rho$-diagram. Let us do this for the equations of state as they are encountered with increasing density. I.e., when a star is born, its density is relatively low, and perhaps radiation pressure is important in its interior. I.e., $P = \frac{a}{3}T^4$ (a being the radiation constant), and isothermals in the $\log P - \log \rho$-diagram have the slope 0. I.e., contracting stars cross the isothermals "from below", which means the temperature increases. For higher density, the ideal gas pressure $P = \frac{\Re}{\mu}\rho T$ (with mean molecular weight μ) may become dominant. Then, isothermals have the slope 1, which is still less than the slope of the evolutionary tracks 4/3. I.e., still the isothermals are crossed "from below" by contracting stars, which again implies an increasing temperature.

Let us point out, at this stage, that real stars of course stop on their evolutionary tracks in the $\log P - \log \rho$-diagram for long time when they burn a nuclear fuel in their core. However, every fuel extincts at one point, and then the contraction of the core goes on! It can only be stopped when the core becomes unstable, i.e. collapses or explodes — note that the virial theorem is not valid any more in this cases —, or when it hits Pauli's exclusion principle, i.e. when the core is extremely degenerate.

Degeneracy of the stellar core implies: the density has grown so much that in order to avoid three fermions (here: electrons; only two with opposite spin are allowed) in the same phase space cell, we must have fermion momenta exceeding those given by the Maxwell-Boltzmann distribution. Remember that the phase space is constructed from the three spatial and the three momentum coordinates. I.e., for a high matter density ρ, a high (Pauli-forbidden) density in the phase space can be avoided by large particle momenta. For a hot gas, this may not be a problem. But for a "cold" gas, it means that it must be degenerate. A gas is called completely degenerate when all energy states are filled up from below. I.e., no single electron can move to a lower energy state. For a degenerate gas, the pressure is dominated by the degeneracy pressure of the electrons, the pressure from the Maxwellian velocities of the ions can be neglected.

Getting back to our contracting star in the ideal gas regime, it may well become degenerate during its further evolution. Whether or not this happens can — again — be seen from Eq. (1.19): Combined with the ideal gas law, $P \sim \rho T$, we obtain

$$\rho \sim T^3/M^2 \, . \tag{1.20}$$

I.e., for a given temperature — or a given thermonuclear burning stage (cf. Section 1.2) — the density is higher in a star/core of lower mass. In other words, if we consider a certain burning stage — as an example: helium burning, which has an associated temperature of $2 \cdot 10^8$ K — there is a critical mass M_{He} (here: of the helium core) such that objects with higher mass reach this stage non-degenerate, while objects with lower mass are significantly degenerate at this point.

Let us assume our star has a low enough mass such that it would indeed become degenerate in its core. From the Fermi-Dirac statistic, it follows that the equation of state of degenerate gas is $P = k_1 \rho^{5/3}$ in case of non-relativistic electron velocities, and $P = k_2 \rho^{4/3}$ for a relativistic degenerate electron gas, where k_1 and k_2 are fundamental constants. For the non-relativistic case, the slope of an isothermal line in the $\log P - \log \rho$-diagram is obviously 5/3, which is *larger* than the slope of the evolutionary tracks. Note that the value of 5/3 is strictly speaking only obtained for vanishing temperature. However, clearly the isothermals of a considerably degenerate, non-relativistic electron gas will be larger than 4/3, which has the consequence that our contracting star does not become hotter any more, but rather cools. Obviously, a critical mass can be assigned to each of the six thermonuclear burning stages (Table 3) such that for a stellar core above that mass the burning stage is reached, but for cores with lower masses the ignition temperature of that burning stage is not reached.

I.e., for the example above, helium cores just below the critical mass M_{He} start cooling just before they are able to ignite helium. During their contraction, they reach a maximum temperature which is just below the helium ignition temperature of $2 \cdot 10^8$ K. More general, stars which evolve into the "trap" of non-relativistic degeneracy are at the end of their evolution. They will not be able to produce any thermonuclear burning, as this would require an increasing temperature. Instead, they continue to cool towards $T = 0$ and obtain a density determined by Pauli's exclusion principle.

The critical core mass required to reach *all* thermonuclear burning stages is called the Chandrasekhar mass. Its value is about $1.4\,M_\odot$, but it depends on the mean molecular weight per free electron μ_e — reflecting the balance between pressure and gravity:

$$M_{Ch} = \frac{5.836}{\mu_e^2}\,M_\odot. \tag{1.21}$$

While $\mu_e \simeq 2$ for hydrogen free gas in most cases, electron capture at high densities can lead to significantly larger values.

We have thus arrived at the fundamental result that the virial theorem together with the equations of state divides the multitude of stars into two classes: those which contract and heat "forever", and those which cool to "zero" temperature and to a final finite density. Real stars which evolve according to the cooling branch are the white dwarfs, and — those with masses below M_H, which never even ignite hydrogen — the brown dwarfs. And real stars which keep heating up evolve through six discrete nuclear burning stages, where each stage burns a fuel which is left as ashes from the foregoing stage (Table 3). When a fuel in the core is extinct, the corresponding burning continues in a spherical shell around the the the core, which gives rise to the so called onion-skin structure of evolved massive stars (cf. Fig. 1). The result of silicon burning are iron group nuclei, which are the most tightly bound nuclei of all. Any nuclear reactions involving them is thus not capable of producing energy. Therefore, silicon burning is the last thermonuclear burning stage.

However, also after core silicon exhaustion, the stellar core continues to contract. Finally, temperatures are reached where all nuclei are photo-disintegrated into protons

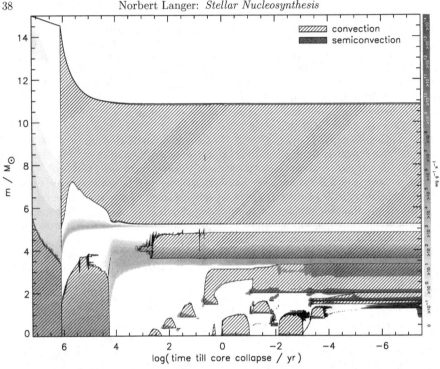

FIGURE 1. So called Kippenhahn-diagram, describing the evolution of the internal structure of a star with an initial mass of $15\,M_\odot$, from core hydrogen ignition until the onset of iron core collapse (cf. Heger *et al.* 2000). The X-axis gives the logarithm of the time (in years) with t=0 defined at the onset of iron core collapse. The Y-axis is a spatial coordinate; however, instead of a radius coordinate, the Lagrangian mass coordinate is used. Hatched areas are convectively unstable, and the convective cores of hydrogen, helium, carbon, neon, oxygen and silicon burning can be identified, successively. Gray shading denotes nuclear energy generation. After core burning, all burning stages develop shell sources, which give rise to the "onion skin model".

and neutrons, which is so energy consuming that the stellar core is destabilised and collapses, forming a neutron star or a black hole, and perhaps a supernova.

2. Nucleosynthesis in Massive Stars

2.1. *Yields from massive star models*

A folding of the stellar metal yields and the stellar initial mass function reveals that stars with initial masses in the range 15 ... $25\,M_\odot$ are responsible for most of the metals in the solar systems and the universe. The yields obtained from hydrostatic and explosive evolutionary models of solar metallicity massive stars on the computer show the very same abundance pattern as found in the solar system for nuclei between oxygen and the iron group. I.e., despite the absolute abundances of isotopes in the mentioned range scatter by several orders of magnitude, a division by the respective solar system abundance results in nearly the same number for all isotopes, with a scatter of only a factor of ~ 2 (Rauscher *et al.* 2002, and references therein). More precisely, an average massive

TABLE 4. Nucleosynthesis results for 15 massive star models computed up to silicon ignition (Langer & Henkel 1995). The symbols have the following meanings: M_i is the initial stellar mass, and Z the metallicity. α_{sc} is the semiconvective mixing parameter, with $\alpha_{sc} = 0$ corresponding to the Ledoux criterion, $\alpha_{sc} = \infty$ to the Schwarzschild criterion for convection. M_f is the final stellar mass, M_{CO} the final CO-core mass and M_{rem} is the assumed remnant mass. M_C and M_O are the total mass of carbon and oxygen ejected by stellar wind mass loss and by the supernova explosion (initially present amounts are *not* subtracted). The values f_{13} ... f_{18} designate production factors for ^{13}C, ^{14}N, ^{17}O, and ^{18}O, and $\Delta Y/\Delta Z$ is the ratio of the *net* yields of helium to metals.

M_i $M_\odot$	Z %	α_{sc}	M_f $M_\odot$	M_{CO} $M_\odot$	M_{rem} $M_\odot$	M_C $M_\odot$	M_O $M_\odot$	f_{13}	f_{14}	f_{17}	f_{18}	$\Delta Y/\Delta Z$
15	2	0.04	14.7	1.8	1.38	0.11	0.34	3.0	3.3	17	32	2.9
20	2	0.04	16.5	2.4	1.45	0.45	0.89	4.2	4.0	12	36	1.5
25	2	0.04	18.3	2.7	1.49	0.93	1.4	2.5	3.5	12	13	1.2
30	2	0.04	18.3	3.7	1.60	1.1	2.1	2.4	3.4	12	22	0.9
40	2	0.04	9.3	4.6	1.69	1.8	3.2	2.0	5.1	11	3.1	1.2
50	2	0.04	6.3	4.8	1.72	3.1	3.5	1.6	5.8	12	5.6	1.3
20	2	∞	15.9	4.2	1.65	0.31	2.3	3.3	4.0	18	2.5	0.3
25	2	∞	13.2	5.2	1.76	0.55	3.5	1.8	3.4	15	0.4	0.1
30	2	∞	13.6	8.4	2.10	1.1	4.8	1.4	3.8	17	0.4	0.1
20	2	∞	15.9	4.2	1.65	0.31	2.3	3.3	4.0	18	2.5	0.3
20	2	0.04	16.5	2.4	1.45	0.45	0.89	4.2	4.0	12	36	1.5
20	2	0.01	16.5	2.3	1.43	0.43	0.70	4.2	3.9	13	53	1.8
20	2	0	16.0	2.2	1.43	0.38	0.68	4.1	3.6	13	57	2.1
20	0.2	0.04	19.1	2.2	1.43	0.43	0.75	2.3	4.3	7.1	40	1.8
25	0.2	0.04	23.3	2.9	1.51	0.78	1.1	2.3	5.3	5.9	21	1.2
40	0.2	0.04	36.8	6.9	1.96	1.4	4.1	2.0	4.2	3.6	9.8	0.7
15	4	0.04	13.0	2.8	1.50	0.90	1.8	2.6	4.3	23	3.7	0.9

star which starts out with a solar abundance distribution on the zero age main sequence ejects, at the end of its life, about 10 times more of each isotope in the considered range than it contained initially. As this concerns some 40 isotopes, this is clearly a success story, as the number of parameters in the underlying stellar and nuclear physics is clearly much less. This is reflected by the general agreement of detailed yields by various groups (cf. Woosley & Weaver 1995, Thielemann et al. 1996, Hoffman et al. 1999, Limongi et al. 2000, Rauscher et al. 2002, Heger et al. 2002)

However, let us point out that in the vast majority of massive star models which are used as a base for nucleosynthesis studies, neither the role of convection nor that of mass loss is investigated. The importance of these issues, at various metallicities, has been analysed by Langer & Henkel (1995), who computed models which are still the most advanced ones at high mass — i.e., such that Wolf-Rayet stages are achieved for the largest considered initial masses — which include effects of stellar wind mass loss.

The dependence of the nucleosynthesis yields of massive stars mass loss, convection and metallicity is summarised in Table 4. Let us emphasise here the effect of mass loss to *increase* the carbon yield at the expense of oxygen, due to the strong stellar wind in the WC stage (cf. also Maeder 1992, Prantzos et al. 1994). The strength of this effect may determine the contribution of massive stars to the total amount of carbon in our Galaxy.

The dependence of the oxygen yield from the assumed convection criterion is even more dramatic and will be discussed in more detail in the next section.

2.2. Oxygen isotopes from massive stars

Roughly half of the mass of all elements heavier than helium in the Galaxy consists of oxygen, which is thus the dominant metal. Most of the oxygen constitutes of the isotope ^{16}O. The solar system isotopic oxygen ratios are $(^{16}O : ^{17}O : ^{18}O) = (2500 : 5.6 : 1)$; cf. Anders & Grevesse (1989). ^{16}O, obtained as a primary isotope by $\alpha(2\alpha)^{12}C(\alpha, \gamma)^{16}O$ during helium burning, is predominantly produced in massive stars — i.e. stars massive enough to evolve to core collapse (cf. Langer & Woosley 1996). They expel most of their C/O-core in a supernova explosion, while it remains locked in the white dwarf remnant left behind by less massive stars (cf. e.g., Maeder 1992). Note that observational evidence for both, the primary production and the high mass origin of ^{16}O, comes from the high [O/Fe] ratios observed in very metal poor stars (e.g., Gratton & Ortolani 1986).

Both stable neutron-rich oxygen isotopes, ^{17}O and ^{18}O, are thought to be of secondary origin, i.e. their production requires heavy elements to be present in the initial composition of the star. ^{17}O is produced during hydrogen burning (cf. Clayton 1968) while ^{18}O is produced by α-capture on ^{14}N. As in the Galaxy today, and more so at earlier epochs, ^{16}O is the dominant metal in the interstellar medium, ^{17}O and ^{14}N are mainly produced from ^{16}O via the CNO cycle, and since ^{18}O is formed from ^{14}N, ^{16}O is the primary seed for both, ^{17}O and ^{18}O. A major difference between these isotopes comes from the fact that ^{18}O is produced by α-capture *and* destroyed by α-capture, i.e. it can reside only in a very narrow temperature range in the stellar interior, while ^{17}O is destroyed by α-capture at similar temperatures as ^{18}O, but it is produced at hydrogen burning temperatures.

The production of both isotopes requires the operation of the CNO cycle, which occurs in stars more massive than $\sim 1.2\,M_\odot$. Therefore, it is presumed that stars with initial masses in the range $1.2\,M_\odot \lesssim M_{ZAMS} \lesssim 10\,M_\odot$ have contributed to the Galactic ^{17}O and ^{18}O abundances (Henkel *et al.* 1994, Prantzos *et al.* 1996). In the case of ^{17}O, this is well supported by stellar models, since ^{17}O is dredged up into the H-rich stellar envelope on the red giant branch (e.g., El Eid 1994), which is subsequently lost due to stellar winds. The contribution of the intermediate-mass stars to the ^{18}O enrichment is much less clear, since ^{18}O lives between the H- and He-burning nuclear shell sources, and its survival depends critically on the still poorly understood mixing events between these shell sources during the so called thermal pulses (e.g., Vassiliadis & Wood 1993). Intermediate mass stars may even effectively destroy rather than produce ^{18}O (cf. Prantzos *et al.* 1996). Therefore, it is clear that intermediate mass and massive stars contribute significantly to the ^{17}O production, while the situation is unclear for ^{18}O.

The temporal evolution of the three stable oxygen isotopes has been considered in models of the Galactic chemical evolution (Timmes *et al.* 1995, Prantzos *et al.* 1996), which requires the knowledge of the isotopic yields as function of the stellar mass and metallicity. In the following, we investigate the production of the oxygen isotopes in massive stars. As in intermediate mass stars, the production of all oxygen isotopes is subject to uncertainties. However, in contrast to the case of intermediate mass stars, which evolve through the complicated thermal pulse phase, we can quantitatively investigate the uncertainties in the case of massive stars.

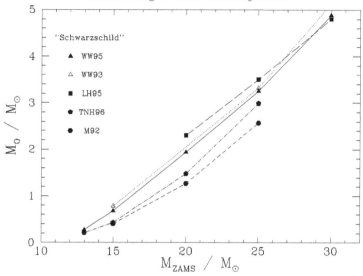

FIGURE 2. Oxygen yield M_O in solar masses as function of the initial stellar mass M_{ZAMS} according to five different sets of stellar models (WW93 = Weaver & Woosley 1993, WW95 = Woosley & Weaver 1995; LH95 = Langer & Henkel 1995; TNH96 = Thielemann *et al.* 1996; and M92 = Maeder 1992) for which the oxygen yield is unaffected by stellar wind mass loss ($M_{ZAMS} \lesssim 30\,M_\odot$) and in which the effect of mean molecular weight gradients on convection is ignored (Schwarzschild criterion for convection) or assumed to be weak.

2.2.1. *The ^{16}O yields in massive star models*

The amount of ^{16}O produced by a massive star depends strongly on **a)** the mass of the C/O-core at the time of iron core collapse and **b)** the C/O-ratio established in the C/O-core at core-He exhaustion. Both factors can be influenced in various ways. While the C/O-core mass is a strong function of the He-core mass and of the convection model, the C/O-ratio of core-He burning is sensitive to the $^{12}C(\alpha,\gamma)^{16}O$ nuclear reaction rate and — again — the employed convection model. While we postpone the discussion of the dependence of the ^{16}O yield on stellar wind mass loss to later, let us in the following compare the results of models where mass loss is unimportant.

Fig. 2 shows the ^{16}O yields M_O obtained by various authors for solar metallicity stars in the initial mass range $13\,M_\odot \le M_{ZAMS} \le 30\,M_\odot$ computed with the Schwarzschild criterion for convection (or similar assumptions). While all models shown agree in the general trend of an almost linear increase of M_O with increasing initial mass, there is a considerable scatter for a given value of M_{ZAMS}. E.g., at $M_{ZAMS} = 20\,M_\odot$ — around which the product of metal mass output times IMF statistical weight is largest — the discrepancy in the oxygen yields of different authors is roughly $1\,M_\odot$, with values ranging from $1.27\,M_\odot$ (Maeder 1992) to $2.3\,M_\odot$ (Langer & Henkel 1995). A similar scatter is found at $M_{ZAMS} = 25\,M_\odot$.

An understanding of the differences of the oxygen yields of different authors is not easily obtained. E.g., one might expect that calculations using a faster $^{12}C(\alpha,\gamma)^{16}O$-rate would obtain a larger oxygen mass; however, Fig. 2 shows that the opposite is the case. While Maeder (1992) and Thielemann *et al.* (1996) used the Caughlan *et al.* (1985) rate which is roughly 1.4 times faster than the rate used by Woosley & Weaver (1995) and Langer & Henkel (1995), they obtained much less oxygen. It is even more surprising that

Maeder (1992), who applied "convective core overshooting" and thus arrives at larger core masses, obtains the smallest oxygen yields.

While the origin of the latter discrepancy could not be resolved on the basis of the published data, the small oxygen yields of Thielemann *et al.* (1996) may be related to the fact that they only compute helium cores of a fixed mass M_{He} rather than complete stars. This appears to have two major drawbacks: firstly, they have to adopt a $M_{He}(M_{ZAMS})$-relation; for a given ZAMS mass their He-core mass appears to be somewhat smaller than the final He-core masses of the models presented in the other papers discussed here. E.g., at $M_{ZAMS} = 25\,M_\odot$: $M_{He}(LH95) = 8.4\,M_\odot$, $M_{He}(WW95) = 9.1\,M_\odot$, and $M_{He}(M92) = 9.7\,M_\odot$, while Thielemann *et al.* (1996) relate a $8\,M_\odot$ helium core with a $25\,M_\odot$ star. Secondly, in a realistic model the He-core mass is not constant but grows with time by substantial amounts. In a $20\,M_\odot$ star, e.g., it grows from $4\,M_\odot$ to more than $6\,M_\odot$, i.e. by more than 50% (e.g., Langer 1991a). This leads to different trajectories in the $\rho_c - T_c$-diagram during He-burning and to different growth rate of the He-burning convective core and consequently to differences in the rate of mixing of α-particles into the He-burning region.

Finally, the oxygen yield is also depending on assumptions on the mass of the compact remnant which remains bound after the supernova explosion. While Woosley & Weaver (1995) and Thielemann *et al.* (1996) evolve their models until iron core collapse and through the supernova explosion and can thus derive a remnant mass best, Maeder (1992) and Langer & Henkel (1995) stop their calculations after carbon and oxygen burning, respectively, and apply a remnant mass C/O-core mass relation. At $M_{ZAMS} = 25\,M_\odot$, the small oxygen yield of Maeder (1992) may be partly related to the large adopted remnant mass in this model — $2.72\,M_\odot$, compared to values between 1.8 and $2.1\,M_\odot$ in the other three cases — (this explanation fails, however, at $M_{ZAMS} = 20\,M_\odot$), while those of Langer & Henkel (1995) are generally somewhat smaller than those of Woosley & Weaver (1995).

The comparison of the ^{16}O yields of models of different authors in Fig. 2 reveals already discrepancies of the order of a factor of 2. Fig. 3 illustrates an uncertainty of the oxygen yield due to a physical effect which has not been considered in the models shown in Fig. 2: in calculations which apply the Schwarzschild criterion for convection the effect of mean molecular weight barriers to slow down convective mixing through these barriers is ignored. On the other extreme, such mixing is practically prevented in case the Ledoux criterion for convection is used. Semiconvection models allow a mixing through these barriers on a certain time scale which, in the mixing scheme of Langer *et al.* (1983), is inversely proportional to an (uncertain) efficiency coefficient $\alpha_{semiconv}$. Fig. 3 shows that the ^{16}O yield is strongly depending on this parameter. In the case of a solar metallicity $20\,M_\odot$ star, the ejected ^{16}O mass varies from $0.68\,M_\odot$ to $2.3\,M_\odot$, i.e. by more than a factor of 3. This effect and its order of magnitude are confirmed in the calculations of Weaver & Woosley (1993). They find the oxygen yields to differ by factors 3.7 and 1.8 in calculations at 15 and $25\,M_\odot$, respectively (cf. also Fig. 4), depending on the efficiency of semiconvective mixing.

We conclude that, despite the ^{16}O yield of massive stars seems — at first glance — to be easily evaluated since it basically depends on the C/O-core mass, it may be as uncertain as by a factor of 3 in the stellar mass range which is most relevant for the ^{16}O budget of galaxies. Next, we show that for $M_{ZAMS} \gtrsim 30\,M_\odot$, the situation becomes even more complex due to effects of stellar mass loss.

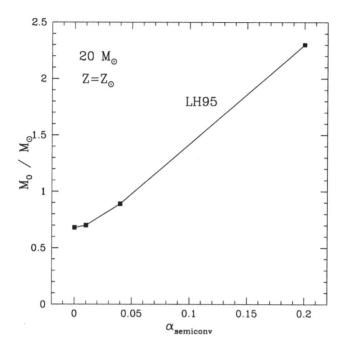

FIGURE 3. Oxygen yield of four solar metallicity $20\,M_\odot$ models of Langer & Henkel (1995; LH95), computed with various values of the semiconvective mixing parameter $\alpha_{\mathrm{semiconv}}$, which determines the time scale of mixing in semiconvective regions (cf. Langer *et al.* 1983). $\alpha_{\mathrm{semiconv}} = 0$ means the time scale is infinite (no mixing in semiconvective regions and Ledoux criterion for convection). The entry for the largest value of the semiconvection parameter corresponds to the Schwarzschild criterion for convection, i.e. $\alpha_{\mathrm{semiconv}} = \infty$; it is arbitrarily placed at $\alpha_{\mathrm{semiconv}} = 0.2$.

2.2.2. *The ^{16}O yield and stellar wind mass loss*

The ^{16}O yield of a massive star remains basically unaffected by stellar mass loss as long as it maintains hydrogen-rich surface layers, because the evolution of the He-burning core is essentially determined by the evolution of the He-core mass (cf. Maeder 1992, Woosley *et al.* 1993). However, stars with $M_{\mathrm{ZAMS}} \gtrsim 30\,M_\odot$ may — at $Z \simeq Z_\odot$ — lose their entire H-rich envelope and evolve into so called Wolf-Rayet (WR) stars. WR stars have such strong mass loss that they can effectively blow off a substantial part of their mass (Langer 1989b) — and thus of their He-core. Consequently, these objects eject much of the helium inside the He-core before it can be burnt into metals, and later much of the carbon which is synthesised during core helium burning before it can be processed further into oxygen. WR mass loss thus increases the helium and ^{12}C yields at the expense of ^{16}O.

This effect can be clearly seen in Fig. 4, where the oxygen yields of stars with and without mass loss are compared (separately for models with slow and fast semiconvective mixing). For the assumption of very efficient mass loss, Maeder (1992) obtained $2.1\,M_\odot$

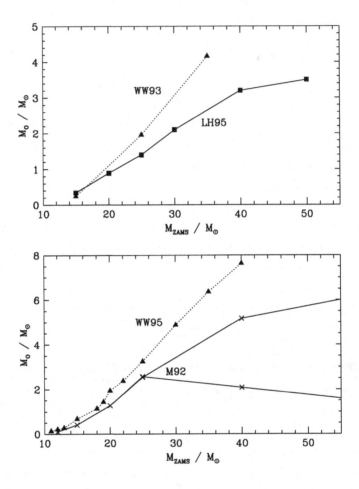

FIGURE 4. Comparison of the oxygen yield M_O as a function of the initial stellar mass M_{ZAMS} for models with and without stellar wind mass loss, and assuming a remnant mass of the order of $2\,M_\odot$. Upper panel: Models with slow semiconvective mixing. Weaver & Woosley's (1993; WW93) models are computed without mass loss, Langer & Henkel (1995; LH95) included mass loss. Lower panel: Models with fast semiconvective mixing resp. Schwarzschild criterion for convection. Woosley & Weaver (1995; WW95) neglected mass loss, Maeder (1992; M92) explored models with standard and excessive mass loss.

of oxygen out of a $40\,M_\odot$ star, while Woosley & Weaver (1995) obtain $7.6\,M_\odot$ for the case of a similar remnant mass as that in Maeder (1992).

As the stellar mass loss rates are likely to be higher at larger metallicity Z, the quenching of the ^{16}O yield due to mass loss is strongly Z-dependent. This is comprehensively shown in the models of Maeder (1992) (cf. also Meynet *et al.* 1994), and its significance in models of Galactic chemical evolution has been demonstrated by Prantzos *et al.* (1994).

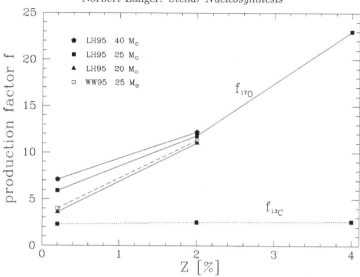

FIGURE 5. Dependence of the ^{17}O production factor (ratio of the ejected versus the initial amount of ^{17}O) on the initial stellar metallicity Z, according to models of Langer & Henkel (1995; LH95) and Woosley & Weaver (1995; WW95). The steep increase with Z is atypical for secondary isotopes. This is demonstrated by comparison with the ^{13}C production factor of the 25 M$_\odot$ LH95 models (dotted line), which shows no dependence on Z.

2.2.3. *The production of ^{17}O as function of metallicity*

As stated above, ^{17}O is produced as a secondary isotope in the CNO cycle, and its final abundance depends primarily on the initial ^{16}O abundance in the star. However, the latter statement holds only for a fixed burning temperature. That the situation of ^{17}O is more complex can be seen in Fig. 5, which displays the ^{17}O production factor for various stellar masses as function of the initial stellar metallicity Z. Note that the initial metal distribution in the models of Langer & Henkel (1995) is the solar distribution scaled according to the actual metallicity.

Since the initial amount M_{in} of a metal isotope in a star is proportional to Z, and the produced and expelled amount M_{out} of a secondary isotope is also proportional to Z, its production factor $f = M_{out}/M_{in}$ should be independent of Z. This is in fact so for many secondary isotopes as can be seen at the example of ^{13}C in Fig. 5. However, this figure shows also that — in contrast to the expectation — the ^{17}O production factor increases strongly with metallicity.

Let us compare the behaviour of ^{17}O to that expected for a primary isotope. In this case, again $M_{in} \propto Z$, but the production does not depend on the initial metal content, i.e. $M_{out} \simeq const.$, and thus $f \propto \frac{1}{Z}$. I.e., the production factor of a primary isotope decreases with increasing Z. This is just opposite to the behaviour of ^{17}O. ^{17}O is thus not behaving like a primary nor like a secondary isotope; its production factor f is roughly proportional to Z, and thus $M_{out} \propto Z^2$. I.e., ^{17}O may be called a "super-secondary" isotope.

What is special about ^{17}O, in particular in comparison to all other secondary CNO isotopes which show, as ^{13}C in Fig. 5, a clean secondary behaviour? The overproduction of secondary CNO nuclei is basically determined by the CNO equilibrium abundance distribution (cf. Clayton 1968). An inspection of the CNO equilibrium abundances of the secondary CNO isotopes in Figure 3 of Arnould & Mowlavi (1993) shows that the ^{17}O equilibrium concentration — in contrast to that of all other secondary isotopes — is extremely temperature sensitive. In particular, it is strongly decreasing for increasing temperature. Since the nuclear energy production rate in the CNO cycle is proportional to the metallicity (Clayton 1968) but the luminosity of a main sequence star is mainly fixed by its mass (Kippenhahn & Weigert 1990), the H-burning temperature in massive main sequence stars of a given mass is larger for lower metallicity. $25 M_\odot$ stars in the Z-range considered in Fig. 5 (0.2% ... 4%) vary in their central temperature at a time where the central hydrogen mass fraction is 50% in the range $(42 ... 36) \times 10^6$ K. We conclude that the decreased temperature in massive H-burning stars of larger metallicity results in a larger ^{17}O equilibrium abundance and thus in a larger ^{17}O production.

We should note that the absolute value of the ^{17}O production factor depends on the still uncertain $^{17}O+p$ nuclear cross section (cf., Aubert *et al.* 1996); however, the general trend discussed here is likely to be independent of that.

2.2.4. How to produce ^{18}O ?

The production of ^{18}O is perhaps even more complex than that of the other oxygen isotopes. First, it is always destroyed during hydrogen burning with the CNO cycle. As stated in Section 1, it is produced by α-capture on ^{14}N. However, it is also destroyed by α-capture, through $^{18}O(\alpha,\gamma)^{22}Ne$. Therefore, ^{18}O can exist only in a very narrow temperature interval in the stellar interior. Moreover, it is destroyed in all layers which are convectively coupled to the main helium burning energy source, i.e. in the He-burning convective core and in convective He-shells. This is the reason why the production of ^{18}O depends critically on the semiconvective mixing parameter $\alpha_{semiconv}$ (cf. Section 2). For larger values of this parameter, the mentioned convection zones have larger spatial extents and destroy more ^{18}O. Fig. 6 shows that this effect is extremely strong. While, in a $20 M_\odot$ star at $Z = Z_\odot$, 57 times the initial amount of ^{18}O is ejected in the SN explosion in the case the Ledoux criterion for convection is used without semiconvective mixing ($\alpha_{semiconv} = 0$), almost nothing is produced in case the Schwarzschild criterion for convection is applied ($\alpha_{semiconv} = \infty$). Note that this effect has also been found in the models of Weaver & Woosley (1993).

Perhaps, this extreme sensitivity of the ^{18}O production on the convective mixing criterion can turn out to be very useful. In case massive stars are the main producers of ^{18}O, even order of magnitude chemical evolution arguments might be used to constrain the uncertain but extremely important parameter $\alpha_{semiconv}$ in stellar models. According to present day yields of low and intermediate-mass stars (cf. Prantzos *et al.* 1996), these don't produce any ^{18}O but rather destroy it. Some of the massive star models of Langer & Henkel (1995) computed with the Schwarzschild criterion also are net destructors of ^{18}O. Thus, the consideration of hindering effects of mean molecular weight barriers on convective mixing seems to be required to understand the production of ^{18}O to the extent it is presently observed in the Galaxy.

2.2.5. Clues from oxygen

We have seen that the production of all three stable oxygen isotopes in massive stars is affected by at first glance unexpected physical effects. The ^{16}O yield can currently, for a given stellar mass and metallicity, not be predicted better than within a factor of 2 or 3. Further, it depends — despite of its primary character — on the metallicity through the Z-dependence of stellar winds. A decreasing oxygen yield with increasing metallicity should lead to a negative slope of the [O/H] vs. [Fe/H] relation (assuming here a clean primary behaviour of iron) which is on fact observed (cf. Timmes *et al.* 1995). Though

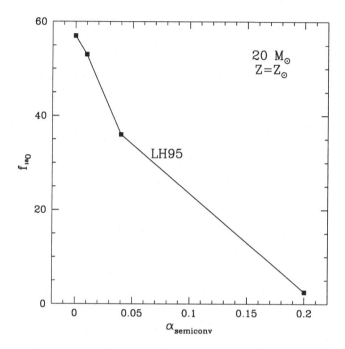

FIGURE 6. ^{18}O production factor (cf. Fig. 5) in solar metallicity $20\,M_\odot$ models of (Langer & Henkel 1995; LH95), as function of the semiconvective mixing parameter $\alpha_{\mathrm{semiconv}}$ (cf. Fig. 3; note especially that the entry for the largest value of $\alpha_{\mathrm{semiconv}}$ corresponds to $\alpha_{\mathrm{semiconv}} = \infty$).

this observation is usually explained with the onset of iron production due to Type Ia supernovae (cf. also Edmunds *et al.* 1991, for an alternative explanation), there may be an impact of the Z-dependent oxygen yields on this feature.

For ^{17}O, its component produced in massive stars behaves in a "super-secondary" way, i.e. its yield increases stronger than linear with Z, for $0.1 \lesssim Z/Z_\odot \lesssim 2$. And ^{18}O is hard to produce at all; it may be mainly destroyed in low and intermediate mass stars, while in massive stars its production depends on assumptions on convection. There are no low metallicity halo star observations to test the chemical evolution of these isotopes, but radio observations of molecular clouds give the intriguing result that the ^{18}O/^{17}O ratio is constant ($\simeq 3.5$) throughout the Galactic disk, with the solar system having a different ratio of $\simeq 5.5$ (Henkel & Mauersberger 1993, Wilson & Rood 1994). Prantzos *et al.* (1996) studied the Galactic evolution of the CNO isotopes and conclude that "there is simply no universal nucleosynthesis mechanism capable to explain" this result. I.e., especially the evolution of ^{18}O remains far from being understood, and its exploration involves at the same time the physics of evolving galaxies and of mixing processes in massive stars.

2.3. *Pre-supernova surface abundances*

Table 2.3 illustrates the progression of enrichments at the surface of massive stars at the time of their collapse with increasing initial mass. While stars with initial masses below

TABLE 5. Surface mass fractions of various isotopes in stellar evolution models in the initial mass range $15 M_\odot \leq M_{initial} \leq 50 M_\odot$ (Langer & Henkel 1995) at the pre-SN stage. The pre-SN configuration is also indicated, where RSG means red supergiant and WC stands for Wolf-Rayet star of the carbon sequence. The last column gives the initial abundances used in the stellar evolution calculations.

isotop	$15 M_\odot$ RSG	$20 M_\odot$ RSG	$25 M_\odot$ RSG	$30 M_\odot$ RSG	$40 M_\odot$ WC	$50 M_\odot$ WC	initial
^{1}H	$6.75 \cdot 10^{-1}$	$6.48 \cdot 10^{-1}$	$6.31 \cdot 10^{-1}$	$6.16 \cdot 10^{-1}$	0.0	0.0	$7.00 \cdot 10^{-1}$
^{4}He	$3.06 \cdot 10^{-1}$	$3.32 \cdot 10^{-1}$	$3.49 \cdot 10^{-1}$	$3.65 \cdot 10^{-1}$	$7.22 \cdot 10^{-1}$	$1.49 \cdot 10^{-1}$	$2.80 \cdot 10^{-1}$
^{12}C	$2.40 \cdot 10^{-3}$	$2.00 \cdot 10^{-3}$	$2.05 \cdot 10^{-3}$	$2.01 \cdot 10^{-3}$	$2.07 \cdot 10^{-1}$	$4.94 \cdot 10^{-1}$	$3.48 \cdot 10^{-3}$
^{13}C	$1.04 \cdot 10^{-4}$	$2.08 \cdot 10^{-4}$	$1.35 \cdot 10^{-4}$	$1.39 \cdot 10^{-4}$	$4.70 \cdot 10^{-7}$	0.0	$3.87 \cdot 10^{-5}$
^{14}N	$3.04 \cdot 10^{-3}$	$4.17 \cdot 10^{-3}$	$4.42 \cdot 10^{-3}$	$4.83 \cdot 10^{-3}$	$4.51 \cdot 10^{-3}$	0.0	$1.03 \cdot 10^{-3}$
^{15}N	$2.27 \cdot 10^{-6}$	$2.05 \cdot 10^{-6}$	$1.69 \cdot 10^{-6}$	$1.62 \cdot 10^{-6}$	0.0	0.0	$3.77 \cdot 10^{-6}$
^{16}O	$9.09 \cdot 10^{-3}$	$8.21 \cdot 10^{-3}$	$7.85 \cdot 10^{-3}$	$7.42 \cdot 10^{-3}$	$4.81 \cdot 10^{-2}$	$3.32 \cdot 10^{-1}$	$9.98 \cdot 10^{-3}$
^{17}O	$5.69 \cdot 10^{-5}$	$6.04 \cdot 10^{-5}$	$6.46 \cdot 10^{-5}$	$6.95 \cdot 10^{-5}$	$8.33 \cdot 10^{-6}$	0.0	$3.80 \cdot 10^{-6}$
^{18}O	$1.18 \cdot 10^{-5}$	$3.33 \cdot 10^{-6}$	$8.67 \cdot 10^{-6}$	$8.22 \cdot 10^{-6}$	$8.92 \cdot 10^{-4}$	0.0	$2.00 \cdot 10^{-5}$
^{19}F	$3.26 \cdot 10^{-7}$	$3.44 \cdot 10^{-7}$	$2.73 \cdot 10^{-7}$	$2.53 \cdot 10^{-7}$	$0.03 \cdot 10^{-9}$	$0.05 \cdot 10^{-9}$	$3.74 \cdot 10^{-7}$
^{20}Ne	$1.70 \cdot 10^{-3}$	$1.70 \cdot 10^{-3}$	$1.70 \cdot 10^{-3}$	$1.70 \cdot 10^{-3}$	$1.68 \cdot 10^{-3}$	$1.66 \cdot 10^{-3}$	$1.70 \cdot 10^{-3}$
^{21}Ne	$3.93 \cdot 10^{-6}$	$3.90 \cdot 10^{-6}$	$3.28 \cdot 10^{-6}$	$3.05 \cdot 10^{-6}$	$1.38 \cdot 10^{-5}$	$3.61 \cdot 10^{-5}$	$4.14 \cdot 10^{-6}$
^{22}Ne	$1.10 \cdot 10^{-4}$	$1.00 \cdot 10^{-4}$	$9.77 \cdot 10^{-5}$	$9.43 \cdot 10^{-5}$	$1.21 \cdot 10^{-2}$	$1.84 \cdot 10^{-2}$	$1.24 \cdot 10^{-4}$
^{23}Na	$5.10 \cdot 10^{-5}$	$5.56 \cdot 10^{-5}$	$6.53 \cdot 10^{-5}$	$6.98 \cdot 10^{-5}$	$1.78 \cdot 10^{-4}$	$1.76 \cdot 10^{-4}$	$3.46 \cdot 10^{-5}$
^{24}Mg	$5.38 \cdot 10^{-4}$	$5.38 \cdot 10^{-4}$	$5.38 \cdot 10^{-4}$	$5.38 \cdot 10^{-4}$	$4.11 \cdot 10^{-4}$	$2.26 \cdot 10^{-4}$	$5.38 \cdot 10^{-4}$
^{25}Mg	$6.52 \cdot 10^{-5}$	$6.22 \cdot 10^{-5}$	$5.92 \cdot 10^{-5}$	$5.70 \cdot 10^{-5}$	$2.12 \cdot 10^{-4}$	$9.34 \cdot 10^{-4}$	$6.81 \cdot 10^{-5}$
^{26}Mg	$7.86 \cdot 10^{-5}$	$8.08 \cdot 10^{-5}$	$8.34 \cdot 10^{-5}$	$8.52 \cdot 10^{-5}$	$3.90 \cdot 10^{-4}$	$2.03 \cdot 10^{-3}$	$7.50 \cdot 10^{-5}$
^{27}Al	$6.05 \cdot 10^{-5}$	$6.06 \cdot 10^{-5}$	$6.07 \cdot 10^{-5}$	$6.10 \cdot 10^{-5}$	$7.75 \cdot 10^{-5}$	$8.02 \cdot 10^{-5}$	$6.00 \cdot 10^{-5}$
^{28}Si	$6.80 \cdot 10^{-4}$	$6.80 \cdot 10^{-4}$	$6.80 \cdot 10^{-4}$	$6.80 \cdot 10^{-4}$	$5.59 \cdot 10^{-4}$	$3.63 \cdot 10^{-4}$	$6.80 \cdot 10^{-4}$
^{29}Si	$3.44 \cdot 10^{-5}$	$3.44 \cdot 10^{-5}$	$3.44 \cdot 10^{-5}$	$3.44 \cdot 10^{-5}$	$1.06 \cdot 10^{-4}$	$1.49 \cdot 10^{-4}$	$3.44 \cdot 10^{-5}$
^{30}Si	$2.28 \cdot 10^{-5}$	$2.28 \cdot 10^{-5}$	$2.28 \cdot 10^{-5}$	$2.28 \cdot 10^{-5}$	$7.86 \cdot 10^{-5}$	$2.43 \cdot 10^{-4}$	$2.28 \cdot 10^{-5}$
^{56}Fe	$1.32 \cdot 10^{-3}$	$1.32 \cdot 10^{-3}$	$1.32 \cdot 10^{-3}$	$1.32 \cdot 10^{-3}$	$1.32 \cdot 10^{-3}$	$1.32 \cdot 10^{-3}$	$1.32 \cdot 10^{-3}$
^{26}Al	$0.25 \cdot 10^{-9}$	$5.22 \cdot 10^{-9}$	$9.07 \cdot 10^{-8}$	$2.99 \cdot 10^{-7}$	$4.09 \cdot 10^{-7}$	$3.60 \cdot 10^{-7}$	0.0

about $35 \, \mathrm{M}_\odot$ end their lives as hydrogen-rich red supergiants, more massive stars evolve into Wolf-Rayet stars which may be completely hydrogen free and show the products of helium burning at their surface.

We want to emphasise the relevance of this for the interpretation of supernova observations and related topics: the most massive stars give rise to Type I (hydrogen-free) supernovae — so called Type Ib/c; cf. Langer & Woosley (1996) —, while the majority of massive stars forms the Type II (hydrogen-rich) supernovae. The mass dependence of the final surface composition seems also to be well reflected in the energy dependence of isotopic abundances in Galactic cosmic rays (Biermann *et al.* 2001). Possible changes of the surface abundances due to effects of rotational mixing are discussed in the next section.

2.4. *Effects of Rotation*

In general, the surface abundances in models calculated without rotation remain intact until the first dredge-up of H-burned material on the red giant branch (e.g., Schaller *et al.* 1992). Since H-burning is dominated by the CNO-cycle in this mass range, then the surface abundances of carbon, nitrogen, and oxygen are predicted to change in a discernible way. However, there is a growing amount of evidence to suggest that the abundances of CNO do not retain their natal values until the stars reach the red giant branch.

So called OBN stars, among them main sequence stars, have been found to be extremely enriched in CNO-cycled material (Schönberner *et al.* 1988, cf. also Walborn 1988). Gies & Lambert (1992) determined CNO abundances in unevolved B-stars and suggested that their spread is larger than expected from random errors in the analysis. In particular, they point out that the range in oxygen abundances is much less than in nitrogen or carbon, and that the nitrogen abundances tend to be slightly enhanced while the opposite is true of carbon, as expected from slight mixing with CNO-processed layers.

Venn (1995) showed that the abundances of CNO in 5 to 20 $M_\odot$ Galactic A-type supergiants are also not the pristine values, but appear to be the result of partial mixing with CNO-cycled layers. Since the A-supergiants do not have first dredge-up values of CNO, they are not evolving on blue-loops after having visited the red giant branch. Finally, Lennon (1994) also finds that B-type supergiants in the Galaxy show evidence for mixing with CNO-cycled gas. Examination of Lennon's C/N ratios suggests that the amount of mixing increases with the mass/luminosity of the star.

After the early attempt of Endal & Sofia (1978), several models including rotationally induced mixing as proposed by Zahn (1983, 1992) have been presented. According to Maeder (1987), stars with $M \gtrsim 25\,M_\odot$ might evolve through a bifurcation, i.e. evolving as chemically homogeneous for rotation speeds above a critical value. Langer (1991b) found rotational mixing able to explain the CNO anomalies of the SN 1987A progenitor. Langer (1992) showed that rotational mixing might be responsible for the helium enrichments found in main sequence O stars (Herrero *et al.* 1992). Finally, Denissenkov (1994) has shown that rotational mixing in $10\,M_\odot$ stars may change the CNO surface abundances as they evolve off the main sequence.

The evolutionary models including the physics of rotation can account for many of the observed surface abundance effects mentioned above. As an example, Fig. 7 shows the evolution of the main CNO isotopes and of helium at the surface of stars in the range $12\,M_\odot$ to $20\,M_\odot$, for various initial rotation rates, according to Heger & Langer (2000).

However, rotation is a fundamental parameter of massive stars, not only affecting their surface abundances, but their whole evolution as strongly as initial mass and metallicity (Langer *et al.* 1997b). Fig. 8 shows how rotationally induced mixing changes the evolution in the HR diagram during core hydrogen burning. Although the reduced effective gravity due to the centrifugal force leads to a smaller luminosity initially, rotational mixing between core and envelope results in a larger core mass and thus in a higher luminosity later on, compared to the non-rotating case.

An increased core mass has consequences for all subsequent burning stages, as the core mass determines which temperatures and densities are achieved in the course of stellar evolution (cf. Sect. 1.2). Since the cores contain the ashes of hydrostatic nuclear burning stages, the corresponding chemical yields of massive stars — e.g. that of oxygen — are increased by rotation (cf. Langer *et al.* 1997a). However, while this effect can become of the order of tens of percent, also qualitative changes can be caused by increased core masses. An example is shown in Fig. 9, which compares the evolution of a non-rotating and a rotating $10\,M_\odot$ star in the $\rho_c - T_c$-diagram. While the latter evolves straight through all nuclear burning stages to become a Type II supernova, the non-rotating star ignites carbon in a degenerate core off-center and then cools and may become an ONeMg white dwarf (cf. Ritossa *et al.* 1996).

The physical processes responsible for the mixing of chemical species — besides convection predominantly Eddington-Sweet circulations and shear instabilities — do also

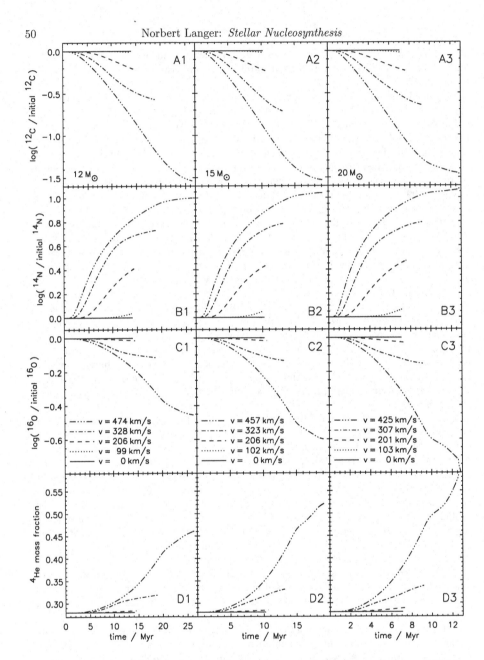

FIGURE 7. Evolution of the surface abundances during central hydrogen burning of $12\,M_\odot$ (left column) $15\,M_\odot$ (middle column), and $20\,M_\odot$ stars for different ZAMS equatorial rotational velocities (see legend) as a function of time (Heger & Langer 2000). Displayed are ^{12}C, ^{14}N, ^{16}O relative to their initial abundance, and the ^{4}He mass fraction.

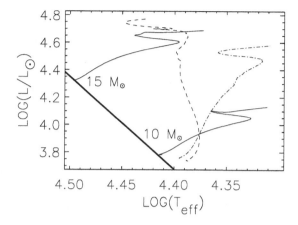

FIGURE 8. Evolutionary tracks of three $10\,M_\odot$ stars with different initial equatorial rotation velocities ($v_{eq} = 0$, 280, and 400 km/s; solid, dash-dotted, and dashed lines, respectively) during the core hydrogen burning phase. The track of a non-rotating $15\,M_\odot$ star is also shown. The thick solid line marks the ZAMS position of non-rotating stars. (From Fliegner *et al.* 1996).

transport angular momentum (cf. Pinsonneault 1997). As, in the course of stellar evolution, the cores contract and thus rotate more rapidly, the direction of angular momentum transport is usually from the core into the envelope. Furthermore, massive stars can lose angular momentum through stellar winds (Langer 1998). All of these processes need to be considered in order to obtain an estimate of the rotation rate of the core prior to its collapse. They were included in the recent study of the evolution of a $20\,M_\odot$ star from the zero age main sequence to iron core collapse by Heger *et al.* (2000), who found that after core hydrogen exhaustion — i.e. after the ignition of the first nuclear shell source — the specific angular momentum of the core remains roughly constant (cf. Table 6). This has the consequence that the collapsing iron core (assuming local angular momentum conservation during the collapse) will almost reach critical rotation. This would, of course, have severe consequences for the core collapse supernova mechanism (e.g., Yamada & Sato 1994, Aksenov *et al.* 1997).

However, the pre-supernova calculations of Heger *et al.* (2000) are the first of its kind, and they should be considered as preliminary. Major uncertainties are due the poorly known influence of molecular weight gradients on the shear instability (cf. Maeder 1995, 1997) and our ignorance of the effect of magnetic fields in the stellar interior on the transport of chemicals and angular momentum (cf. Spruit 1998). If confirmed, the results above would have strong implications for Type II and Ib/c supernovae and the resulting neutron stars (cf. Langer *et al.* 1997a).

In contrast to helium and CNO, the light elements are destroyed relatively close to the stellar surface. For both stable boron isotopes, ^{10}B and ^{11}B, the life time against proton capture, the dominant destruction mechanism, is equal to the main sequence life time of a $10\,M_\odot$ star (10^7 yr) at a temperature of roughly $7\,10^6$ K. This temperature occurs sufficiently deep inside the stellar envelope (i.e. roughly $1\,M_\odot$ below the surface; cf. Fig. 10) that its surface abundance can not be altered due to mass loss alone on the main sequence in the B star regime. Thus, the boron abundance in B stars is a critical

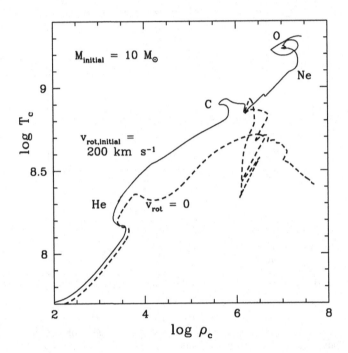

FIGURE 9. Evolutionary tracks of a rotating (solid line) and a non-rotating (dashed line) $10\,M_\odot$ star in the $\log\rho_c - \log T_c$–diagram. The various core burning stages are indicated along the track of the rotating model. (From Langer *et al.* 1997a).

test of mixing processes in the upper stellar envelope, while CNO and helium abundances additionally trace the mixing in deeper layers.

For main sequence secondary stars in massive close binary systems, a strong boron depletion must also be expected due to the transfer of nuclear processed matter; however, a very noticeable CNO-signature is always present in this case (cf. e.g. de Loore & De Greve, 1992). Therefore, the existence of boron depleted stars with a roughly normal nitrogen abundance would be a clear indication of rotational mixing.

Let us reconsider Fig. 8, which shows the stellar evolution tracks of non-rotating models of 10 and $15\,M_\odot$, and of a moderately ($v_{eq} \simeq 280$ km/s at the ZAMS) and a rapidly rotating ($v_{eq} \simeq 400$ km/s) $10\,M_\odot$ star in the HR diagram. Assuming that stars are born with a distribution of initial rotation rates, there is no clear trend of changes in surface composition due to rotational mixing as a function of the proximity to the zero age main sequence. For example, a non- or slowly rotating star of $15\,M_\odot$, which keeps its initial surface abundances during the main sequence evolution, leads to the same point in the HR diagram as a rapidly rotating $10\,M_\odot$ star which has changed surface abundances. This may explain the lack of a correlation of the N-overabundance with stellar age found by Gies & Lambert (1992).

However, due to the different sampling depths of boron and (e.g.) nitrogen, the mod-

TABLE 6. Properties of a $20 \, M_\odot$ star at the mass coordinate $M_r = 1.7 \, M_\odot$ (the mass of the iron core) at various times during the evolution (cf. Heger *et al.* 2000).

evol. stage	j cm^2/s	r cm	ρ g/cm^3	ω s^{-1}	ω/ω_c
ZAMS	10^{17}	$4 \, 10^{10}$	3.4	$7 \, 10^{-5}$	0.03
H-exh.	$2 \, 10^{16}$	$3 \, 10^{10}$	8.0	$2 \, 10^{-5}$	0.008
He-exh.	10^{16}	$3 \, 10^9$	$8 \, 10^3$	10^{-3}	0.01
collapse	10^{16}	$6 \, 10^7$	10^9	3	0.08
n-star	(10^{16})	10^6	$4 \, 10^{14}$	10^4	0.7

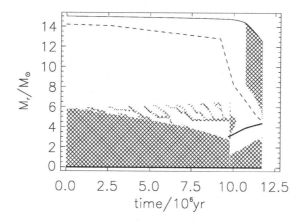

FIGURE 10. Internal structure of a $15 \, M_\odot$ star during core hydrogen and helium burning (Fliegner *et al.* 1996). The solid line on top indicates the total mass of the star as function of time. Hatched areas designate convectively unstable mass zones in the star. The full drawn line at $M_r \simeq 4 \, M_\odot$ and $t \gtrsim 10^7$ yr designates the location of the H-burning shell during core helium burning. The dashed line indicates the threshold temperature for boron destruction.

els of Fliegner & Langer (1995) predict a distinct evolution in the boron-depletion vs. nitrogen-enhancement. The boron vs. nitrogen predictions are remarkably similar for models of different masses and rotation rates (Fig. 11). This is because the evolution proceeds through three different stages of surface abundances: i) An initial phase of unaltered abundances, which lasts roughly $1.5 \, 10^6$ yr. ii) Boron depletion occurs at the stellar surface without a change in nitrogen, since boron survives in only the outermost $1 \, M_\odot$ and nitrogen is synthesised at much deeper layers. iii) After another time span of about $7 \, 10^6$ yr, nitrogen starts to be enriched with only a small additional boron depletion.

Due to the degeneracy of the HRD positions during the main sequence evolution (see Fig. 8), a comparison of the observed stellar boron abundances and effective temperatures

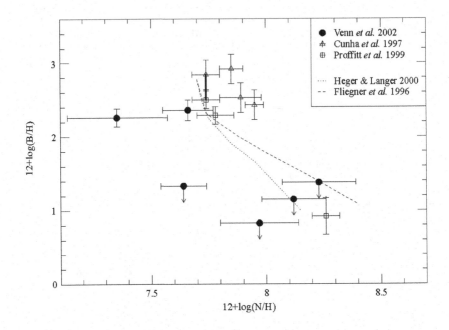

FIGURE 11. Boron versus nitrogen in Galactic B-stars. One star, HD 35299, has been analysed by both Cunha *et al.* (1997) and Proffitt *et al.* (1999) from the B II and B III resonance lines, respectively; these two measures are connected by a *thick line.* Theoretical results are shown, with (*dotted line*) and without (*dashed line*) μ-gradients for rotational velocities $\sim 200\,\mathrm{km\,s^{-1}}$, a typical equatorial velocity for B-stars.

with theoretical results may be spurious. However, a distinct correlation of the boron depletion with the nitrogen enhancement can be expected. Venn *et al.* (1996) have already looked into the observational boron versus nitrogen dependence, considering eight stars, among them two A supergiants. They found only a weak trend.

Venn *et al.* (2002) re-investigated this problem. Fig. 11 shows that the prediction from the rotating stellar models and the observed boron versus nitrogen abundances agree well. As noted above, this is not trivial since the thermonuclear processes which affect the boron and the nitrogen abundances occur in completely different stellar layers. In particular, the existence of boron depleted stars with a normal nitrogen abundance (i.e. the stars HD 886, HD 29248, and HD 216919; cf. Venn *et al.* 1996) appears to specifically support the idea of rotational mixing, since such abundance ratios are not considered a possible result of close binary evolution.

In summary, boron abundances appear to be an extremely important diagnostic for the evolution of B stars and cooler luminous stars since rotational mixing on the main sequence deeply affects the post-main sequence evolution. Therefore, a comprehensive study of this element with highest quality data (HST GHRS) for hot stars is highly desirable. This, as well as the analysis of boron in supergiants (cf. Venn *et al.* 1996) may allow us to considerably reduce persisting uncertainties in the theory of rotational mixing, which is not only fundamental for studying stars within the mass range considered here, but which is also required for a thorough understanding of more massive stars (Maeder 1987, Fliegner *et al.* 1996).

3. The s-Process

3.1. *Low mass stars*

It is known since long that thermally pulsing Asymptotic Giant Branch (TP-AGB) stars provide a site for the so called s-process, i.e., the slow neutron capture process which forms neutron-rich isotopes heavier than iron (Clayton 1968). Heavy elements primarily produced by the s-process are overabundant at the surface of AGB stars (Smith & Lambert 1990; Lambert, this volume), including technetium (Little *et al.* 1987) which has no stable isotope and which is produced as ^{99}Tc ($\tau_{1/2} = 2.1\,10^5$ yr) in the s-process. In particular the roughly solar magnesium isotopic pattern found in s-process enriched AGB stars has demonstrated that the ^{13}C(α,n) rather than the ^{22}Ne(α,n) neutron source is likely to operate the s-process in AGB stars (Guélin *et al.* 1995, Lambert *et al.* 1995).

Evidence for *in situ* s-processing is found exclusively in carbon stars (Smith & Lambert 1990), which correspond to a late evolutionary stage on the TP-AGB where the stars have large ^{12}C enrichments in their envelopes (Iben & Renzini 1983, Wallerstein & Knapp 1998). The ^{12}C enrichment implies that these stars contain, at certain times, a region at the bottom of their hydrogen-rich envelope where ^{12}C is abundant. This region where protons and ^{12}C coexist may then perhaps form ^{13}C through ^{12}C(p,γ)^{13}N($\beta^+\nu$)^{13}C. Although this scenario is unrivalled, the formation of a layer which is rich in protons *and* ^{12}C in TP-AGB models has proven to be difficult, and its existence had hitherto to be assumed *ad hoc* in all s-process calculations (Gallino *et al.* 1998). Iben & Renzini (1982) found a ^{13}C layer in low metallicity ABG models. Recently, Herwig *et al.* (1997) have obtained a ^{13}C-rich layer in TP-AGB models of solar metallicity, by invoking a diffusive overshoot layer at convective boundaries. In the following we shall discuss the finding of Langer *et al.* (1999) that mixing induced by differential rotation is capable of selfconsistently producing a ^{13}C-rich layer in TP-AGB models.

Even though in the example of a $3\,M_\odot$ star of Langer *et al.* (1999) the star evolves close to rigid rotation on the main sequence, the extreme expansion of the hydrogen-rich envelope and the compactness of the C/O-core on the AGB create an enormous shear at the core-envelope interface. While this remains without drastic consequences as long as active nuclear shell sources separate both regions well in terms of entropy, this is different during the thermal pulses where both nuclear shell sources are periodically switched off.

Fig. 12 shows the evolution of the internal structure during and after the 25th thermal pulse of the rotating model. It shows that the tip of the pulse-driven convection zone leaves after its decay a region of strong rotational mixing. This mixing becomes even stronger when the convective envelope extends downward during the third dredge-up event. The reason is that convection enforces close-to-rigid rotation (cf. Langer *et al.* 1999), with an envelope rotation rate which is many orders of magnitude smaller than that of the core. The resulting strongly differential rotation allows the Goldreich-Schubert-Fricke instability, and to a lesser extent the shear instability and Eddington-Sweet circulations, to produce a considerable amount of mixing between the carbon-rich layer and the hydrogen envelope.

By applying the concept of rotationally induced mixing as it has been developed for massive stars during the last years without alteration to a $3\,M_\odot$ TP-AGB model sequence, Langer *et al.* (1999) obtained conditions which appear favourable for the development of the s-process, i.e. a ^{13}C-rich layer which produces a considerable neutron flux later-on.

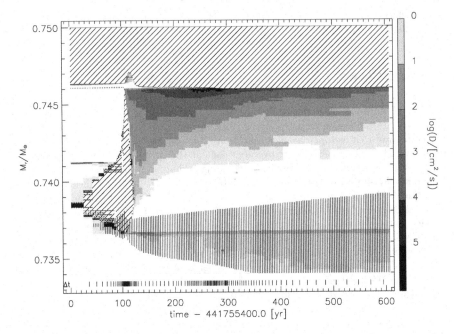

FIGURE 12. Section of the internal structure during and after the 25th thermal pulse of a rotating $3\,M_\odot$ sequence (Langer *et al.* 1999). Diagonal hatching denotes convection. The convective envelope extends down to $M_r \simeq 0.746\,M_\odot$. The pulse driven convective shell is located at $0.737\,M_\odot \lesssim M_r \lesssim 0.746\,M_\odot$ and $30\,\mathrm{yr} \lesssim t \lesssim 120\,\mathrm{yr}$. Vertical hatching denotes regions of significant nuclear energy generation, i.e., the hydrogen burning shell (at $M_r \simeq 0.746$ and $t \lesssim 100\,\mathrm{yr}$) and the helium burning shell ($0.734\,M_\odot \lesssim M_r \lesssim 0.739\,M_\odot$ and $t \gtrsim 40\,\mathrm{yr}$). Gray shading marks regions of significant rotationally induced mixing (see scale on the right side of the figure). Vertical marks at the bottom of the figure denote the time resolution of the calculation, where every fifth time step is indicated.

Although this model develops only a very late and weak third dredge-up, the mechanism which diffuses the protons into the carbon layer and ^{12}C into the envelope must occur with a similar magnitude in all TP-AGB stars which develop a third dredge-up. The reason is that the huge specific angular momentum jump at the hydrogen/carbon interface — five orders of magnitude — is independent of the depth of the third dredge-up.

The maximum ^{13}C abundance and its distribution in the $3\,M_\odot$ model is, at first, similar to that found due to diffusive convective overshooting by Herwig *et al.* (1997). However, the rotational mixing spreads the ^{13}C peak out before the neutrons are produced (cf. Fig. 13), which is not the case in the models of Herwig *et al.* (1997). At the present time one can not discriminate which of these scenarios would agree better with empirical constraints. However, we want to stress that both mechanisms of ^{13}C production, rotation and overshooting, do not exclude each other, and that it is possible that they act simultaneously in AGB stars.

Finally, we want to emphasise that, although stars of less than $\sim 1.3\,M_\odot$ lose 99% of their angular momentum due to a magnetic wind during their main sequence evolution, it can not be excluded that the proposed mechanism of ^{13}C-production due to differential

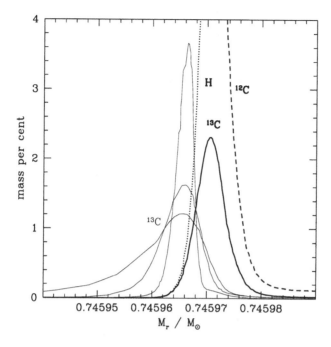

FIGURE 13. Chemical profiles at the location of the maximum depth of the convective envelope during the 25th thermal pulse (cf. Fig. 12) of a rotating $3\,M_\odot$ AGB sequence (cf. Langer *et al.* 1999). The dotted and dashed lines mark the hydrogen and the ^{12}C mass fractions at $t = 1704\,\text{yr}$, with $t = 0$ defined as in Fig. 12. The fat solid line denotes the ^{13}C mass fraction at the same time. The three thin solid lines represent the ^{13}C mass fractions at $t = 2016\,\text{yr}$, $t = 4155\,\text{yr}$, and $t = 5139\,\text{yr}$, with a later time corresponding to a smaller peak abundance. The maximum ^{13}C mass fractions of 3.6% occurs at $t = 2016\,\text{yr}$. The ^{13}C peak moved inwards in the time interval from $t = 1704\,\text{yr}$ to $t = 2016\,\text{yr}$ due to continued proton captures on both, ^{12}C and ^{13}C.

rotation also works for them. Certainly, the sun's core will spin-up and the envelope will further spin down during its post-main sequence evolution, which may result in a specific angular momentum jump of similar magnitude. The investigation of the mass and metallicity dependence of the production of ^{13}C due to rotation is an exciting task for the near future.

3.2. *Massive stars*

Massive stars achieve a natural neutron production via the ^{22}Ne(α,n) reaction since, as almost all initially present CNO nuclei in the core are converted into ^{14}N during hydrogen burning, which in turn is fused into ^{22}Ne via ^{14}N$(\alpha,\gamma)^{18}$F$(e^+\nu)^{18}$O$(\alpha,\gamma)^{22}$Ne at the beginning of helium burning (Peters 1968). Detailed models have shown that only stars above $\sim 20\,M_\odot$ achieve the burning of ^{22}Ne during core helium burning, while ^{22}Ne is consumed during core or shell carbon burning in lower mass stars. The general conclusion is that the solar system distribution of the weak s-process component is well reproduced by current massive star models (Lamb *et al.* 1977, Langer *et al.* 1989, Prantzos *et al.* 1990, Arcoragi *et al.* 1991, Rayet & Hashimoto 2000).

However, as pointed out by Rayet & Hashimoto (2000), the metallicity dependence of the weak s-process production is not yet well understood — which may hamper an understanding of observed s-abundances in metal poor stars — due to nuclear physics and stellar structure uncertainties (cf. also Baraffe & Takahashi 1993).

4. Nucleosynthesis in Binary Systems

Most stars appear to be members of binary or multiple systems. The fraction of massive stars being members of *close* binaries — i.e. such in which mass overflow is expected to occur — is estimated to be of the order of 20 ... 40% (cf. Garmany *et al.* 1980, Podsiadlowski 1997, Mason *et al.* 1998). Thus, it appears necessary to investigate the effect of binary mass transfer on the overall massive star nucleosynthesis yields: if the mass transfer would increase the yield of an isotope only by a factor of ~ 3, then massive close binaries might be the dominant source of this isotope. Clearly, most isotopes are not affected by binaries at this level (cf. De Donder & Vanbeveren 2002).

Wellstein *et al.* (2001) have studied the nucleosynthesis processes in typical massive close binaries. Fig. 14 gives an example of the evolution of a close $20 + 18\,M_\odot$ pair of Case A — i.e. mass transfer occurs during the core hydrogen burning phase of the primary (the initially more massive star) — in the HR diagram. Due to the transfer of most of the hydrogen-rich envelope of the primary to the secondary component, the primary becomes a helium star (cf. point D in Fig. 14) while the secondary evolves into a luminous blue supergiant. In both cases, the core masses evolve differently compared to single stars of the same initial mass (i.e., 20 or $18\,M_\odot$ in our example); the primaries' core masses are smaller, that of the secondaries larger.

Then, the primary may, as a helium star, develop strong Wolf-Rayet type winds mass loss (cf. Langer 1989ab, Woosley *et al.* 1995), further reducing its helium core mass. This can affect the chemical yields of primary isotopes substantially, e.g. the carbon yield is enhanced at the expense of oxygen (cf. Langer & Henkel 1995). As long as the primary's helium core mass remains above $\sim 2\,M_\odot$ it will develop a collapsing iron core and become a supernova of Type Ib or Ic (Woosley *et al.* 1995).

As for the secondary star, which accretes the envelope of the primary during its core hydrogen burning evolution, it is a common assumption that it evolves after accretion exactly like a single star of the corresponding new mass ($\sim 32\,M_\odot$ in our example). However, Braun & Langer (1995) found that secondaries may retain significant structural differences compared to single stars, e.g. with the consequence that the supernova explosion occurs in the blue supergiant stage (cf. Fig. 14) — as in the case of the progenitor of SN 1987A — rather than in the red supergiant stage.

However, independent of this phenomenon, Wellstein *et al.* (2001) found that substantial differences in the synthesis of secondary CNO isotopes do occur in the secondary components of massive close binaries, due to the interplay between CNO burning and so called thermohaline mixing. This mixing process does occur in the whole hydrogen-rich envelope of the secondary from the surface down to the convective H-burning shell since the star accretes helium enriched matter from the primary component (i.e. matter with a higher mean molecular weight is lying above matter with lower mean molecular weight). As the time scale for thermohaline mixing — which is for the first time treated as a time dependent process by Wellstein *et al.* (2001) — and CNO processing at the bottom of the mixed zone are comparable, the production particularly of ^{13}C and ^{14}N is boosted in the whole H-rich envelope, increasing the yield of these isotopes by factors of the order of 2 ... 3 (cf., Braun 1997).

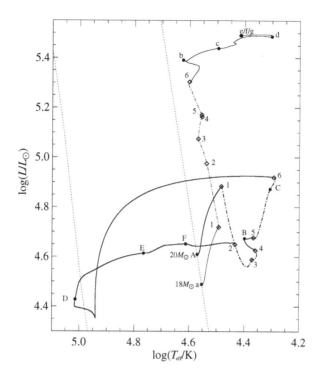

FIGURE 14. Evolutionary tracks in the HR diagram of the components of a 20+18 $M_\odot$ case A close binary system with a metallicity of $Z_\odot/4$ and an initial period of 2.5 days. The path of the primary component (initial mass 20 $M_\odot$) is marked by the thick line and upper case letters, that of the secondary by the thin line and lower case letters. Mass transfer stages correspond to the dot-dashed parts of the lines. The thin dotted lines designate the zero age main sequence and the location of pure helium stars (helium main sequence). The letters designate beginning and end of nuclear burning stages, i.e. core hydrogen burning (a/A – b/B), core helium burning (c/C – d/D), core carbon burning (e/E – f/F). g/G marks the beginning of core neon burning. Numbers designate mass transfer events for both stars. 1: beginning of Case A mass transfer, 2: maximum of mass transfer rate, 3: start of slow phase of Case A mass transfer, 4: end of Case A mass transfer, 5: start of Case AB mass transfer, 6: end of Case AB mass transfer. The final masses of the primary and secondary are 3 and 32 $M_\odot$, respectively.

4.1. ^{26}Aluminium in Massive Binaries

In massive stars, ^{26}Aluminium is produced during core hydrogen burning according to
$$^{24}\text{Mg}(p,\gamma)^{25}\text{Al}(\beta^+\nu)^{25}\text{Mg}(p,\gamma)^{26}\text{Al}\nearrow(p,\gamma)^{27}\text{Si}(\beta^+\nu)\searrow$$
$$\searrow(\beta^+\nu)^{26}\text{Mg}(p,\gamma)\nearrow^{27}\text{Al}(p,\gamma)^{24}\text{Mg}$$
The synthesis of ^{26}Al in secondary components of close binaries was found to deserve special attention. The reason is that for ^{26}Al, as it is β-unstable with a mean life time of $1.03\,10^6$ yr ($\tau_{1/2} = 7.2\,10^5$ yr), not only the amount which is synthesised matters, but also the time of the synthesis. The γ-ray line emission from the decay of ^{26}Al in the Galaxy is observed. However, we can only see the decay of ^{26}Al nuclei in the interstellar medium; the decay inside stars is unobservable. Therefore, the ^{26}Al which is observed

should either be produced during supernova explosions or shortly before. From the spatial distribution of the γ-ray line emission (Prantzos & Diehl 1996) we know that it originates from massive stars. Supernovae are in fact the currently favoured production site, although the corresponding yields are very uncertain (Weaver & Woosley 1993, Thielemann *et al.* 1996).

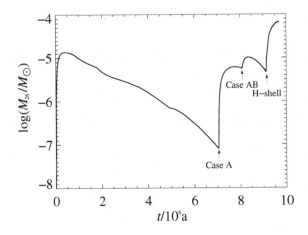

FIGURE15. Evolution of the total mass of ^{26}Al inside the secondary component of a $20+18\,M_\odot$ case A close binary system (cf. Fig. 14) as a function of time (in 10^6 yr). The beginning of the Case A and Case AB mass transfer phase is indicated, as well as the beginning of hydrogen shell burning which marks the end of the core helium burning evolutionary phase. When the star explodes as a supernova of Type SN 1987A ($t = 9.68\,10^6$ yr) it contains almost $10^{-4}\,M_\odot$ of ^{26}Al. Note that the initial metallicity of the stars is $Z_\odot/4$.

However, ^{26}Al is also produced during hydrostatic hydrogen burning, by proton capture on ^{25}Mg. Although very massive stars, through extremely strong stellar winds, can eject ^{26}Al generated during H-burning and contribute to the Galactic ^{26}Al (Langer *et al.* 1995, Meynet *et al.* 1997), the hydrogen burning contribution of less massive stars (say 10 ... $30\,M_\odot$) is not considered as important (though see Langer *et al.* 1997a) since the major fraction of the ^{26}Al decays inside the star before it is released in the course of the supernova explosion.

Wellstein *et al.* (2001) found this to be different in massive close binary secondaries. Fig. 4.1 shows the time dependence of the amount of ^{26}Al generated by hydrogen burning in the interior of the secondary component of the $20+18\,M_\odot$ system shown in Fig. 14. Obviously, the ^{26}Al mass, although $10^{-5}\,M_\odot$ initially, would be of the order of $10^{-8}\,M_\odot$ at the end of the evolution if no mass accretion would occur. However, since fresh fuel — i.e. also fresh ^{25}Mg — is mixed into the core due to the mass accretion process (cf. Braun & Langer 1995) the amount of ^{26}Al is increased by orders of magnitude at that time.

This would already be sufficient to produce as much as $\sim 10^{-6}\,M_\odot$ ^{26}Al from this star of initially $18\,M_\odot$ — two orders of magnitude more than expected from single star calculations at $Z_\odot/4$ (cf. Langer *et al.* 1995). However, the H-burning shell source in the secondary component is much more efficient than in corresponding single stars. This leads to the coupling of an extended convection zone to the hydrogen burning shell, and consequently to the enrichment of the whole convection zone with ^{26}Al. In the end, the secondary star contains almost $10^{-4}\,M_\odot$ of ^{26}Al which will be liberated during the supernova explosion. As the chosen example appears to be a rather typical case, the Galactic ^{26}Al production may in fact be dominated by massive close binary systems.

4.2. Progenitors of Type Ia Supernovae

Close binaries consisting of a degenerate and a non-degenerate star — e.g. of an unevolved star and one which has completed its evolution — are not as rare as one might expect. E.g., we know more than 300 Cataclysmic Variables (Downes *et al.* 2001). Such systems are rather frequent since stars of different masses have vastly different life times. I.e., while the initially more massive star in a binary may already be "dead", its companion may still be at the very beginning of its evolution. And as stars spend most of their life on the main sequence, the pairing of a main sequence star and a white dwarf is the most common variety of single degenerate binaries. We will concentrate on such systems in the following.

The combination of a main sequence star and a white dwarf in a close binary system can result in a large number of highly spectacular and astrophysically interesting phenomena. We need a *close* binary, as the spectacle begins when the main sequence star, which expands due to its nuclear evolution — core hydrogen burning —, fills its critical volume and starts to transfer mass onto the white dwarf. Depending on the rate at which the white dwarf receives hydrogen-rich matter, hydrogen burning and subsequently helium burning may occur on top of its C/O-body, either steadily or — more likely — thermally unstable or even explosive. Such systems have the potential to explode the whole white dwarf, i.e. to produce a Type Ia supernova.

4.2.1. Supersoft X-ray Sources

The regime of steady nuclear burning in the outer layers of an accreting C/O-white dwarf is essential to produce Chandrasekhar-mass white dwarfs — the currently favoured Type Ia supernova progenitors (Livio 2000). In order to achieve this, the white dwarf accretion rate needs to remain in a very limited range (Nomoto & Kondo 1991), which — for a main sequence donor star — translates into a very limited range of donor star masses: roughly $1.5\,M_\odot \ldots 2.5\,M_\odot$ (Hachisu *et al.* 1996; Li & van den Heuvel 1997; Langer *et al.* 2000). The lower limit comes from the constraint that the donor star should be more massive than the white dwarf in order to obtain high enough accretion rates. I.e., $\dot{M} \simeq M_d/\tau_{th}$, M_d being the mass of the donor star and τ_{th} its thermal time scale. Once the white dwarf becomes more massive than the donor, the mass transfer leads to a widening of the orbit and the mass transfer rate drops by many orders of magnitude (cf. Fig. 15).

The single degenerate SN Ia progenitor scenario has the advantage over other scenarios that it may have observed counterparts: the supersoft X-ray sources. Indeed, steady nuclear burning in a white dwarf results in effective temperatures of $0.5 \ldots 1 \times 10^6$ K, and a luminosity of the order of the white dwarf's Eddington luminosity — which roughly fits to the observed supersoft sources (Greiner 2000). About half of them are also binaries with orbital periods of the order of one day. However, the following two problems with this idea have been discussed recently. First of all, so far neither the white dwarf nor the main sequence companion have ever been unambiguously identified in any of the sources. And secondly, some of the systems have orbital periods which appear too short to allow for a main sequence star fitting into the system. We believe that both problems are connected and may have their root in the fact that the main sequence donor stars are

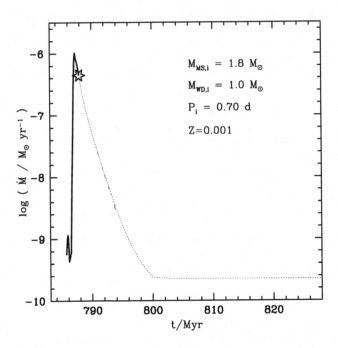

FIGURE 15. Evolution of the mass transfer rate as function of time, for a binary consisting initially of a $1.8\,M_\odot$ main sequence star and a $1\,M_\odot$ white dwarf in a 0.7 d orbit. The line is drawn fully until the white dwarf reaches $1.4\,M_\odot$ (star symbol). The calculation is continued assuming the compact star can accrete further (dotted part of the line; cf. Langer *et al.* 2000). The first phase of high mass transfer rates may correspond to the supersoft X-ray sources. After its decay ($t > 800\,\mathrm{Myr}$), the system is CV-like; only the beginning of this second phase of low mass transfer rates is shown.

quite extraordinary: unlike normal main sequence stars, their interior grossly deviates from thermal equilibrium.

Fig. 16 shows the evolution of a $2.1\,M_\odot$ donor star in the HR diagram, from hydrogen ignition until and throughout much of the mass transfer evolution. Its luminosity can be seen to drop by as much as a factor of 30, mostly due to the mechanical work which the expanding envelope has to provide against the stellar gravitational potential. Only if the envelope expansion time were large against its thermal time scale could the mechanical energy loss be compensated by the core's nuclear fusion; this is not the case here. Thus, the donor star is extremely underluminous.

The thermal time scale mass transfer is initiated by the shrinkage of the orbit as a consequence of angular momentum conservation, as long as $M_d > M_{WD}$: the Roche lobe filling main sequence star is squeezed into a continuously shrinking volume. The stellar radius can fall much below the zero-age main sequence mass-radius relation. Consequently, orbital periods can be achieved which are smaller by a factor f than periods corresponding to unperturbed main sequence stars filling their Roche lobes. For conser-

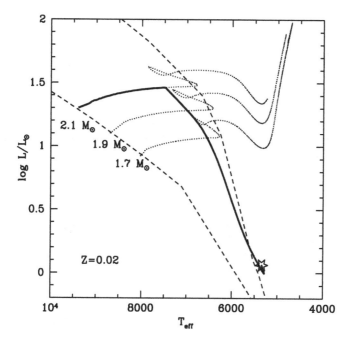

FIGURE 16. Evolution in the HR diagram of a $2.1\,M_\odot$ donor star having an $0.8\,M_\odot$ white dwarf in an orbit with a period 1.23 d (thick solid line). The track ends at a time when the white dwarf mass reaches $1.4\,M_\odot$ (star symbol). For comparison, evolutionary tracks for single stars with $2.1\,M_\odot$, $1.9\,M_\odot$, and $1.7\,M_\odot$ (dotted lines), and the zero and terminal age main sequence (dashed lines) are shown.

vative systems, this factor is given by

$$f = P_{\min}/P_{\mathrm{i}} = \left(\frac{4q_i}{(q_i+1)^2}\right)^3, \tag{4.22}$$

with $q_i = M_{\mathrm{WD,i}}/M_{\mathrm{MS,i}}$, and the index i referring to the initial situation, i.e. to the time before the mass transfer. I.e., for reasonable initial mass ratios, say $q_i > 0.4$, periods of up to a factor of 2 smaller than with unperturbed donors are achieved. Thereby, orbital periods of 4 hours and smaller are conceivable with the thermal time scale mass transfer scenario (Langer *et al.* 2000).

Applying the constraints of steady hydrogen and helium shell burning as function of the accretion rates (cf. Li & van den Heuvel 1997) determines which MW+WD binary has the chance to produce a Type Ia supernova, as only for mass transfer rates in the narrow range allowed by these constraints can the white dwarf mass actually grow. Fig. 17 shows, for two metallicities, which white dwarf masses could actually be achieved is such systems if the white dwarf would be allowed to grow even beyond the Chandrasekhar mass. It confirms that such binary systems may indeed constitute a major SN Ia progenitor channel, as, for solar metallicity, a wide range of initial parameters leads to "white dwarf masses" well above the Chandrasekhar mass, indicating that even were the mass accretion process somewhat inefficient, there would be enough mass flowing to

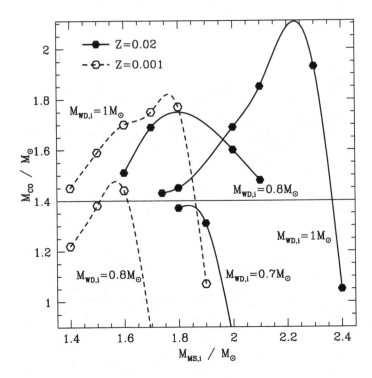

FIGURE 17. Maximum achievable CO-core masses as function of the initial mass of the main sequence star, for various initial white dwarf masses and for the two metallicities considered here, as indicated

the white dwarf with the right accretion rate to achieve the Chandrasekhar limit, for main sequence star masses in the range $1.6 \dots 2.3 \, M_\odot$.

4.2.2. *Helium shell flashes*

As discussed above, supersoft X-ray sources may be compatible with the scenario of accreting white dwarfs which burn hydrogen steadily. However, not for all accretion rates for which hydrogen burns steadily does helium burning proceed steadily as well. As discussed by Cassisi *et al.* (1998) and Kato & Hachisu (1999), helium shell burning may instead become unstable. The corresponding instability can be modelled by computing helium accreting C/O-white dwarfs — which simulates continuous hydrogen shell burning. Helium shell burning can become unstable due to two ingredients. Firstly, the helium-rich layers may become somewhat degenerate, especially for small accretion rates. And secondly, the larger the white dwarf mass the smaller is the radius of the degenerate C/O-core, and the smaller is the geometrical thickness of the helium burning shell. And thin helium burning shells undergo the same thermal instability which drives the thermal pulses in evolved AGB stars — even if the helium shell is non-degenerate.

Fig. 18 gives an example: a C/O-white dwarf of initially $0.8 \, M_\odot$ which accretes at a rate of $5 \times 10^{-7} \, M_\odot \, \mathrm{yr}^{-1}$ undergoes stable helium shell burning for more than $10^5 \, \mathrm{yr}$, accreting thereby $0.05 \, M_\odot$. During this stage, the system might represent a supersoft

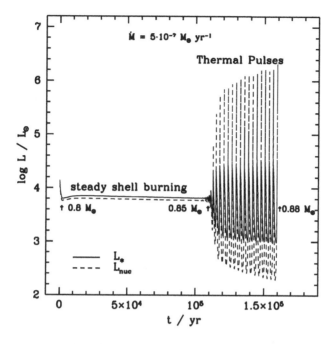

FIGURE 18. Stellar and nuclear luminosity as function of time for a C/O-white dwarf model starting at $0.8\,M_\odot$, which accretes helium at a constant rate of $5 \times 10^{-7}\,M_\odot\,\text{yr}^{-1}$ (cf. Scheithauer 2000).

X-ray source. However, the helium shell becomes violently unstable once the white dwarf mass grows further. And clearly, the amplitude of the instability grows the more massive the white dwarf becomes.

The helium shell burning instability has two consequences. As Fig. 18 implies, not only the nuclear luminosity $(\int_M \varepsilon_{nuc} dM)$ becomes large at the peak of the instability, but also the surface luminosity. Quickly, values exceeding the white dwarf's Eddington luminosity are achieved. This will lead to radiation driven mass loss from the white dwarf (cf. Kato & Hachisu 1999), which questions whether it can ever reach the Chandrasekhar limit.

Secondly, even though the helium shell instability proceeds on a time scale which is long compared to the hydrodynamic time scale — i.e. there is no mass ejection due to the inertia of the expanding envelope — the white dwarf may be driven to obtain a large radius. Being in a compact binary, this may result in a contact or common envelope situation in which the least hazardous thing happening to the white dwarf would again be: it may lose its extended envelope and fail to reach the Chandrasekhar mass.

4.2.3. *White dwarf spin-up*

As we have seen above, it is not easy for a white dwarf to accrete in such a way that it can grow to the Chandrasekhar limit. However, there is one more problem which has not been mentioned yet: the angular momentum problem. Most white dwarfs accrete via an accretion disk. Therefore, the specific angular momentum of the accreted matter may

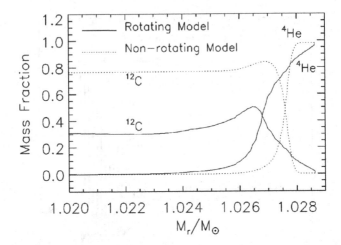

FIGURE 19. Comparison of carbon and oxygen profiles in helium accreting CO-white dwarfes with and without effects of rotation considered (Yoon & Langer 2002). Rotational mixing of α-particles into the helium buring shell results in a much reduced carbon abundance in the case with rotation.

correspond to that of matter rotating at Keplerian speed at the white dwarf equator, i.e.

$$j = 3 \times 10^{17} \sqrt{\frac{M}{M_\odot} \frac{R}{0.01\,R_\odot}} \quad \text{cm}^2\,\text{s}^{-1} \quad . \tag{4.23}$$

It may be a good approximation to neglect the angular momentum in the white dwarf before it starts accreting — both, observations (Heber *et al.* 1997; Koester *et al.* 1998) and stellar evolution models (Langer *et al.* 1999) find small spin rates of isolated white dwarfs. However, assuming the white dwarf always rotates as a solid body implies that it approaches critical rotation as

$$\Omega = \frac{\omega}{\omega_{\text{crit}}} = \frac{3}{4k^2} \left(1 - \left(\frac{M_{\text{WD,i}}}{M_{\text{WD}}} \right)^{4/3} \right), \tag{4.24}$$

with ω and ω_{crit} being its rotation and critical rotation frequency (Langer *et al.* 2000). With a gyration constant of $k = 0.4$ (Ritter 1985) this means that white dwarfs with an initial mass of $M_{\text{WD,i}} = 0.6\,M_\odot$, $0.8\,M_\odot$, and $1.0\,M_\odot$ reach critical rotation at $0.71\,M_\odot$, $0.96\,M_\odot$, and $1.20\,M_\odot$, respectively.

The real situation may be worse, as the time scale to transport the angular momentum from the white dwarf surface into its degenerate core need to be considered. Clearly, any finite angular momentum redistribution time makes the white dwarf envelope reach critical rotation earlier (cf. Yoon & Langer 2002, Yoon *et al.* 2002). Without any doubt, if Chandrasekhar mass white dwarfs are responsible for Type Ia supernovae, there must be a way to remove angular momentum from the accreting stars or from the accreted material. At present, it is an open question of how this can be achieved (cf. Livio & Pringle 1998).

4.2.4. *Outlook*

Type Ia supernovae exist, and their spectra appear most consistent with models of exploding Chandrasekhar-mass white dwarfs (Höflich *et al.* 1998). I.e., there is likely a solution to the problem of unstable shell burning and spin-up in accreting white dwarfs. There may be one promising line of thaught which combines both: to investigate the effect of rotation on the behaviour of the burning shells. In preliminary models, Yoon & Langer (2002) find that rapidly rotating white dwarfs can stabilise their otherwise unstable helium shells. In the near future, we will investigate this effect in detail in order to understand whether it has the potential to alleviate the problem of forming Chandrasekhar-mass white dwarfs.

A nucleosynthesis consequence of the accreting white dwarf models with rotation of Yoon & Langer (2002) is a strongly reduced carbon abundance in the part of the white dwarf which was added due to helium shell burning (cf. Fig. 19). This might not only reduce the carbon yield of Type Ia supernovae, but influence the energy production and the light curve of the exploding white dwarf, as the buring of oxygen produces less energy than that of carbon.

5. The Most Massive Stars

5.1. *The Eddington limit*

The Eddington limit is not only important in determining the upper mass limit of stars, but it is likely to induce strong mass loss once a massive star comes close to it. I.e., the Eddington limit may shape the initial-final mass relation for the most massive star, with severe consequences for nucleosynthesis of may species.

The most spectacular phenomenon in connection with the Eddington limit, associated with Luminous Blue Variables (LBVs), are certainly their giant eruptions, which can last several years and may be recurrent on a time scale of centuries (cf. Humphreys & Davidson 1994), η Carinae's A.D. 1840–1860 and P Cygni's A.D. 1600 outbursts being the major examples. Circumstellar shells, which exist around virtually all LBVs (Nota *et al.* 1995), may be considered as fossil records of such outbursts (García-Segura *et al.* 1996). No fully self-consistent models for the giant eruptions do yet exist; however, among other ideas, the evolution of massive stars towards the Eddington limit has been put forward as an explanation by various authors (cf. Humphreys & Davidson 1994, and references therein). In the following we show that, in case the Eddington limit is involved in LBV eruptions, in fact not the Eddington limit itself but rather critical rotation defines the hydrostatic stability limit, no matter what the actual rotation rate of the star is.

5.1.1. *Does the Eddington limit apply in the stellar interior?*

The Eddington limit, in brief, is a limit to hydrostatic stability defined by the condition that the outwards directed force induced by the radiation momentum balances the inwards directed gravity. We want to emphasise that, as a stability limit, the Eddington limit applies only to hydrostatic situations. This does not mean that radiation forces are unimportant in hydrodynamic situations (e.g stellar pulsations); however, a stability limit makes of course no sense in an unstable situation.

The "classical Eddington limit" $L > L_{\text{edd}} := 4\pi c G M / \kappa_e$, with L and M being stellar luminosity and mass, and κ_e the electron scattering opacity, is derived from the momentum equation as $a_{\text{rad}} + g > 0$, where $a_{\text{rad}} = \kappa_e L / (4\pi R^2 c)$ is the radiative acceleration and the gravity $g = GM/R^2$, with R being the stellar radius. Instead of the above definition of the Eddington limit, one may introduce the dimensionless Eddington factor

$\Gamma := L/L_{\mathrm{edd}}$, and by replacing κ_e by the true flux-mean opacity coefficient κ one gets

$$\Gamma = \frac{\kappa}{4\pi cG}\frac{L}{M} \qquad (5.25)$$

and instability for $\Gamma > 1$.

Eq.(5.25) can be generalised to be evaluated in the stellar interior, i.e. $\Gamma(r) = \frac{\kappa(r)}{4\pi cG}\frac{L(r)}{M(r)}$, and some authors used the condition $\Gamma(r) > 1$ as stability criterion. However, this is wrong since for $\Gamma(r) \to 1$ convection must set in. This can be seen for the case of LBVs, where the equation of state of gas and radiation applies as $P := P_{\mathrm{gas}} + P_{\mathrm{rad}} := \frac{\Re}{\mu}\rho T + \frac{a}{3}T^4$, even when we neglect ionization which would reduce the adiabatic temperature gradient ∇_{ad} and make convection more likely. With $\beta := P_{\mathrm{gas}}/P$, we have

$$\nabla_{\mathrm{ad}} = \frac{8 - 6\beta}{32 - 24\beta - 3\beta^2} \qquad (5.26)$$

(Kippenhahn & Weigert 1990). The radiative temperature gradient

$$\nabla_{\mathrm{rad}} := \frac{3}{16\pi acG}\frac{\kappa(r)L(r)P}{M(r)T^4} \qquad (5.27)$$

can be written as

$$\nabla_{\mathrm{rad}} = \frac{\Gamma(r)}{4\,(1 - \beta)}. \qquad (5.28)$$

The Schwarzschild criterion for convection $\nabla_{\mathrm{rad}} \geq \nabla_{\mathrm{ad}}$ can thus be expressed as

$$\Gamma(r) \geq (1 - \beta)\frac{32 - 24\beta}{32 - 24\beta - 3\beta^2}. \qquad (5.29)$$

Since, for $\beta > 0$, the r.h.s. of (4) is always smaller than 1, the Schwarzschild criterion is fulfilled for $\Gamma(r) \to 1$.

Since the convective flux does not contribute to the radiative acceleration one might think of a modified Eddington limit in the stellar interior according to

$$\Gamma'(r) := \frac{\kappa(r)}{4\pi cG}\frac{L(r) - L_{\mathrm{conv}}(r)}{M(r)} > 1, \qquad (5.30)$$

where the convective luminosity $L_{\mathrm{conv}}(r)$ is subtracted. Especially in the deep stellar interior where convection is almost adiabatic, almost arbitrarily large luminosities can be transported through convection. Therefore, even for $\Gamma(r) \gg 1$, $\Gamma'(r) < 1$ is the result and no instability occurs.

However, even for convection zones close to the stellar surface, where convection becomes strongly non-adiabatic and $L_{\mathrm{conv}} \to 0$, $\Gamma'(r) = \Gamma(r) > 1$ is *not* a criterion for instability. This can be seen when the Schwarzschild criterion is written in entropy formulation as $\frac{ds}{dr} \leq 0$. Since $\frac{ds}{dr} \propto \frac{1}{a^2}\frac{dP}{dr} - \frac{d\rho}{dr}$ (with a being the adiabatic sound speed), we have for adiabatic convection $(\frac{ds}{dr} = 0)$ $\frac{d\rho}{dr} = \frac{1}{a^2}\frac{dP}{dr}$. As in a hydrostatic situation the total pressure decreases outwards, the density must do the same. But for strongly non-adiabatic convection $\frac{ds}{dr}$ is large and negative, and thus a *positive* density gradient may be established. This leads to a positive gas pressure gradient, which can stabilise the stratification even for $\Gamma'(r) > 1$.

In summary, the Eddington limit, in whichever form, can not be applied in the stellar interior to predict instability. It only applies at the stellar surface, where density inversions can not occur and the convective flux is negligible. In the next section, however, we show that the Eddington limit by itself should also not applied at the stellar surface.

5.1.2. *The Ω-limit*

In a hydrostatic rotating star, the momentum equation has to include the centrifugal acceleration $a'_{\text{cen}}(r,\theta) = a_{\text{cen}}(r)cos(\theta)$, where θ is the latitudal coordinate. To obtain a stability criterion, we follow the previous section and only evaluate the momentum equation at the stellar surface, where density inversions and convective flux contributions are negligible. Furthermore, we concentrate on the equatorial plane, since there the destabilising centrifugal term is largest. Defining $v_{\text{rot}}^2/R := a_{\text{cen}}(R,0°)$, we are left with a momentum balance as $a_{\text{rad}} + g + v_{\text{rot}}^2/R = 0$ or $\frac{1}{R}\left(\frac{\kappa L}{4\pi Rc} + \frac{GM}{R} + v_{\text{rot}}^2\right) = 0$. We can thus define

$$v_{\text{crit}}^2 := \frac{GM}{R}(1 - \Gamma) \qquad (5.31)$$

and obtain as criterion for instability

$$\Omega := \frac{v_{\text{rot}}}{v_{\text{crit}}} > 1. \qquad (5.32)$$

It is crucial to realize that for $\Gamma \to 1$ Eq. (5.31) gives $v_{\text{crit}} \to 0$. Therefore, if we assume a star to evolve towards the Eddington limit ($\Gamma \to 1$), no matter what its rotation rate may be, it will arrive at critical rotation well *before* $\Gamma = 1$ is actually reached. Therefore, one may rather speak of the "Ω-limit" instead of the Eddington limit. Note that in this simplified approach gravity darkening is neglected; however, Maeder (1999) found this to not change our conclusions for the most massive stars qualitatively.

As a consequence, the location of the occurrence of the instability in the HR diagram depends on the stellar rotation rate. In contrast to the Eddington limit, the Ω-limit is not — for a fixed metallicity and L/M-relation — a single line in the HR diagram. Furthermore, while the most massive stars may arrive at the Ω-limit independent of their rotation speed, it may happen that two less massive single stars start with the same initial mass and metallicity, but only the faster rotating one arrives at the Ω-limit while the slower one does not and evolves further into the red supergiant region. This scenario might apply to the less luminous LBVs, which do have red supergiant counterparts of equal luminosity.

A second consequence is that the outflow which occurs due to a star hitting the Ω-limit is predicted to be highly bipolar. This is illustrated by a hydrodynamical simulation of the giant eruption of η Car by Langer *et al.* (1999). The fact that virtually all LBV nebulae are highly bipolar supports the idea that the Ω-limit is actually involved in the LBV instability.

5.1.3. *Rotating very massive stars*

Let us now describe several effects which may happen when massive stars approach the Ω-limit. Striking qualitative features can be derived from simplified models of rotating, mass losing $60\,M_\odot$ stars (Langer 1998).

In calculating these models, it was assumed that the stars remain always rigidly rotating, which is justified given the short time scales of angular momentum redistribution (Zahn 1992, Maeder 1997), and confirmed by more sophisticated models (Heger *et al.* 2000). The angular momentum is carried only as a passive quantity, i.e. any feedback of rotation on the stellar structure is neglected. For stars approaching the Ω-limit, this may appear oversimplified; however, note that Ω becomes close to unity only in a tiny fraction of the the stellar mass due to the iron opacity peak, while it remains small in the rest of the star which always remains perfectly spherical. Much more uncertain is the evaluation of the Eddington factor Γ. In the models presented here, the maximum value occurring in the gray plane-parallel atmosphere down to an optical depth of $\tau = 100$ was adopted; thereby the convective flux and density inversions can be neglected. However, when Ω approaches 1, these layers may expand in the equatorial region of the star. Since the expansion will lead to cooling, the opacity in these layers will increase. Furthermore, viscous coupling would increase their specific angular momentum and bring them even closer the Ω-limit. Therefore, the quoted values of Γ and Ω may in fact be lower limits.

In the non-rotating case, the value of Γ evaluated as described above is 0.78 on the zero age main sequence (log $T_{\rm eff}$ = 50 000 K) and rises to 0.96 at $t = 2.6\,10^6$ yr (log $T_{\rm eff}$ = 25 000 K) due to the iron opacity peak. The core H-burning phase is finished at $t = 3.4\,10^6$ yr and log $T_{\rm eff}$ = 20 000 K with $\Gamma = 0.84$. The evolution of Γ is reflected in the behaviour of the critical rotational velocity (cf. Eq. 5.31) as shown in Fig. 20; it has a pronounced minimum of $v_{\rm crit,min} \simeq 100\,{\rm km\,s}^{-1}$ at $t \simeq 2.6\,10^6$ yr. Since the rotational velocity tends to remain almost constant on the main sequence (Fig. 20) the question arises: what happens to a star with an initial rotational velocity above $\sim 100\,{\rm km\,s}^{-1}$?

As shown by Friend & Abbott (1986), rotating massive stars have an enhanced radiation driven stellar wind mass loss well before they actually hit the Ω-limit, with an enhancement factor

$$\frac{\dot{M}}{\dot{M}(v_{\rm rot} = 0)} = \left(\frac{1}{1-\Omega}\right)^\xi, \quad \xi \simeq 0.43. \tag{5.33}$$

Formally, $\dot{M} \to \infty$ for $\Omega \to 1$ in Eq. (5.33). However, the mass loss remains finite due to the effect that it imposes a loss of angular momentum which results in a (non-magnetic) spin-down of the star: In rigidly rotating stars, the specific angular momentum increases from the center to the surface; due to mass loss, it is continuously transported from the core to the envelope where it is blown off. Therefore, the mass loss rate (due to Eq. 5.33) of the considered stars at the Ω-limit, $\dot{M}_\Omega$, is self-regulated by the condition $\dot{M}_\Omega = \dot{M}(v_{\rm rot} = v_{\rm crit})$.

It can be seen from Fig. 20 that $\dot{M}_\Omega \simeq 10^{-5}\ {\rm M}_\odot\ {\rm yr}^{-1}$ for the considered $60\,{\rm M}_\odot$ sequences. It should be stressed that, although the model is simplified and the analysis only qualitative, the derived mass loss rate at the Ω-limit is expected to be even quantitatively correct, since it is established by the rate of angular momentum loss which is associated with the mass loss. Once the critical rotational velocity is reached, the angular momentum loss rate is set by the expansion time scale of the star (i.e. its speed in the HR diagram) but does not depend on the value of the critical velocity.

A further striking feature shown in Fig. 20 is the convergence of the rotational velocities to the value of $v_{\rm crit,min}$ (i.e., $\simeq 100\,{\rm km\,s}^{-1}$), which is produced by the fact that the time dependence of the critical rotational velocity is almost independent of the mass loss history, in particular its minimum value. Note, however, that although the rotation velocities of the four sequences displayed in Fig. 20 at the end of the main sequence evolution are almost identical, their masses at that time are greatly different, and thus their ensuing post-main sequence evolution is expected to be very different as well.

These main sequence models at the Ω-limit can certainly not be related to LBVs, but perhaps rather to B[e] stars; the material of stars at the Ω-limit which has excess angular momentum can not remain on the star but it is also not clear whether it can be pushed to infinity; some of it may form a ring or disk close to the star. However, as the mass loss rate at the Ω-limit depends on the evolutionary speed of the star in the HR diagram, one can expect values of the order $10^{-5}\ {\rm M}_\odot\ {\rm yr}^{-1} \times \tau_{\rm H}/\tau_{\rm KH}$ for the expansion phase after core H-exhaustion, when the star hits the Ω-limit again due to the helium opacity peak. Here, $\tau_{\rm H}$ is the H-burning time scale and $\tau_{\rm KH}$ the thermal time scale of the star, such that values of $\dot{M}_\Omega$ of the order of $10^{-3} - 10^{-2}\ {\rm M}_\odot\ {\rm yr}^{-1}$ can be expected, which is the order of magnitude observed for LBVs.

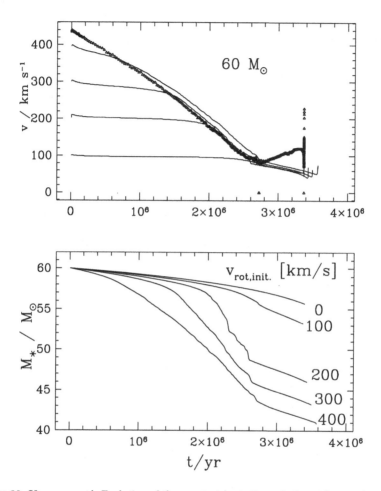

FIGURE 20. **Upper panel:** Evolution of the equatorial rotation velocity with time during the core hydrogen burning phase of four $60\,M_\odot$ sequences with different initial rotation rates (see at $t = 0$). The evolution of the critical rotation velocity (Eq. 5.31) is displayed for the sequence with $v_{\rm rot,init.} = 100\,\mathrm{km\,s^{-1}}$ by the triangles. It is very similar for the other sequences. For $v_{\rm crit} \simeq v_{\rm rot}$, the stars evolve at the Ω-limit. **Lower panel:** Evolution of the stellar mass with time for the same $60 M_\odot$ sequences. The initial equatorial rotation velocities are given as labels. For comparison, the evolution of a non-rotating star is shown in addition.

5.2. *Evolution of very massive stars*

Evolutionary calculations for stars of even higher masses have been presented by Figer *et al.* (1998). Fig. 21 shows their stellar evolutionary tracks with a metallicity of 2 % and initial masses in the range $60...300\,M_\odot$. These models are calculated with a hydrodynamic stellar evolution code described in Langer *et al.* (1994), with an outer boundary condition adequate for spherically symmetric winds of arbitrary optical thickness (Heger & Langer

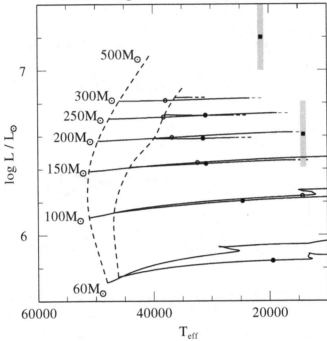

FIGURE 21. Evolutionary tracks of non-rotating models in the initial mass range 60 ... 300 $M_\odot$ and metallicities of Z=0.04 (thick continuous lines) and Z=0.02 (thin continuous lines; Figer *et al.* 1998). Dashed lines mark the corresponding zero age main sequences. Filled and open dots on the tracks mark the first occurrence of nuclear processed material at the stellar surface, for Z=0.04 and Z=0.02, respectively. Two possible positions of the Pistol star are indicated (cf. Figer *et al.*).

1996). Due to the lack of empirical mass loss rates in the regime of the calculations, the radiation driven wind theory according to Kudritzki *et al.* (1989) was applied, with wind parameters $k = 0.085$, $\alpha = 0.657$, $\delta = 0.095$, and $\beta = 1$ (Pauldrach *et al.* 1994). I.e., the dependence of the mass loss rate on the luminosity, effective temperature, mass and surface hydrogen mass fraction was taken into account.

The models lose less mass than, e.g., those of Schaller *et al.* (1992); the 60, 100, and 150 $M_\odot$ sequences have, at core hydrogen exhaustion, 54, 84, and 110 $M_\odot$, respectively. The 200, 250, and 300 $M_\odot$ sequences are terminated at central hydrogen mass fractions of $X_c = 0.16$, 0.22, and 0.29 where the remaining masses are 158, 197, and 244 $M_\odot$, due to a hydrodynamic instability occurring at the stellar surface (cf. below).

The reason for the smaller amounts of lost mass compared to Schaller *et al.* (1992) and even more so compared to Meynet *et al.* (1994) is understood as follows. Langer *et al.* (1994) and Meynet *et al.* (1994) argued that the average amount of lost mass during core hydrogen burning for very massive stars (e.g. $\sim 60\,M_\odot$) should be larger than the values resulting from semi-empirical mass loss formulae (e.g. that of de Jager *et al.* 1988) or current radiation driven wind models (e.g. Kudritzki *et al.* 1989) in order to obtain agreement with observed mean properties of very massive post-main sequence stars, e.g. Wolf-Rayet stars. This apparent contradiction is resolved by considering the influence

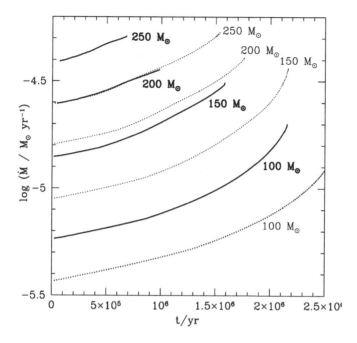

FIGURE 22. The mass loss rate of 100 ... 250 $M_\odot$ stellar evolution models (cf. Fig. 21) — calculated according to Kudritzki *et al.* (1989) — as function of time, for Z=0.04 (continuous lines) and Z=0.02 (dotted lines).

of rotation on the main sequence mass loss of very massive stars. Langer (1998) finds that, depending on their initial rotation rate, $60\,M_\odot$ stars can end their core hydrogen burning with masses in the range 56 ... $41\,M_\odot$, the largest value corresponding to the non-rotating case.

In the models of Figer *et al.* (1998), which were constructed to understand a very luminous but relatively cool star, the so called Pistol Star, it was assumed that the rotation rate is small, since several investigations have shown that rapidly rotating very massive stars, due to rotationally induced internal mixing processes, do not evolve to cool surface temperatures (Maeder 1987, Langer 1992, Fliegner & Langer 1995, Meynet 1998). Consequently, the use of a mass loss in the lower part of the possible range — e.g. that according to Kudritzki *et al.* (1989) — seemed appropriate.

For the same reason, no "convective core overshooting" was invoked. This has been applied in many massive star calculations in the recent years in order to obtain a wider main sequence band (cf., e.g., Schaller *et al.* 1992). However, as rotationally induced mixing has a very similar effect (e.g., Langer 1992, Fliegner *et al.* 1996), the interpretation that the main sequence widening is due to rotation and thus that the convective cores of non-rotating stars are not extended over their sizes predicted by the Schwarzschild criterion was adopted.

Furthermore, massive stars are supposedly becoming unstable when they approach or penetrate the so called Humphreys-Davidson limit in the HR diagram (Humphreys & Davison 1994). We have seen above that, whichever instability mechanism causes the

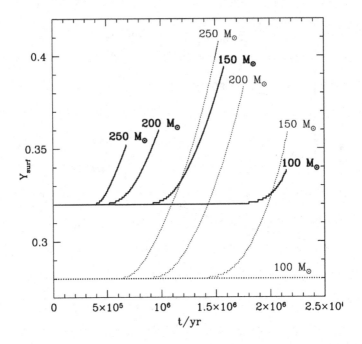

FIGURE 23. Evolution of the surface helium mass fraction Y_{surf} as function of time for the same stellar evolution models as shown in Fig. 21, for Z=0.04 (continuous lines) and Z=0.02 (dotted lines).

existence of the HD-limit, its location in the HR diagram must strongly depend on the rotation rate of the considered stars. I.e., in case the Eddington-limit is the physical cause of the HD-limit, the so called Ω-limit, i.e. the corresponding instability limit for rotating stars (Langer 1997), is located at lower luminosities for more rapidly rotating stars. Or, vice versa, the instability limit for slowly- or non-rotating stars is located at higher luminosity than that for stars rotating with the average rate.

The finding of Figer *et al.* that the most massive stellar models become unstable at around $T_{eff} \simeq 20\,000$ K and $\log L/L_\odot \gtrsim 6.5$, together with the coincidence of the position of the Pistol Star with these figures, lead to the proposition of the following scenario. The Pistol Star is unusually massive and may therefore have an unusual formation history. Apparently, it has obtained very little angular momentum, which perhaps allowed its mass to grow so much. Consequently, it evolved with less-than-average mass loss towards the cool side of the HR diagram. Still during core hydrogen burning, it arrived at its Eddington-limit and strongly increased its mass loss rate. This is its present evolutionary stage. As it is still burning hydrogen in its core, the probability for the star to be found in this stage is not small. According to the evolutionary models, its initial mass is 200 ... 250 $M_\odot$, and its present age is in the range 1.7 ... $2.1\,10^6$ yr. The amount of mass lost prior to the occurrence of the surface instability is 42 ... 53 $M_\odot$.

5.3. *Supermassive stars*

What are the largest possible stellar masses? Even though in practice, the answer to question may be constrained by the physics of star formation, we shall have a look at possible answers from the point of view of stellar structure and stability.

In many places in the literature one can find the statement that the Eddington limit determines the largest possible stellar mass: as $L_{Edd} = 4\pi cGM/\kappa$, and L grows faster than M with increasing stellar mass, the Eddington limit must be hit at a certain stellar mass. However, this argument is flawed. As mentioned above, the more massive a massive star is, the more is the pressure inside the star dominated by radiation pressure. With β being again defined as $\beta = P_{gas}/P = (1 - P_{rad})/P$, homology relations yield

$$\frac{1 - \beta}{\beta} \sim M^2 \tag{5.34}$$

(Kippenhahn & Weigert 1990), implying a strong decrease of β with M. Radiation pressure dominated stars are almost fully convective (see above; Eq. 5.29), and fully convective stars can be well approximated by $n = 3$ polytropes, for which

$$\beta = \frac{4.3}{\mu} \left(\frac{M_\odot}{M} \right)^{1/2} \tag{5.35}$$

(Fuller *et al.* 1986).

Kippenhahn & Weigert (1990) have shown that for $\beta \to 0$ it is

$$L \sim M . \tag{5.36}$$

Besides the immediate consequence that the lifetime of radiation dominated stars becomes independent of the mass, as the amount of fuel is proportional to the rate at which they spend it, this obviously topples the premise in the flawed argument above that L grows faster than M. To be more quantitative, we may combine the hydrostatic equation

$$\frac{dP}{dr} = -\rho \frac{GM_r}{r^2}, \tag{5.37}$$

with the radiative diffusion equation

$$L_r = 4\pi r^2 \frac{ac}{3\kappa\rho} \frac{dT^4}{dr} . \tag{5.38}$$

We may assume radiative diffusion despite radiation dominated stars are almost fully convective, since Eq. (5.28) indicates that

$$\nabla_{rad} \to 0.25 \quad \text{for} \quad \beta \to 0 , \tag{5.39}$$

and Eq. (5.26) shows that also

$$\nabla_{add} \to 0.25 \quad \text{for} \quad \beta \to 0 , \tag{5.40}$$

which means that neglecting the convective energy flux is a good approximation (cf. also Hillebrandt *et al.* 1987). Inserting $P = \frac{1}{1-\beta}\frac{a}{3}T^4$ into Eq. (5.37), and assuming $(1 - \beta)$ to be constant throughout the star, we obtain for the stellar surface

$$L = \frac{4\pi cG}{\kappa}(1 - \beta)M = (1 - \beta)L_{Edd} . \tag{5.41}$$

This confirms not only that indeed L is proportional to M, but also that L remains *smaller* than L_{Edd} however large the mass may be, if only by the factor $1 - \beta$.

However, supermassive stars encounter more stability problems than just the Eddington limit. The almost completely adiabatic structure of radiation dominated stars implies an adiabatic index

$$\gamma_{ad} := \left(\frac{d\ln P}{d\ln \rho} \right)_{ad} \tag{5.42}$$

to be

$$\gamma_{ad} \simeq \frac{4}{3} + \frac{\beta}{6} + O(\beta^2) \tag{5.43}$$

to first order in β (Chandrasekhar 1939). I.e., radiation dominated stars are very close to the critical value of $\gamma_{ad} = 4/3$ and thus "trembling on the verge of instability" as Fowler (1964) expressed it. Any small disturbance may push them over the edge towards instability. It turns out that indeed for stars above $\sim 10^5 \, M_\odot$, general relativistic effects destabilise contracting pre-main sequence stars before they can ignite hydrogen burning. The extra gravity caused by the thermal energy of the star is sufficient to make them collapse (Hoyle & Fowler 1963, Appenzeller & Fricke 1972, Fricke 1974, Fuller *et al.* 1986, Hillebrandt *et al.* 1987), which remains true if rotation is considered (Fricke 1974, Baumgarte & Shapiro 1999a, 1999b, Shibata & Shapiro 2002). As the ensuing explosive hydrogen burning is unable to turn the collapse into an explosion for low metallicity, supermassive stars are discussed as an intermediate stage towards forming supermassive black holes which are found to exist in the cores of galaxies in increasing number (Ferrarese & Merritt 2000).

REFERENCES

AKSENOV, A. G., ZABRODINA, E. A., IMSHENNIK, V. S. & NADEZHIN, D. K., 1997 *Astronomy Letters* 23, 677

ANDERS, E. & GREVESSE, N., 1989, *GeoChim. CosmoChim.*, 53, 197

APPENZELLER, I. & FRICKE, K., 1972, A&A 21, 285

ARCORAGI, J.-P., LANGER, N., ARNOULD, M., 1991, A& A, 249, 134

ARNOULD, M. & MOWLAVI, N., 1993, in: *Inside the Stars*, IAU Colloquium 137, ASP Conference Series, Vol. 40. W. W. Weiss & A. Baglin eds., San Francisco: Astronomical Society of the Pacific, p. 310

AUBERT, O., PRANTZOS, N. & BARAFFE, I., 1996, A&A 312, 845

BARAFFE, I. & TAKAHASHI, K., 1993, A&A 280, 476

BAUMGARTE, T. W. & SHAPIRO, S. L., 1999a, ApJ 526, 937

BAUMGARTE, T. W. & SHAPIRO, S. L., 1999b, ApJ 526, 941

BIERMANN, P. L., LANGER, N., SEO, EUN-SUK, STANEV, T., 2001, A&A 369, 269

BRAUN, H., 1997, PhD thesis, LMU München

BRAUN, H. & LANGER, N., 1995 A&A 297, 483

BURBIDGE, E. M., BURBIDGE, G. R., FOWLER, W. A. & HOYLE, F., 1957, RMP 29, 547

CASSISI, S., IBEN, I. & TORNAMBE, A., 1998, ApJ 496, 376

CASUSO, E. & BECKMAN, J. E., 2000, PASP 112, 942

CAUGHLAN, G. R., FOWLER, W. A., HARRIS, M. J. & ZIMMERMAN, B. A., 1985, Atomic Data and Nucl. Data Tables, 32, 197

CHANDRASEKHAR, S., 1939, *Introduction to the Study of Stellar Structure*, University Press, Chicago

CLAYTON, D. D., 1968, *Principles of Stellar Evolution and Nucleosynthesis*, Univ. of Chicago Press

COWAN, J. J., CAMERON, A. G. W. & TRURAN, J. W., 1985, ApJ 294, 656

CUNHA, K., LAMBERT, D. L., LEMKE, M., GIES, D. R. & ROBERTS, L. C., 1997, ApJ 478, 211

DE DONDER, E. & VANBEVEREN, D., 2002, New Astron. 7, 55

DENISSENKOV, P., 1994, A&A, 287, 113

DOWNES, R. A., *et al.* , 2001, PASP 113, 764

EDMUNDS, M. G., GREENHOW, R. M., JOHNSON, D., KLÜCKERS, V. & VILA, M. B., 1991, MNRAS, 251, 33p

EL EID, M. F., 1994, A&A 285, 915

ENDAL, A. S. & SOFIA, S., 1978, ApJ, 220, 279

FERRARESE, L. & MERRITT, D., 2000, ApJL 539, L10

FIELDS, B. D. & OLIVE, K. A., 1999, ApJ 516, 797

FIGER, D. F., NAJARRO, F., MORRIS, M., MCLEAN, I. S., GEBALLE, T. R., GEHZ, A. M. & LANGER, N., 1998, ApJ 506, 384

FLIEGNER, J. & LANGER, N., 1995, in: *Wolf-Rayet Stars: Binaries, Colliding Winds, Evolution*, Proc. IAU-Symp. No. 163, K.A. van der Hucht *et al.* ed., Kluwer, p. 326

FLIEGNER, J., LANGER, N. & VENN, K. A., 1996 A&A 308 L13

FOWLER, W. A., 1964, Rev. Mod. Phys., 36, 545, 1104E

FRICKE, K., 1974, ApJ, 189, 535

FRIEND, D. B. & ABBOTT, D. C., 1986, ApJ 311, 701

FULLER, G. M., WOOSLEY, S. E. & WEAVER, T. A., 1986, ApJ 307, 675

GALLINO, R., ARLANDINI, C., BUSSO, M., LUGARNO, M., TRAVAGLIO, C., STRANIERO, O., CHIEFFI, A. & LIMONGI, M., 1998, ApJ 497, 388

GARCÍA-SEGURA, G., MAC LOW, M.-M. & LANGER N., 1996, A&A, 305, 299

GARMANY, C. D., CONTI, P. S. & MASSEY, P., 1980 ApJ 242, 1063

GIES, D. R. & LAMBERT, D. L., 1992, ApJ, 387, 673

GORIELY, S., JOSE, J., HERNANZ, M., RAYET, M. & ARNOULD, M., 2002, A&A 383, 27

GRATTON, R. G., & ORTOLANI, S., 1986, A&A 169, 201

GREINER, J., 2000, New Astron. 5, 137

GUÉLIN, M., FORESTINI, M., VALIRON, P., ANDERSON, M. A., CERNICHARO, J. & KAHANE C., 1995, A&A 297, 183

HACHISU, I., KATO, M. & NOMOTO, K., 1996, ApJ 470, L97

HEBER, U., NAPIWOTZKI, R. & REID, I. N., 1997, A&A 323, 819

HEGER, A. & LANGER, N., 1996, A&A, 315, 421

HEGER, A. & LANGER, N., 2000, ApJ, 544, 1016

HEGER, A., LANGER, N. & WOOSLEY, S. E., 2000, ApJ 528, 368

HEGER, A., WOOSLEY, S. E., RAUSCHER, T., HOFFMAN, R. D. & BOYES, M. M., 2002, New Astron. Rev., 46, 463

HENKEL, C., & MAUERSBERGER, R., 1993, A&A 274, 730

HENKEL, C., WILSON, T. L., LANGER, N., CHIN, Y.-N., & MAUERSBERGER, R., 1994, in: *The Structure and Content of Molecular Clouds*, T.L. Wilson, K.J. Johnston, eds., Springer (Lecture Notes in Physics 439), Berlin, p. 72

HERRERO, A., KUDRITZKI, R. P., VILCHEZ, J. M., KUNZE, D., BUTLER, K. & HASER, S., 1992, A&A, 261, 209

HERWIG, F., BLÖCKER, T., SCHÖNBERNER, D. & EL EID, M. F., 1997, A&A 324, L81

HILLEBRANDT, W., THIELEMANN, F.-K. & LANGER, N., 1987, ApJ 321, 761

HOFFMAN, R. D., WOOSLEY, S. E., FULLER, G. M. & MEYER, B. S., 1996, ApJ 460, 478

HOFFMAN, R. D., WOOSLEY, S. E., WEAVER, T. A., RAUSCHER, T. & THIELEMANN, F.-K., 1999, ApJ 521, 735

HÖFLICH, P., WHEELER, J. C. & THIELEMANN, F.-K, 1998, ApJ 495, 617

HOYLE, F. & FOWLER, W. A., 1963, MNRAS, 125, 169

HUMPHREYS, R. M. & DAVIDSON, K., 1994, PASP, 106, 1025

IBEN, I. JR. & RENZINI, A., 1982, ApJ 263, L23

IBEN, I. JR. & RENZINI, A., 1983, ARAA 21, 271

DE JAGER, C., NIHEUVENHUIJZEN, H. & VAN DER HUCHT, K. A., 1988, A&AS, 72, 259 M., Busso, M., Gallino, R., Limongi, M.

KATO, M. & HACHISU, I., 1999, ApJL 513, L41

KIPPENHAHN, R. & WEIGERT, R., 1990, *Stellar Structure and Evolution*, Springer, Berlin

KOESTER, D., DREIZLER, S., WEIDEMANN, V. & ALLARD, N. F., 1998, A&A 338, 617

KUDRITZKI, R. P., PAULDRACH, A., PULS, J. & ABBOTT, D. C., 1989, A&A, 219, 205

LAMB, S. A., HOWARD, W. M., TRURAN, J. W. & IBEN, I. JR., 1977, ApJ 217 213

LAMBERT, D. L., SMITH, V. V., BUSSO, M., GALLINO, R. & STRANIERO, O., 1995, ApJ 450, 302

LANGER, N., 1989a A&A 210, 93

LANGER, N., 1989b A&A 220, 135

LANGER, N., 1991a, A&A 252, 669

LANGER, N., 1991b, A&A 243, 155

LANGER, N., 1992, A&A 265, L17

LANGER, N., 1997, in *Luminous Blue Variables: Massive Stars in Transition*, A. Nota, H.J.G.L.M. Lamers, eds, ASP Conf. Ser., p. 83

LANGER, N., 1998 A&A 329 551

LANGER, N., ARCORAGI, J.-P. & ARNOULD, M., 1989 A&A 210 187

LANGER, N., BRAUN, H. & FLIEGNER, J., 1995 *Astrophys. Space Sci.* 224 275

LANGER, N., DEUTSCHMANN, A., WELLSTEIN, S. & HÖFLICH P., 2000, A&A 362, 1046

LANGER, N., FLIEGNER, J, HEGER, A. & WOOSLEY, S. E., 1997a *Nucl. Phys.* A621 457c

LANGER, N., HAMANN, W.-R., LENNON, M., NAJARRO, F., PULS, J. & PAULDRACH, A., 1994, A&A, 290, 819

LANGER, N., HEGER, A. & FLIEGNER, J., 1997b *Proc. IAU-Symp. No. 189 on Fundamental Stellar Properties* eds. T R Bedding *et al.*(Dordrecht: Reidel) p 343

LANGER, N., HEGER, A., WELLSTEIN S. & HERWIG, F., 1999, A&A 346, L37

LANGER, N. & HENKEL, C., 1995 *Space Sci. Rev.* 74 343

LANGER, N., SUGIMOTO, D. & FRICKE, K. J., 1983, A&A, 126, 207

LANGER, N. & WOOSLEY, S. E., 1996, in: *From Stars to Galaxies — The Impact of Stellar Physics on Galaxy Evolution*, eds. C. Leitherer, U. Fritze-von Alvensleben, & J. Huchra, ASP Conf. Series Vol. 98, p. 220

LENNON, D. J., 1994, Space Sci. Rev., 66, 127

LI, X.-D. & VAN DEN HEUVEL, E. P. J., 1997, A&A 322, L9

LIMONGI, M., STRANIERO, O. & CHIEFFI, A., 2000, ApJS 129, 625

LITTLE, S. J., LITTLE-MARENIN, I. R. & HAGEN BAUER, W., 1987, AJ 94, 981

LIVIO, M., 2000, in: *Type Ia Supernovae, Theory and Cosmology*, J. C. Niemeyer and J. W. Truran, eds., Cambridge University Press, p. 33

LIVIO, M. & PRINGLE, J. E., 1998, ApJ 505, 339

MAEDER, A., 1987, A&A 178, 159

MAEDER, A., 1992, A&A 264, 105

MAEDER, A., 1995, A&A 299 84

MAEDER, A., 1997, A&A 321 134

MAEDER, A., 1999, A&A 347, 185

MARIGO ,P., 2001, A&A 379, 194

MASON, B. D., GIES, D. R., HARTKOPF, W. I., BAGNUOLO, W. G. JR., BRUMMELAAR, T. T. & MCALISTER, H. A., 1998 ApJ 115 821

MENEGUZZI, M., AUDOUZE, J. & REEVES, H., 1971, A&A, 15, 337

MEYNET G., 1998, *Boulder-Munich II: Properties of Hot, Luminous Stars*, ASP Conference Series, Vol. 131. I. Howarth ed. San Francisco: Astronomical Society of the Pacific, p. 96

MEYNET, G., ARNOULD, M., PRANTZOS, N. & PAULUS, G., 1997 A&A 320 460

MEYNET, G., MAEDER, A., SCHALLER G., SCHAERER, D. & CHARBONNEL, C., 1994, A&AS, 103, 97

MOWLAVI, N., JORISSEN, A. & ARNOULD, M., 1998, A&A 334, 153

NOMOTO, K. & KONDO Y., 1991, ApJ 367, L19

NOTA, A., LIVIO, M., CLAMPIN, M. & SCHULTE-LADBECK, R., 1995, ApJ, 448, 788

PAULDRACH, A. W. A., KUDRITZKI, R. P., PULS, J., BUTLER, K. & HUNSINGER, J., 1994, A&A, 283, 525

PETERS, J. G., 1968, ApJ 154, 225

PINSONNEAULT, M., 1997 ARA&A 35 557

PODSIADLOWSKI, P., 1997 in: *Evolutionary Processes in Binary Stars* eds. R A M J Wijers *et al.*NATO ASI Ser. C Vol. 477 (Dordrecht: Reidel) p 181

PRANTZOS, N., AUBERT, O. & AUDOUZE, J., 1996, A&A, 309, 760

PRANTZOS, N. & DIEHL, R., 1996 *Phys. Rep.* 267 1

PRANTZOS, N., HASHIMOTO, M. & NOMOTO, K., 1990, A&A, 234, 211

PRANTZOS, N., VANGIONI-FLAM, E., & CHAUVEAU, S., 1994 A&A, 285, 132

PROFFITT, C. R., JÖNSSON, P., PICKERING, J. C. & WAHLGREN, G. M., 1999, ApJ, 516, 342

QIAN, Y.-Z., 2000, ApJ 534, 67

RAUSCHER, T., HEGER, A., HOFFMAN, R. D. & WOOSLEY S. E., 2002, ApJ 576, 323

RAYET, M., ARNOULD, M., HASHIMOTO, M., PRANTZOS, N. & NOMOTO, K., 1995, A&A 298, 517

RAYET, M. & HASHIMOTO, M., 2000 A&A 354, 740

REEVES, H., FOWLER, W. A. & HOYLE, F., 1970, Nature, 226, 727

RITOSSA, C., GARCIA-BERRO, E. & IBEN, I. JR., 1996 ApJ 460, 489

RITTER, H., 1985, A&A 148, 207

SCHALLER, G., SCHAERER, D., MEYNET, G. & MAEDER, A., 1992, A&AS, 96, 269

SCHEITHAUER, S., 2000, Master Thesis, Potsdam University

SCHÖNBERNER, D., HERRERO, A., BUTLER, K., BECKER, S., EBER, F., KUDRITZKI, R. P. & SIMON, K. P., 1988, A&A, 197, 209

SHIBATA, M. & SHAPIRO, S. L., 2002, ApJ 572, 39

SMITH, V. V. & LAMBERT, D. L., 1990, ApJS 72, 387

SPRUIT, H.C., 1998 A&A 333, 603

THIELEMANN, F.-K., NOMOTO, K. & HASHIMOTO, M., 1996 ApJ 460 408

TIMMES, F. X., WOOSLEY, S. E. & WEAVER T. A., 1995, ApJS, 98, 617

VAN DEN HOEK, L. B., GROENEWEGEN, M. A. T., 1997, A&AS 123, 305

VASSILIADIS, E. & WOOD, P. R., 1993, ApJ, 413, 641

VENN, K. A., 1995, ApJ, 449, 839

VENN, K. A., BROOKS, A. M., LAMBERT, D. L., LEMKE, M., LANGER, N., LENNON, D. J., KEENAN, F. P., 2002, ApJ 565, 571

VENN, K. A., LAMBERT, D. L., LEMKE, M., 1996, A&A, 307, 849

WALBORN, N., 1988, in IAU-Symp. No. 108 on *Atmospheric Diagnostics of Stellar Evolution*, K. Nomoto, ed., p. 70

WALLERSTEIN, G. & KNAPP, G. R., 1998, ARA&A 36, 369

WASSERBURG, G. J. & QIAN, Y.-Z., 2000, ApJ 529, 21

WEAVER, T. A. & WOOSLEY, S. E., 1993 *Phys. Rep.* 227 65

WELLSTEIN, S., LANGER, N. & BRAUN, H., 2001, A&A, 369, 939

WILSON, T., & ROOD, R., 1994 ARAA, 32, 191

WOOSLEY, S. E., LANGER, N. & WEAVER, T. A., 1993, ApJ 411, 823

WOOSLEY, S. E., LANGER, N. & WEAVER, T. A., 1995 ApJ 448, 315

WOOSLEY, S. E. & WEAVER, T. A., 1995 ApJS 101 181

YAMADA, S. & SATO, K., 1994 ApJ 434 268

YOON, S.-C. & LANGER, N., 2002, *The Physics of Cataclysmic Variables and Related Objects*, ASP Conference Proceedings, Vol. 261. B. T. Gänsicke, K. Beuermann, & K. Reinsch eds. San Francisco: Astronomical Society of the Pacific, p. 79

YOON, S.-C., LANGER, N. & SCHEITHAUER, S., 2002, A&A, in preparation

ZAHN, J. P., 1983, in 13th Saas-Fee Course on *Astrophysical Processes in Upper Main Sequence Stars*, B. Hauck, A. Maeder, eds.

ZAHN, J. P., 1992, A&A, 265, 115

Observational Aspects Of Stellar Nucleosynthesis

By D A V I D L. L A M B E R T

Department of Astronomy, University of Texas, Austin, TX 78712, USA

1. Introduction

The origins of the chemical elements must rank highly in any intelligent citizen's list of questions about the natural world. Thanks to the efforts of observers and theoreticians over the last half-century, the citizen may now be provided with answers to 'Where, when, and how were the elements made?' This remarkable achievement of astrophysics provides one focus for this set of lectures. It is impossible to tell in the available space the complete story of nucleosynthesis from hydrogen to uranium (and beyond) with full justice to the observational and theoretical puzzles that had to be addressed.†

Nucleosynthesis began with the Big Bang (see Steigman's contribution to this volume). According to the standard model of this event, nucleosynthesis completed in the first few minutes of the Universe's life resulted in gas composed of ^{1}H, and ^{4}He with ^{1}H/^{4}He $\simeq$ 0.08 by number of atoms, and trace amounts of ^{2}H, ^{3}He, and ^{7}Li. The inability of the rapidly cooling low density Big Bang to synthesise nuclides beyond mass number 7 is due to the fact that all nuclides of mass number 5 and 8 (i.e., potential products from ^{1}H + ^{4}He and ^{4}He + ^{4}He) are highly unstable.

Ashes of the Big Bang cooled. The photons of the cosmic microwave background radiation were set free to roam the Universe. Then came what is known as 'The Dark Ages' before galaxies were formed. One view is that nucleosynthesis did not resume until stars appeared in galaxies. An alternative view is that pre-galactic objects – very massive or supermassive stars – seeded the primordial gas from which galaxies formed. Evidence for these views and the nucleosynthesis achieved by pre-galactic objects is not in my remit (see contributions by Langer and Pettini).

Nucleosynthesis by stars habiting regions in galaxies is referred to as stellar nucleosynthesis, a label covering a variety of sites and nuclear processes. Much of the nucleosynthesis is accomplished by three main types of stars: (i) Massive stars (here, initial mass $M \geq 8M_\odot$ but $M \leq 100M_\odot$) perform nucleosynthesis up to and through their death as a supernova of Type II (SN II) or Type Ib; (ii) Intermediate and low mass stars ($M \leq 8M_\odot$) make their principal contributions – the products of H- and He-burning and the s-process – as asymptotic red giant branch stars and feed these contributions to the interstellar medium via a wind; (iii) White dwarfs gaining mass from a companion star and attaining the Chandrasekhar mass explode as a supernova of Type Ia (SN Ia) and experience explosive nucleosynthesis. Under different circumstances, the white dwarf may lose mass in novae outbursts, whose contributions to nucleosynthesis are very modest.

Limited nucleosynthesis occurs outside stars in gas exposed to very energetic particles. One such site is the interstellar medium of a galaxy permeated by cosmic rays. High-energy collisions between a nucleus among the cosmic rays and one in the interstellar

---

† I dedicate my contribution to the memory of Sir Fred Hoyle whose book *Frontiers of Astronomy*, when received as a school prize in 1956, transformed my passion for matters aerodynamical to matters astrophysical in general and matters cosmochemical in particular.

medium result in a spallation of the nuclei; for example, $p + {}^{16}O \to X + x$ where X might be one of the Li, Be, or B nuclides, and x represents other fragments and products. In what is called a direct reaction, p is a cosmic ray proton and ${}^{16}O$ is an interstellar resident. Inverse reactions – p is interstellar and ${}^{16}O$ is a cosmic ray – also occur. Necessarily, $\alpha + \alpha$ reactions occur, as α-particles are the second most abundant species in both the cosmic rays and the interstellar medium, and contribute the Li isotopes through the fusion reactions $\alpha + \alpha \to {}^{6}Li$ and ${}^{7}Li$.

In the following pages, I offer one observer's perspective on the principal sites of stellar nucleosynthesis. I stress observations of stable stars. My choice of topics and their presentation is surely idiosyncratic but other numerous reviews are available for digestion by eager students demanding a thorough and balanced education. If my lectures persuade a student or two to read widely and, in particular, to explore beyond the 'standard' texts and papers, I shall be happy. After several introductory sections, I discuss two issues of nucleosynthesis not addressed by the other lecturers: the origins of Li, Be, and B; the s-process.

2. Stellar nucleosynthesis - A site survey

2.1. *Sites*

Nucleosynthesis results from reactions between a nucleus and either a charged particle, a neutron, or a photon. The Coulomb barrier requires that synthesis resulting from charged particle collisions takes place at the high energies found in hot stellar interiors or in Galactic cosmic rays and particle beams from other cosmic particle accelerators. Photodisintegration builds lighter nuclei from heavier ones but requires energetic photons found in hot stellar interiors or in the inner regions of an accretion disk around a massive compact object. Of course, reactions between a nucleus and a neutron may occur at low energies, but activation of astrophysical neutron sources demands the high temperatures of a stellar interior. Stellar nucleosynthesis in stable and exploding stars is the principal origin of the chemical elements. Synthesis driven by collisions between energetic and ambient particles in interstellar space, also circumstellar environments or stellar surfaces, makes a sensible contribution for a few nuclides of special astrophysical interest.

Gravity's tendency to collapse stars controls a star's structure.† Consider a main sequence star. Gravitational collapse is opposed by a pressure - temperature gradient. The temperature gradient directs the energy flow outwards to the surface. Lost energy is replaced in the interior by burning H to He through either the pp-chains or the CNO-cycle. Main sequence life ends when the core is He-rich rather than H-rich, as it was at the outset. Gravity requires that the He core begin a collapse to the higher temperature required for He burning by the 3α-process. Hydrogen may continue to burn in a shell around the He core. In this process, the structure of the outer layers is changed; the star becomes a red giant or supergiant. The He core of the evolved star ignites and He is transmuted to a mixture of C and O. This phase ends when He in the core is exhausted. At this point, stellar evolution experiences a bifurcation by mass.

In low mass stars the C-O core is supported against gravitational collapse by the pressure of degenerate electrons. This is not the case for high mass stars whose C-O core collapses to the ignition point of carbon. Division between low and high mass stars is put at about $8M_\odot$. A massive star's core cycles through the following stages: fuel ignition, burning ended by fuel exhaustion, and gravitational collapse to ignite the spent fuel. The fuel sequence begins with H burning, proceeds through He, C, Ne, O, to Si

---

† A reader new to stellar structure is directed to the text by Phillips (1993).

burning after which the core is composed of Fe at which point the cycle is broken with a catastrophic collapse leading to a supernova with massive amounts of material ejected at high velocity and the star reduced to a compact remnant (a black hole or a neutron star). Massive stars are key factories for the manufacture and distribution of products of nucleosynthesis occurring in the phases of hydrostatic burning prior to the explosion and in the explosive event itself. Langer's lectures discuss massive stars, their structure and contribution to cosmochemistry.

Low mass stars with their electron degenerate C-O core evolve as mass-losing AGB stars. The mass loss trims the star to a mass less than the Chandrasekhar limit of $1.4M_\odot$. The slimmed star evolves as a post-AGB star to a white dwarf. In most AGB stars, energy to offset gravity is released through either H-burning in a thin layer just below the base of the convective envelope or He-burning in the thin He-rich shell separating the H-rich layers from the C-O core. Alternation of H-burning and He-burning occurs with a long episode of the former followed by a brief phase of the latter. The initial phase of He-burning is often referred to as a thermal pulse or a He-shell flash. The AGB stars are the dominant contributors of the neutron capture products from the s-process, specifically the main s-process. Additionally, they may be significant contributors of certain light nuclides (e.g., ^{7}Li, ^{13}C, ^{14}N, and ^{19}F).

A white dwarf safely below the Chandrasekhar limit is exempt from making contributions to stellar nucleosynthesis unless it gains mass. A white dwarf in a binary system may accrete mass from its companion. The fate of the white dwarf is dependent on the rate of accretion and the total mass accreted. Accretion may lead to nova explosions in which mass is shed; the net gain in mass may be slight. Such a white dwarf may avoid attaining the critical maximum mass. Novae explosions - examples of explosive nucleosynthesis - have been greatly studied theoretically and observationally. It is fair to say that their collective contribution to Galactic chemical evolution is slight except, perhaps, for a few rare nuclides, such as ^{15}N which is a product of the hot CNO-cycles which power a nova explosion.

If mass accummulates to drive the white dwarf's mass to the Chandrasekhar limit, degenerate electrons can no longer thwart gravity. A catastrophe ensues in the form of a thermonuclear explosion - the white dwarf becomes a supernova of type Ia. Explosive nucleosynthesis in a SN Ia is a vigorous hot field with many questions remaining about the probable progenitors, the nature (detonation or deflagration?) of the explosion and its fruits. This research, which is driven now in part by the use of these supernovae as standard candles to probe the acceleration of the Universe, is critical to understanding the nucleosynthesis of Fe-peak nuclides, the principal product of SN Ia.

2.2. *Surveying tools*

In assessing the various sites, a surveyor now has available a wonderful array of tools to gather vital observations. An initial classification of observations can be made by the mode of analysis: chemical or spectroscopic. Chemical denotes those observations in which atoms or nuclei are handled and sorted. Spectroscopic observations refer, of course, to acquisition and analysis of astronomical spectra. One of the joys of spectroscopic exploration today is the almost complete access to the electromagnetic spectrum allowing observations from the γ-lines of nuclear deexcitation and decay to the infrared lines of the rotational transitions of molecules.

The following examples of chemical analyses merit mention:

• Carbonaceous chondrites. These meteorites are presumed to be a record of the early solar system's composition, and even of the Sun's natal interstellar cloud. Volatile elements including some of most abundant and astrophysically interesting (H, He, C, N, and O - for example) are not surprisingly underabundant in these meteorites. In abundance tables such as that used to compile the data in Fig. 1, the abundance of a volatile is obtained from a solar, stellar, or other astronomical measurement.

• Interstellar grains in meteorites. Isolation of very small grains from chondrites has provided a wealth of data. These grains securely identified as interstellar grains that were present in the solar nebula and – somehow – survived entry incorporation into meteorites variously sample environments as diverse as the circumstellar envelopes of AGB stars and the ejecta of supernovae. For example, SiC grains from C-rich AGB stars provide data on the range of $^{12}C/^{13}C$ ratios for carbon stars for comparison with spectroscopic estimates from field carbon stars. Earlier discoveries of isotopic anomalies in inclusions in meteorites, including excess ^{26}Mg from ^{26}Al decay, have stimulated much work in stellar nucleosynthesis.

• Galactic cosmic rays. The composition of observed Galactic cosmic rays differs from their composition at their point of origin: source abundances are obtained by correcting for the changes caused by spallation en route through the interstellar medium from source to Earth. The cosmic rays for example have a higher than standard ratio of B (relative to C) but this is entirely due to spallation of C, N, and O by interstellar protons and α-particles. The sources of the cosmic rays remain uncertain.

For the spectroscopic analyses, I identify three kinds of environments: stellar atmospheres, stellar ejecta and interstellar gas, and stellar systems.
• Stellar atmospheres. Stellar spectroscopy, most particularly stellar photospheric spectroscopy, is a major source of the abundance data used to address questions of nucleosynthesis.

• Stellar ejecta and interstellar gas. Products of stellar nucleosynthesis may be observed directly in ejecta from various stars. Obvious examples include supernovae, novae, planetary nebulae, and massive young stars such as the Wolf-Rayet stars with local nebulae.

• Stellar systems. Here, I place analyses of stellar atmospheres that provide data for collections of stars in one system. The system might be a self-contained one like a globular cluster or an open one like the solar neighbourhood. Observations of individual stars are now being extended beyond the Magellanic Clouds to more distant galaxies. In the case of very distant galaxies, the integrated starlight is observed and analysed to obtain important though restricted data on elemental abundances. Abundance data with other data (e.g., kinematics) may be used to trace the chemical evolution of the system. Since nucleosynthesis is but of many factors influencing the chemical evolution, detailed insights into operation of stellar nucleosynthesis sites are limited. On the contrary, it is the case that understanding of stellar nucleosynthesis may be used to advance understanding of a system such as our Galaxy and its bulge, disks, and halo.

3. Numbers and Notation

Stellar spectroscopists have evolved a standard notation for specifying the chemical composition of a stellar atmosphere. By 'abundance of the element E' is meant – usually – the ratio of the number density n(E) to that of hydrogen, i.e., A(E) = n(E)/n(H). The

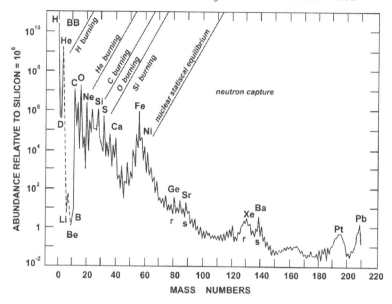

FIGURE 1. The standard or solar system abundance distribution of nuclides, normalised to 10^6 ^{28}Si atoms. From Pagel (1997).

abundance A(E) is often quoted on the logarithmic scale where $\log \epsilon(E) = \log A(E) + 12.0$.

Frequently, the stellar and solar abundance of an element may be compared using 'the square-bracket notation', i.e., $[E/H] = \log \epsilon(E)_{Star} - \log \epsilon(E)_{Sun}$. For example, we may speak of Galactic metal-poor stars as those with $[Fe/H] \leq -1$, that is stars with less than one-tenth the Fe abundance of the Sun. The notation is readily extended to a ratio of two elements E_1 and E_2; for example, a star with $[Si/Fe] = +0.5$ has a Si/Fe abundance ratio that is three times that of the Sun. Two notes of caution: (i) when [X] is quoted check that the Sun was adopted as the reference object, and (ii) A few authors may define [X] quite differently!

The student new to the field may not appreciate the distribution of the elements in the Sun, a 'typical' star. Fig. 1 shows solar abundances compiled largely by combining spectroscopic estimates for the solar photosphere with chemical determinations from carboncaeous chondrites, and supplemented in a few cases with entries from other sources. In this figure, abundances are given relative to n(Si) = 10^6, a carry over from the analyses of the chondrites. Clearly, H and He are the dominant elements. Notable too are the low abundances of Li, Be, and B. Abundances decline from C to Ca and Sc but then comes the iron-peak. Elements with mass numbers beyond the iron peak show a drop in abundance with evident structure, here labelled by the letters r and s. Annotations above the curve denote nuclear processes primarily responsible for synthesis of the elements. The total range in elemental abundance is about 10^{12} from the most abundant H to the rarest of the heavy elements (Au, for example). Surface abundances show star-to-star variations on the solar distribution ranging from minor to major, common to rare, and understood to puzzling. Across this parameter space are sought the clues to stellar nucleosynthesis and competing processes.

On occasions, a stellar composition will be expressed in terms of mass fractions. Stellar structure calculations commonly use mass fractions. It is the universal custom to refer to the H and He mass fractions by the letters X and Y, respectively. The mass fraction of all other elements - 'metals' in astronomical parlance - is denoted by Z. Today, Z may be broken into parts: Z(C), Z(O), etc. Of course, X + Y + Z = 1. Chemical evolution of the Universe from the Big Bang to the Sun could be summarized as the change from $(X,Y,Z) \simeq (0.76, 0.24, 0.00)$ immediately following the Big Bang to (0.73, 0.25, 0.02), which is representative of the Sun and the present solar neighbourhood – a few per cent change in 10 billion years. Tales of stellar nucleosynthesis are coded in the few per cent rise of Z.

4. Pioneering Tales

Casual reading of the literature may give the impression that secrets of stellar nucleosynthesis are revealed only by detailed quantitative analysis of a stellar spectrum involving model atmospheres and spectrum synthesis with perhaps attention to non-LTE and other effects. Exacting but not exciting work may be the verdict. Although there is important information in abundances that can come only from a detailed and precise analysis, there have been and surely remain startling results which can be unearthed from quite simple analyses and even from mere inspection of a spectrum. I discuss two such examples.

4.1. *Technetium in S Stars*

Merrill's (1952) discovery of Tc I resonance lines in spectra of S-type long-period variable red giant stars was, despite Merrill's initial reservations, taken as proof positive that these stars were factories for heavy element production. S-type stars are noted for their strong bands of ZrO and LaO, and enhanced lines of heavy elements such as Zr and Ba. Technetium is not just another heavy element but one of only two (promethium is the other) between H and Bi with no stable nuclides. With a half-life of only about 10^5 years for the longest lived isotope (^{99}Tc), it was apparent that Tc (and other heavy elements) must be synthesised in these red giants. Hoyle (1994) and Cameron (1999) convey the excitement generated by this spectroscopic discovery and the impact on their ground-breaking surveys of stellar nucleosynthesis (Burbidge et al. 1957; Cameron 1957).

How synthesis of Tc and friends might occur by a chain of neutron captures and β-decays was first mapped out by Cameron (1955). Following Burbidge et al., we refer to the chain as the *s*-process. In contemporary language, the S-type red giants are examples of asymptotic red giant branch (AGB) stars. AGB stars have an electron degenerate C-O core and an extensive and convective H-rich envelope with a thin He-shell sandwiched between the core and envelope. It is in the He-shell that the *s*-process operates. It is the convective envelope that dredges the *s*-products including Tc to the spectroscopically accessible surface. Understanding of *s*-process operation in AGB stars began in 1952 with an astute observer searching photographic stellar spectra for the strongest lines of Tc I then recently reported for the first time in a laboratory spectrum obtained and analysed at the National Bureau of Standards.

4.2. *The Spite Plateau*

Today, the concept of chemical evolution of the Galaxy – old stars are metal-poor and young stars are metal-rich – is embedded so firmly in our thinking that it may be hard to imagine a time when it was supposed that all normal stars had a very similar ('cosmic') composition. Yet, this was once the accepted idea or prejudice. The first real challenge came from Chamberlain & Aller's (1951) abundance analysis of two subdwarfs showing them to be significantly metal-poor relative to the Sun. From this beginning sprang a long and continuing suite of investigations of the compositions of metal-poor stars. A key result soon became apparent: an abundance E/H declines with decreasing Fe/H among normal stars down to the present observational limit of [Fe/H] $\sim$ -4. (Relative abundances [E/Fe] in unevolved stars may show small variations about zero. These variations are grist to a chemical evolutionist's mill.) The implication is that the first generations of Galactic stars formed from Big Bang debris only slightly contaminated by products from cosmic cauldrons be they pre-Galactic 'stars' or very early generations of Galactic stars. Stellar nucleosynthesis gradually enriched the interstellar medium in elements E and Fe and raised the abundances over time to their present values.

Spite & Spite's (1982) remarkable discovery that lithium has a constant abundance in warm subdwarfs† showed that, in contrast to all other elements, lithium had an origin that predated the formation of Galactic stars. The Li abundance in these subdwarfs with [Fe/H] runing from $\simeq$ -4 up to about -1.5 is $\log \epsilon(\mathrm{Li}) \simeq 2.0$, a factor of about 1.3 dex below the Li abundance in the solar system. The fact that the Li abundance is almost unchanged as other elements, the products of stellar nucleosynthesis, rise 300-fold in abundance is essentially evident by inspection of a collection of spectra of warm subdwarfs of similar temperature; the strength of the 6707 Å Li I resonance doublet varies very little from star-to-star but metal lines vary greatly in strength. Spite & Spite drew the conclusion that the lithium in warm subdwarfs was a product of the Big Bang, and identified the Li abundance with that provided by the Big Bang. Today, the tools of model atmospheres and synthetic spectroscopy are brought to bear in order to satisfy the cosmologists' desire for a precise estimate of the Li abundance in subdwarfs. A parallel debate ensues about the relationship between the present surface Li abundance and a star's initial abundance: has the surface Li been altered over the more than 10^{10} years of a subdwarf's life?

5. An Assumption and a Warning

In applications of stellar abundances to questions of cosmochemistry, an assumption is invoked. Put in its strongest form, this states that the star's present surface composition is identical to that of its natal interstellar cloud. When applicable and in conjunction with corresponding assumptions about interstellar clouds, this assumption allows the chemical evolution of the Galaxy to be traced using stars of different ages or more realistically stars of different metallicity.

Although this assumption is behind the identification of the Li abundance of the Spite plateau with the yield from the Big Bang. Lithium in subdwarfs should also stand as a warning not to take the assumption for granted. In the cool subdwarfs, lithium is obviously depleted; the plateau Li abundance is maintained for stars with effective temperatures hotter than about 5800 K, but Li in cooler stars shows an increasing depletion with decreasing effective temperature. This depletion is attributed to destruction of Li by protons at the base of the star's convective envelope. Can we be sure that Li in plateau stars is not also reduced (or increased) from its initial abundance? Diffusion, rotationally-induced mixing, and other processes have been examined theoretically. Observers tend to the point of view that the small or non-detectable scatter in Li abundances among plateau stars sets a tight constraint on depletion achieved by the proposed

† Lithium is quite obviously depleted in cool subdwarfs. Hence, the warm subdwarfs are used to establish the constant abundance.

processes, for example, rotational mixing due to star-to-star differences in rotational velocity and angular momentum should result in a star-to-star scatter in Li abundance. Surface depletion of Li, which is commonly observed among main sequence stars of all metallicities, continues to present puzzles to observers and theoreticians. This failure of the assumption for main sequence stars may be limited to Li and possibly Be and B, but see 'chemically peculiar' stars below. If Li, Be, and B are the only elements affected by surface depletion, all other elements in main sequence stars may be included in an observer's armoury when attacking issues in Galactic chemical evolution.

The assumption fails more broadly for red giants. Technetium is a glaring reminder. Thanks to the deep convective envelope, red giants dredge up nuclear-processed material to the atmosphere. Giants evolving to He-core ignition dredge-up CN-cycled material. Among the changes in surface composition are a reduction in the ^{12}C abundance, an increase in the ^{14}N abundance and a lowering of the ^{12}C/^{13}C ratio. Surface Li, Be, and B abundances are reduced by dilution with the Li, Be, and B-free gas from the interior. Clearly, abundances of the light nuclides – Li to N and possibly O – are a convolution of the initial (natal) abundances and the changes occurring in the giant. Unravelling this convolution and reconciling the observed effects of the convective mixing with theoretical models remains a challenge. Yet despite the dredge-up, it is reasonable to identify the abundances of heavier elements with those of the natal cloud. After He-core ignition and exhaustion, the red giant evolves to the AGB and becomes a major player in cosmochemistry. Many more elements have surface abundances affected by internal nucleosynthesis and dredge-up to the surface. Indeed, AGB stars are primarily analysed for what they tell us about internal nucleosynthesis.

I continue this section with brief remarks on some stars whose compositions defy direct connections to matters of cosmochemistry (here, understood to include nucleosynthesis). Many main sequence stars of spectral types B, A, and F have anomalous chemical compositions. The label 'chemically peculiar' (Preston 1974) is attached to these stars. An example must suffice. Among the hotter CP stars are the HgMn stars with optical spectra marked by strong lines of Mn II and Hg II. Chi Lupi analysed thoroughly by Leckrone et al. (1999) using optical and ultraviolet spectra has a Hg abundance 10^6 in excess of the solar abundance, and overabundances also of the elements Pt, Au, and Tl bracketing Hg in atomic number. Oddly, Mn has a near solar abundance. Other iron-peak elements are normal, but Zn is at least a factor of 10^4 underabundant. This range of at least a factor of 10^{10} variation with respect to solar abundances is remarkable. Although quantitative understanding of abundance anomalies of CP stars has not been achieved, it is broadly supposed that anomalies arise through diffusion of elements into or out of the atmosphere. CP stars are warning lights that stellar compositions do not always provide data simply relatable to questions of nucleosynthesis. Of historical interest only (?) are the early theoretical ideas linking the abundance anomalies of chemically peculiar stars to surface nuclear reactions† or to deposition of ejecta from a companion which exploded as a supernova.

A listing of other failures of the unrestricted assumption would be lengthy. Some offer opportunities to the cosmochemist but others thwart the search for the origins of the elements. Examples must suffice. Severe mass loss may expose interior regions previously experiencing nuclear processing (Wolf-Rayet stars, for example); mass transfer across a binary system may expose a stellar core (Algols) or make processed material available in

---

† Some CP stars have strong surface magnetic fields which might spawn flares capable of particle acceleration - an old idea (Fowler et al. 1955) deserving of reexamination?

a more spectroscopically friendly way (Barium stars and CH subgiants). These examples expose nuclear-processed to spectroscopic scrutiny.

Counter examples exhibiting severe abundance anomalies unrelated to nuclear processing include those post-AGB stars with an atmosphere exhibiting the selective accretion (or retention) of gas but not dust onto a star with a thin convective envelope. A wonderful example is the bright star HR 4049 with [Fe/H] = -4.8 but near-solar abundances of C,N, O, S, and Zn (Lambert et al. 1988; Takada-Hidai 1990; Waelkens et al. 1991; Takeda et al. 2002). This star with other Fe-poor post-AGB A-type stars is a spectroscopic binary (Van Winckel et al. 1995). The abundance pattern (e.g., high S and Zn but low Ca and Fe) for the stellar photosphere resembles that found in diffuse interstellar gas clouds (Savage & Sembach 1996). HR 4049 posesses a strong infrared excess indicative of warm dust near the star. The speculation is that the post-AGB star has accreted gas but not dust from a circumstellar or circumbinary cloud. Elements with a 'high' condensation temperature condense into and onto grains leaving the gas depleted in those elements. Winnowing of dust from gas occurs as dust grains are driven away from the star by radiation pressure exerted by starlight. If the dust-to-gas density ratio is sufficiently low, dust will be expelled leaving behind a cloud of gas depleted in those elements which condense readily into dust. Accretion of this gas by the star provides the observed composition - at least, qualitatively. Certainly, this picture suggests why Mg, Ca, Fe etc. are severely underabundant at the stellar surface but C, N, O, S, and Zn are not. As the drive to find extremely metal-poor stars (say, [Fe/H] < -4 for the natal cloud) gathers momentum, the example of HR 4049 should be remembered; it is a peculiar star with an initial metallicity near-solar. Indeed, identification of the metallicity [Fe/H] ~ -5 with that of the star's natal cloud would –because the star is bright and in a very rapid phase of evolution– imply that main sequence and red giant stars with [Fe/H] ~ -5 should be common - they are decidedly not.

Many questions remain about how dust-gas winnowing works. Although some questions may dissolve under theoretical scrutiny, observers must suppose that a search for other types of stars with similar abundance anomalies will yield clues to the answers. The RV Tauri variables, possibly post-AGB or post-RGB stars, are cooler than stars like HR 4049 but with similar although milder abundance anomalies. Many RV Tauri stars have strong infrared excesses from dust. A few are known to be spectroscopic binaries and possibly all are binaries (Van Winckel et al. 1999). Beginning with the unexpected discovery of severe dust-gas winnowing for IW Car (Giridhar et al. 1994), we have surveyed a large number of northern RV Tauri stars and begun to define the parameters controlling the winnowing (Giridhar et al. 2000). To illustrate the range of the abundance anomalies, I show in Fig. 2 results for [Zn/Fe] from our survey (Giridhar et al. in preparation): the most severe anomaly [Zn/Fe] ≃ 2 is less extreme than that for HR 4049.

6. Black Boxes and Black Magic

Novice practitioners of the art of quantitative stellar spectroscopy are likely early in their studies to encounter a series of 'black boxes'. One such box will contain grids of model atmospheres bearing a certificate such as 'Made at Harvard' or 'Made in Uppsala'. Into a second box one puts a model atmosphere and pulls out a synthetic spectrum. Then follows an iterative process through which one achieves a satisfactory match of synthetic to observed spectrum consistent with other information about the star including photometric and astrometric data. It is possible to derive a stellar chemical composition using the black boxes and have minimal appreciation of the physics of model atmospheres

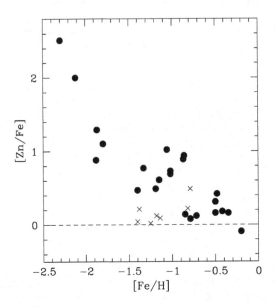

FIGURE 2. Zinc and iron abundances for a sample of RV Tauri variables. Normal stars have [Zn/Fe] $\simeq 0$ for the illustrated range of [Fe/H]. The filled circles refer to variables belonging to the Galactic disk for which [Fe/H] ~ 0 was their initial metallicity; depletion of Fe but not Zn moves the star in this diagram to high [Zn/Fe] at low [Fe/H]. The crosses denote high-velocity stars for which the intrinsic metallicity was most probably sub-solar. These stars fall close to the line [Zn/Fe] = 0 indicating little if any distortion of atmospheric abundances by dust-gas winnowing. A likely explanation is that the dust-to-gas ratio for intrinsically metal-poor gas leads to severe drag on the grains such that little dust-gas separation is achieved.

and photospheric absorption line formation.† An appealing aspect of black boxes is that abundances are obtainable quickly and time is saved for delving into the astrophysical problems opened up by the abundance data. A danger inherent in ready access to black boxes is that their limitations may not be understood; the possibility of deriving erroneous results then looms large.

These lectures are not the vehicle for thorough discussion of atmospheres and line formation. I assume that the student has had some exposure to the rudiments of observational and theoretical aspects of stellar spectra. Rather, I have chosen to pose and comment on three topics which experience tells me may bemuse the tyro.

6.1. *Why are abundance analyses incomplete?*

Accurate abundances for all elements (and isotopes) comprise the ideal database with which to challenge or promote theoretical ideas on nucleosynthesis. Ideals are unattainable. Why this is so for abundances may puzzle the unworldly theoretician, but the

† In these lectures, I shall not discuss spectra of extended atmospheres – chromospheres, coronae, winds and circumstellar shells – and the limited information they may yield about elemental and isotopic abundances.

practising spectroscopist soon understands that several basic barriers impede the drive to complete the database. I list the most important of them:

• The elemental abundance is so low that not even the strongest resonance lines of the dominant stage of ionization rise above the detection limit. This limit for an absorption line obviously depends on the quality of the spectrum (signal-to-noise, resolution). In practice, blends may intervene to set the limit.

• Potentially detectable lines fall outside the observed spectral window. The corollary is obvious: exploration of other windows may give lines from which an abundance is determinable. Boron is an outstanding illustration of this corollary. Its low abundance demands access to the strong resonance lines which happen for reasons of atomic physics to be in the ultraviolet. The lines B I 2497 Å, B II 1362 Å, and B III 2066 Å have all been observed in appropriate stars with the *Hubble Space Telescope* and provided novel data on the Galactic evolution of boron, the isotopic ratio, and internal destruction or dilution of boron in low and high mass stars. (He-like B^{3+} and H-like B^{4+} with resonance lines deep in the UV are undetectable now and forever.)

• An element may be abundant but the dominant stages of ionization in the atmosphere do not provide detectable lines. Helium stands as the quintessential example. In normal cool stars, helium is abundant ($n(He)/n(H) \sim 0.1$) but no photospheric lines are seen. Helium is present in the atmosphere as neutral atoms. The He I leading resonance line is inaccessible at 304 Å. Lines from excited states do occur at accessible wavelengths but these states are very thinly populated. First excited states are at about 21 eV above the ground state with a population $10^{-\chi\theta} = 10^{-21\theta}$ of the total He abundance, where $\theta = 5040/T$. For the Sun, $\theta \sim 1$, and a He I line strength is governed by not the abundance of 0.1 but an effective abundance of 10^{-22}, which is a factor of 10^{10} less than required to provide a detectable photospheric line.

Lack of an abundance estimate for a trace element like boron has no effect on the accuracy of the abundance analysis for other elements but merely restricts astrophysical interpretations involving B. On the other hand, helium is an abundant elements with effects on the atmospheric structure and through this on the derived abundances of other elements. Although rarely stated explicitly, abundance analyses of cool stars are dependent on an assumption about the He/H ratio; the assumption enters both into the model atmosphere and synthetic spectrum calculations. For normal stars, ignorance about the He/H ratio is mitigated by the fact that the He/H ratio is surely constrained within tight limits (Y = 0.24 to 0.26, see above).

6.2. Why does a line of E yield $A(E) = n(E)/n(H)$?

Absorption lines are defined relative to the continuum. In the case of a resolved line, one may describe the line in terms of its depth relative to the local continuum across the line, say $R(\Delta\lambda) = F_l(\Delta\lambda)/F_c$ where $\Delta\lambda$ is the wavelength measured from line centre, F_c is the flux in the continuum, and F_l is the flux at a wavelength within the line. The total absorption by the line obtained by integrating $R(\Delta\lambda)$ over the line profile is known as the equivalent width W_λ. When a line profile is not resolved yet unaffected by blending from neighbouring lines, the equivalent width is independent of the resolution even though the line profile is set by the instrumental profile and not by the intrinsic stellar profile.

The equivalent width like the line depth is, of course, determined by the ratio of line to continuum fluxes. This ratio is a key to understanding how the equivalent of a line of an atom or ion of element E provides the abundance E/H without the measurement of a line of hydrogen. Consider a normal atmosphere in which temperature increases with increasing depth. Suppose the line-absorbing atoms (or ions) and continuum-absorbing particles are similarly distributed with depth. If the number density of line absorbers is increased, the flux in the line will decrease and the line in the emergent spectrum will strengthen. If the number of continuum absorbers is decreased, the flux in the continuum will increase but the line will also get stronger; the continuum photons will on average come from deeper in the atmosphere, traverse a larger column density of line absorbing particles, and, hence, experience a higher probability of being absorbed within the line. This crude argument suggests that the equivalent width is likely controlled by the ratio of line (κ_l) to continuous (κ_c) opacity. It is rarely controlled solely by the line opacity. It will be appreciated that the temperature gradient in the atmosphere is factor too; the steeper the gradient, the larger is the contrast between the flux in the line and continuum. For weak lines (see next section for the definition of 'weak'), it can be shown that the line depth and equivalent width depend under limited but often realistic conditions on the product of κ_l/κ_c and the temperature gradient (see Böhm-Vitense 1989).[†]

It is through the ratio of line to continuous opacity that the abundance as E/H may emerge as the natural quantity. One example must suffice. In a solar temperature star, iron is predominantly singly ionized: $n(Fe^o) << n(Fe^+) >> n(Fe^{2+})$, and continuous opacity at all wavelengths except the UV is dominated by the trace quantity of H^- ions. Consider a line of Fe I. Ignoring temperature dependent factors, the line to continuous opacity ratio is set by $n(Fe^o)/n(H^-)$. Saha ionization equilibrium gives $n(Fe^o)/(n(Fe^+)n(e)) = A(T)$, and $n(H^-)/(n(H)n(e)) = B(T)$ where $A(T)$ and $B(T)$ are familiar functions of temperature, partition functions, and ionization potential. Combining the Saha equations with the line to opacity ratio, we find that $\kappa_l/\kappa_c \propto N(Fe^+)/N(H)$ $\simeq N(Fe)/N(H) = A(Fe)$ from Fe I lines alone. The same exercise for Fe II lines is left to the reader, who may also wish to extend exploration to hotter stars where photoionization of H atoms is the leading contributor of continuous opacity.

There are situations where the strength of a line is not directly relatable to E/H. In the UV of cool stars of solar metallicity, photoionization of excited states of neutral atoms not the photodetachment of H^- ions may dominate the continuous opacity. For example, in the case of the B I 2497 Å resonance lines in cool stars of solar metallicity, Mg atoms contribute significant opacity with the result that the B I lines depend in part on the adopted B/Mg ratio.

6.3. The Curve of Growth

To black box users, the phrase 'curve of growth' may denote an outdated method of stellar abundance analysis. I prefer to regard a curve of growth (CoG) as an inherent property of a set of photospheric absorption lines and would stress that recognition of basic features of the CoG may ease design of an abundance analysis.

Consider a uniform cold cloud through which starlight passes to the observer. The cloud is seeded with atoms absorbing at the wavelength λ_0 with an absorption coefficient given by the transition's f-value and a profile $\phi(\Delta\lambda)$. Assume the cloud to be transparent at all other wavelengths. Around λ_0, the transmitted intensity of starlight is $I(\Delta\lambda) = I_0 \exp(-\tau(\Delta\lambda))$, where I_0 is the intensity in the local continuum, and $\tau(\Delta\lambda)$ is the cloud's optical depth at $\Delta\lambda$ from λ_0.

At very small $\tau(\Delta\lambda)$, the normalized line depth $I(\Delta\lambda)/I_0 = \tau(\Delta\lambda)$ when integrated gives the equivalent width

$$W_\lambda = \frac{\pi e^2}{mc^2} N_L H f \lambda^2 \tag{6.1}$$

† A counterexample. Suppose the line and continuum forming regions are non-overlapping, as in the case of an interstellar cloud projected in front of a star. In this case, the line's equivalent width is independent of the continuum flux, and controlled by the line opacity alone.

where $N_L H$ is the column density of absorbing atoms in the lower state L, f is the absorption f-value, and other symbols represent the usual atomic constants. In this example, $N_L H$ is the abundance sought by observers. In this limit known as the weak-line approximation, also referred to as the linear part of the CoG, the equivalent width (and line depth) and abundance are linearly related. The fractional error in the abundance $N_L H$ is that in the measured W_λ. The need for atomic data is confined to the line's f-value.

The mathematical origin of the weak-line limit lies quite obviously in the expansion $\exp(-\tau) \simeq 1 - \tau$ for small τ which necessarily fails as the product $f N_L H$ and, therefore, τ increases. The physical origin of the failure is readily understood. Atoms in the lower state L wherever they are in the cloud have an equal propensity to remove photons from the beam of starlight impinging on the atoms. When $f N_L H$ is very small, all atoms see the same intensity I_0 and remove the same number of photons. As $f N_L H$ is increased, atoms furthest from the entrypoint of the starlight into the cloud see a reduced intensity $(I < I_0)$ and, hence, remove fewer photons than they did when $f N_L H$ was smaller. The effect on the CoG is obvious: W_λ increases at a less than linear rate with increasing $f N_L H$.

Development of the $W_\lambda - f N_L H$ relation, the CoG, beyond the weak-line limit depends on the profile of the absorption coefficient $\phi(\Delta\lambda)$. An extreme form for the profile is effective at illustrating this point. Suppose $\phi(\Delta\lambda) = a$ for $|\Delta\lambda| = \Delta\lambda_D$ and 0 for $\Delta\lambda \geq \Delta\lambda_D$. Normalization of ϕ provides the relation connecting the constant a, the width $\Delta\lambda_D$, and the f-value - the derivation is left as an exercise for the student! With increasing $f N_L H$, $I(\Delta\lambda)/I_0$ falls within the line to its minimum value of zero. At which point, the equivalent width has saturated at $W_\lambda = 2\Delta\lambda_D$. Note that, unlike W_λ in the weak-line limit, the CoG beyond the weak-line portion depends on the shape of the line absorption coefficient – here, the width $\Delta\lambda_D$. This dependence means that conversion of a measured W_λ to $f N_L H$ for realistic absorption coefficient profiles demands observational or theoretical knowledge of the absorption coefficient's profile. This requirement plus the reduced sensitivity of W_λ to $f N_L H$ make this part of the CoG less attractive for abundance determinations. This stretch of the CoG is variously referred to as the flat, Doppler or saturated part.

Real lines do not have 'square' absorption coefficients. In general, the absorption coefficient will be a Voigt profile arising from the convolution of a Gaussian core from the motions (thermal and nonthermal) of the atoms with a Lorentzian resulting from radiative and collisional damping. An effect of the Gaussian is a rounding of the core (relative to the square profile) which adds a slight slope to the flat portion of the CoG: $W_\lambda \propto \ln(f N_L H)$ approximately. When $f N_L H$ is so large that the Lorentzian wings begin to contribute to the line profile, W_λ increases again with increasing $f N_L H$. As inner regions of the line profile saturate, there 'always' remains the outer wings in which the absorption is weak. Thus, increasing $f N_L H$ leads to increasing W_λ: $W_\lambda \propto \sqrt{f N_L H \Gamma}$ in the case of the interstellar cloud, where Γ is the damping constant characterizing the Lorentzian. Hence, this final extension of the CoG is called the damping or square-root portion.

Stellar photospheres are not interstellar clouds but the run of W_λ with abundance follows a CoG with linear, flat, and damping portions. Lines on the linear portion are highly desirable for abundance analysis because (i) the sensitivity of the abundance to equivalent width errors is least for this portion, and (ii) no information about the line absorption coefficient except its integrated strength (i.e., the f-value) is needed. Of course, high quality spectra are required to measure weak lines accurately and to resolve them from neighbouring lines. In addition, the W_λ corresponding to the (gradual) termination of the linear part of the CoG must be determined. Lines on the flat portion of the CoG are poor abundance indicators. The profile of the absorption coefficient must be determined from the spectrum. Customarily, a Gaussian is assumed; i.e., $\Delta\lambda_D$ has to be estimated. This width contains a component from the atom's thermal motions and another - the microturbulence - from the atmospheric velocity fields. Then, there is the steep magnification of an error in W_λ into an error in abundance. This error is reduced for lines on the damping portion but here knowledge of the damping parameter Γ is needed. This parameter is partly intrinsic to the line (radiative damping) and partly dependent on the atom's environment (collisional damping). While one may have fair estimates of radiative damping, there remain lacunae in our ability to compute collisional damping parameters. Rarely are lines with damping wings used in stellar abundance analyses. If they are, the W_λ – abundance relation is calibrated using weak lines, or a differential analysis of a collection of similar stars is undertaken to obtain relative abundances. An advantage of strong lines is that they may be detected and measured in low resolution spectra of faint stars.

7. Lithium, Beryllium, and Boron

7.1. *Observational Constraints*

Spectroscopists seeking Li, Be, and B abundances in order to answer questions of nucleosynthesis must recognize several constraints in obtaining abundances and some limitations in interpreting the results. The trio are trace elements detectable only through resonance lines†. Accessible lines include Li I at 6707 Å, Be I at 2348 Å, Be II at 3130 Å, and B I at 2497 Å, B II at 1362 Å and B III at 2066 Å.‡ Li atoms being of low ionization potential are detectable only in cool stars. Resonance lines of Be I are irretrievably blended, but the Be II lines are detectable in warm to cool stars. Only boron is potentially detectable from hot stars via B III lines to cool stars via B I lines, but UV spectra are required. The opportunity to obtain the ^{6}Li/^{7}Li and ^{10}B/^{11}B ratios can be exploited for several kinds of stars. Interpretations of Li, Be, and B abundances must address the question: Does the measured abundance reflect the star's initial abundance or has it been modified in the course of the star's evolution? The answer determines whether the abundance is primarily a tracer of Galactic chemical evolution or of processes occurring in the course of stellar evolution.

7.2. *Theoretical Proposals*

In their classic paper, Burbidge, Burbidge, Fowler, & Hoyle (1957) assigned Li, Be, and B to the x-process described quite generally as "mechanisms which may synthesize deuterium, lithium, beryllium, and boron". It was noted that these light elements are destroyed by protons in stellar interiors and, hence, x-process operation outside stars was considered. Spallation reactions - the break-up of larger nuclides in high energy collisions with protons and α-particles - were cited as a possibility at diverse potential sites including stellar atmospheres, supernova outburst, supernova remnants, and nonthermal radiosources. Cameron (1957) said of D, Li, Be, B that they were "not formed in stellar interiors. Possibly made by nuclear reactions in stellar atmospheres". Today, $x =$ the unknown has been changed to $x =$ the almost known - perhaps!

Present theoretical ideas about LiBeB encompass sites as diverse as the Big Bang,

---

† Excited lines of Li I are detectable in certain stars.
‡ Resonance lines of H-like (Li^{2+}, Be^{3+}, and B^{4+}) and He-like (Li$^+$, Be^{2+}, and B^{3+}) ions are in the inaccessible far-UV.

stars of various types, and the interstellar gas (Table 1). After a brief commentary on these sites, I conclude the section with selective remarks on observers' contributions to understanding of LiBeB synthesis from abundance analyses of stable stars.

The Big Bang. In what is generally known as the standard family of Big Bang (Friedmann) models, ^{7}Li is the only LiBeB nuclide synthesised in observable amounts. This Li in full or in part is seen in warm very metal-poor stars, as the Spite plateau. Nonstandard Big Bang models in a wide variety of forms have been proposed. Often, the consequences for the primordial nucleosynthesis are a focus of these proposals.

Stars - SN II. The neutrino torrent created at core-collapse of a massive star is so intense that, despite their very small cross-sections, interactions between neutrinos and nuclei in the outer layers of a supernova synthesise several trace species from abundant targets at levels that suggest Type II SN ejecta may be a major influence on the Galactic chemical evolution of these trace species. Notably, ^{11}B is synthesised from ^{12}C in the C-shell: ^{12}C$(\nu, \nu'n)^{11}$C$(e^+\nu_e)^{11}$B and ^{12}C$(\nu, \nu'p)^{11}$B. Also, ^{7}Li may be made in the He-shell via ^{4}He$(\nu, \nu'n)^3$He followed by ^{3}He$(\alpha, \gamma)^7$Be$(e^-, \nu_e)^7$Li. No other LiBeB nuclide is predicted to be synthesised in interesting amounts. In particular, ^{9}Be, the sole stable Be isotope, is not synthesized.

Red Giants - Tapping the ^{3}He reservoir. Main sequence stars have a reservoir of ^{3}He outside the H-burning core. A part of the reservoir is provided by the star's primordial ^{2}H and ^{3}He; ^{2}H is burnt to ^{3}He during pre-main sequence evolution. Additional ^{3}He is synthesised by main sequence stars from protons by the initial steps of the *pp*-chain. This ^{3}He supplement is significant only in low mass stars with their long main sequence lifetimes; the initial step in the *pp*-chain is very slow being controlled by the weak interaction. Between the surface and the interior of a main sequence star, where ^{3}He is destroyed by protons, lies the ^{3}He reservoir. The significance for lithium synthesis is that ^{3}He may be converted to ^{7}Li under appropriate conditions by the chain ^{3}He$(^4$He,$\gamma)^7$Be$(e^-, \nu_e)^7$Li known as the ^{7}Be-transport mechanism (Cameron & Fowler 1971). Conversion occurs not in the main sequence star but in the red giant. Even quite inefficient conversion of ^{3}He to ^{7}Li and return of ^{7}Li to the interstellar medium would ensure that the ^{3}He reservoir dominated Galactic lithium production.

Very luminous (i.e., highly-evolved) intermediate mass AGB stars are predicted to develop a convective envelope with a base temperature (a hot-bottomed convective envelope or HBCE) sufficiently high to sustain CN-cycling and associated nuclear reactions. Among the latter is the conversion of ^{3}He to ^{7}Li. The ^{7}Be-transport mechanism is initiated at the base of the convective envelope. If convection operates efficiently, ^{7}Be and ^{7}Li escape complete destruction by hot protons and are swept to the surface. Mass loss from the Li-enriched envelope puts the Li into the interstellar medium. If a HBCE is maintained for long periods, the convective envelope will ensure that not only will the reservoir of the irreplaceable ^{3}He be exhausted but that the surface ^{7}Li abundance will decline through repeated exposure to hot protons. Calculations suggest that lithium production via a HBCE is effective only in intermediate mass AGB stars, say $M \sim 5M_\odot$. Observations (see below) confirm the calculations.

Not only may very luminous giants tap the ^{3}He reservoir but there is evidence that much less luminous giants do so. In evolving to become a red giant, the convective envelope dilutes a main sequence star's thin Li-containing outer layer by a factor of about 50. Thus, red giants are expected to be very Li-poor relative to main sequence counterparts. Many indeed are. However, a fraction of low mass red giants at an evolutionary stage well prior to the AGB phase show ^{7}Li at levels exceeding that in most red giants and in a few cases at levels approaching or exceeding the star's presumed initial Li abundance. Lithium has probably been replenished from inside and not from outside by accretion of planets or brown dwarfs. Replenishment must, as in AGB stars, come from conversion of ^{3}He to ^{7}Li. Early consumption of ^{3}He and mixing of ^{7}Li to a red giant's surface may be a natural occurrence for red giants. Charbonnel & Balachandran (2000) suggested that the observed luminosity of a Li-rich red giant coincides with the predicted luminosity at which a giant's H-burning shell encounters a molecular weight gradient left from the deepest extent of the convective envelope at an earlier time. Palacios et al. (2001) argued that consumption of ^{3}He results in a rapid release of energy (a Li-flash) which serves to mix ^{7}Li to the surface. Although the reservoir of ^{3}He in a generation of red giants would suffice for these stars to be players in the Galactic production of ^{7}Li, there are theoretical and observational arguments suggesting that the ^{7}Li is subsequently destroyed by the red giant before it is ejected into the interstellar medium. The fact that the Li-rich giants have a preferred luminosity and do not appear at all luminosities above that at which Li first surfaces strongly suggests the fresh Li is efficiently destroyed. An alternative scenario is suggested by the observations showing the most Li-rich giants are rapid rotators with a strong infrared excess from circumstellar dust (Drake et al. 2002). By an event not identified (violent He-core flash?), a rapidly rotating giant is imagined to experience mixing leading to conversion of ^{3}He to ^{7}Li, and ejection of Li-rich gas. Were this to be a common event, these stars might be players in overall Galactic production of Li.

The ^{3}He reservoir of a main sequence star may also be tapped during a nova outburst. The ^{3}He is in the gas accreted by the white dwarf from its main sequence companion. Although the reaction chain known as the ^{7}Be-transport mechanism is important, other reactions influence the predicted ^{7}Li yields from the explosions on C-O and O-Ne white dwarfs. José (2002) estimates that the contribution from novae to the Galactic ^{7}Li abundance is 'rather small (i.e., less than 15%), even if the ^{7}Li yield from the most favorable case, a massive CO nova, is considered.'

Interstellar Gas. LiBeB synthesis in interstellar (or circumstellar) gas occurs as cosmic rays collide wth ambient nuclei. Spallation reactions of the type (C, N, and O) + (p or α) result in LiBeB production. Fusion reactions $\alpha + \alpha$ contribute to the synthesis of ^{6}Li and ^{7}Li.

These theoretical contributions to the x-process in 2002 are summarized in Table 1. A key point to note is that, unless a site has been overlooked, ^{9}Be is exclusively a product of cosmic ray spallation reactions. Except for finagling of the cosmic ray energy spectrum around the threshold energies of the principal spallation reactions, the relative yields of LiBeB nuclides from spallation are almost energy independent. Therefore, Be abundances may be used to infer the abundances of ^{6}Li, ^{7}Li, ^{10}B, and ^{11}B attributable to spallation. Measured against solar system abundances, spallation can account for the abundances of ^{6}Li, ^{9}Be, and ^{10}B, but fails by about an order of magnitude to explain the observed amount of ^{7}Li, and by about a factor of two the amount of ^{11}B. The latter seems to demand a contribution from Type II SN. Although the same supernovae should contribute ^{7}Li, yet additional ^{7}Li may come from mass loss from those red giants which tap their ^{3}He reservoir. Table 1 also offers qualitative hints on the expected run of LiBeB abundances with metallicity. Absent a contribution from a non-standard Big Bang or a pre-galactic site of nucleosynthesis, one expects ^{6}Li, Be, and B to show a positive correlation with a measure of stellar nucleosynthesis such as the O or Fe abundance: i.e.,

TABLE 1. The x-process

| | Big Bang[a] | Stars | | | | Cosmic Rays |
		AGB	SN II[b]	Novae	$\alpha + \alpha$[c]	Spall.[d]
^{6}Li	...	...	...	...	$\checkmark$	$\checkmark$
^{7}Li	$\checkmark$	$\checkmark$	$\checkmark$	$\checkmark$	$\checkmark$	$\checkmark$
^{9}Be	...	...	...	...	...	$\checkmark$
^{10}B	...	...	...	...	...	$\checkmark$
^{11}B	...	...	$\checkmark$	...	...	$\checkmark$

[a] Standard model assumed.
[b] ^{7}Li/^{11}B $\sim$ solar system ratio?
[c] ^{7}Li/^{6}Li ~ 1
[d] ^{6}Li/^{9}Be/^{10}B $\approx$ solar. Too little ^{7}Li and ^{11}B.

Be/H $\rightarrow 0$ as Fe/H $\rightarrow 0$. The exception is ^{7}Li/H which tends to a constant value (the Spite plateau) at low metallicities because of the Big Bang's ^{7}Li contribution.

7.3. *Observed Abundances*

A suite of observational tests have been performed on the proposals summarized in Table 1. A personal selection is discussed here. Emphasis is on tests using abundances obtained from main sequence and red giant stars. These fall into two categories: (i) direct tests of ^{7}Li synthesis via the ^{7}Be-transport mechanism through observations of lithium in red giants, and (ii) observations of LiBeB in main sequence stars of different metallicity in order to use the Galactic evolution of these nuclides as a (blunt) tool to quantify the relative contributions of the sites in Table 1. A third possible category - observational tests of astration - is noted but not pursued here.

7.3.1. *Li-rich red giants*

Discovery of Galactic Li-rich red giants (McKellar 1940) came years before either the proposal of the ^{7}Be-transport mechanism or the identification of a luminous AGB star with a HBCE as a plausible site. Quantitative checks on production in such AGB stars (Sackmann & Boothroyd 1992) were made possible with the discovery of Li-rich AGB stars in the LMC and SMC (Smith & Lambert 1989, 1990; Plez et al. 1993) – see Fig. 3. These are being extended through observations of AGB stars in Magellanic Cloud clusters. Reliable distances to the Clouds enable the stellar luminosity and mass to be estimated, a task impossible with any precision for Galactic field stars. Confrontation of theory and observation confirmed that Li-production occurs in AGB stars of about 4-6 $M_\odot$ at about the predicted level. Several factors stand between this result and an assessment of AGB stars as ^{7}Li factories effective on the galactic scale. Uncertainties surround both the production of ^{7}Li via the ^{7}Be-transport mechanism and the return of the ^{7}Li via the stellar winds.

7.3.2. *Lithium from Novae?*

In some recipes describing the galactic evolution of the Li abundance, synthesis by novae is an important ingredient: for example, Romano et al. (2001) ascribe 18 % of the solar system's lithium to production in novae. Novae again tap the ^{3}He reservoir of the low mass star which feeds the white dwarf whose surface is the site for the novae explosions. A feature of Li-synthesis by novae is that the ^{7}Be is ejected before it decays

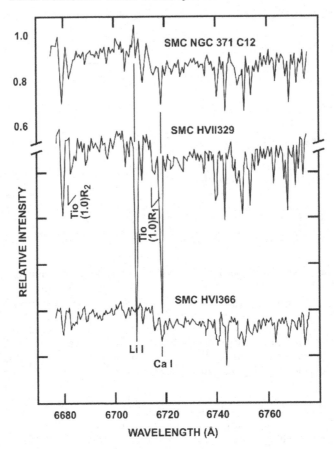

FIGURE 3. Spectra near the Li I resonance doublet of three red giants in the SMC. Lithium is not detectable in either NGC 371 C12 (top), a core-burning supergiant with a luminosity M_{bol} = -7.8, or HV 1366, an AGB star with the relatively low luminosity corresponding to M_{bol} = -5.5. Lithium is very obviously detectable in the higher luminosity (M_{bol} = -6.5) AGB star HV11329. Lithium is strong in the spectra of all AGB stars like HV 11329 near the luminosity limit for intermediate-mass AGB stars.

to ^{7}Li; the half-life of ^{7}Be is 53 days. This means that synthesised Li will not be visible until a couple of months after the thermonuclear runaway.

A corollary deserves a mention in this era of multi-wavelength astrophysics: presence of ^{7}Be may be detectable by the β-decay of this unstable nucleus - ^{7}Be$(e^-\nu_e)^7$Li with 10% of the decays providing a γ-ray at 478 keV. Several searches for the 478 keV γ-ray line have been reported. The most recent is by Harris et al. (2001) who report upper limits on the 478 keV flux at least an order of magnitude higher than predictions for the several well observed novae - 'only weak limits can be placed on the key parameters in the nucleosynthesis and ejection of ^{7}Be'. Prospects for detection of nearby novae using *INTEGRAL* appear quite good.

A tentative identification of the Li I doublet in emission in the spectrum of the fast nova Nova V382 Vel 1999 was advanced by Della Valle et al. (2002). Weak broad emission at 6700 - 6715 Å was present for a few days reaching maximum intensity about 10 days after maximum light. The feature is part of a stronger unidentified emission feature immediately to the blue. The authors remark that 'there is no unambiguous identification for this feature' but suggest that the line strength corresponds 'approximately' to a theoretical prediction of lithium by a fast nova. Perhaps, the principal lesson from this set of spectra is that continued spectroscopic monitoring of novae promises novel rewards.

7.3.3. *Galactic Evolution of Lithium*

Main sequence stars not red giants must be used to map the rise of the Galactic Li abundance from the Spite plateau at low metallicities to its present value in interstellar gas and young stars. Red giants must be discarded because their convective envelope reduces the surface Li abundance below its initial value by an amount that is not yet predictable quantitatively with certainty. Even many main sequence stars have a Li abundance below their initial value. We say lithium has been astrated. Various proposals for astration in main sequence stars have been made but none are developed to the quantitative level that would allow an astrated/reduced Li abundance to be revised upward with the precision desired. Therefore, the upper envelope in a plot of Li/H vs Fe/H is taken to define the Galactic evolution. This shows that the Li abundance remains close to its Spite plateau for $[Fe/H] \leq -1.5$ but then rises to about $\log \epsilon(Li) = 3.3$ at $[Fe/H] = 0$.

Two recent papers of a much larger set attempt to model the evolution of the Li abundance, one more successfully than the other. Scrutiny of the papers may convince the reader that modeling Galactic chemical evolution is as arcane art. As a partial indicator of the imprecision of the art, I draw on the authors' assessments of the various contributors to the solar system's Li abundance. For example, Romano et al. (2001) conclude that the luminous AGB stars contribute a mere 0.5% of the Li, but, in sharp contrast, Travaglio et al. (2001) put the AGB contribution at 40%. Clearly, there is much for a student to do!

7.3.4. *Galactic Evolution of Beryllium*

Beryllium, as noted above, is solely a product of spallation reactions. Observational definition of the run of the Be abundance from the most metal-rich to the most metal-poor unevolved stars is a measure of the spallation activity over the lifetime of the Galaxy: the Be abundance declines steadily with decreasing abundance of a metal such as O or Fe. A question of extragalactic or even primordial relevance is – Does Be/H $\rightarrow$ 0 as (say) O/H $\rightarrow$ 0? or Is there a Be plateau analogous to the Spite lithium plateau? A plateau would suggest that Be was synthesised prior to the formation of the first supernovae in the Galaxy. Beryllium also serves as a calibrator of the yield of boron from spallation. Although an *ab initio* calculation of the evolution of the Galactic Be abundance is fraught with uncertainties, the observed run of the Be versus O or Fe abundances together with the relative predicted yields from spallation of Be and B serves to predict fairly reliably the evolution of the boron abundance in the event that spallation is the sole source of B. Beryllium is a less satisfactory monitor of the Li yield from spallation because in low metallicity gas significant production of Li but not Be may occur from $\alpha + \alpha$ collisions.

Measurements of the Be abundance in very metal-poor dwarf stars via the 3130 Å Be II doublet are challenging observations even for the new generation of large telescopes. Atmospheric extinction is not negligible at 3130 Å. The most metal-poor stars are faint. High-spectral resolution and high signal-to-noise spectra are required to detect and resolve the weak Be II lines from neighbouring lines. To a young enthusiastic observer the following two reports of exposure times may be sobering. Boesgaard et al.'s (1999) Be abundance for a [Fe/H] = -3.0 star is based on 11 hours of integration with the HIRES at the Keck telescope. Primas et al. (2000) consumed 6 hours of time on the VLT with UVES to measure the Be abundance in a [Fe/H] = -3.3 star. Since the chosen stars at [Fe/H] ~ -3 level are among the brightest known examples, it is clear that amassing a large sample of Be abundances in the most metal-poor stars will be a difficult process in face of competing claims for telescope time.

Beryllium abundances are most fruitfully compared with the O abundances because the primary spallation process is expected to be collisions between protons (or αs) and O nuclei. Nitrogen and carbon nuclei can also be spallated to Li, Be, and B but one expects their abundances to be lower than that of O. At this juncture, one must confront the 'oxygen problem'. Various atomic and molecular indicators of the oxygen abundance in metal-poor stars give conflicting results for the O/H abundance. The reader is referred to a series of articles in volume 45 of *New Astronomy Reviews* from an IAU Joint Discussion on the oxygen abundances of old metal-poor stars, and to papers published since then (e.g., Nissen et al. 2002; Meléndez & Barbuy 2002). As a temporary measure to bypass the oxygen problem, another Type II SN product may replace O in quantifying the growth of the Be (and B) abundance. Iron is not the best candidate because the iron yield from SN II is sensitive to the mass cut between the remnant neutron star or black hole and the ejecta. Also, at [Fe/H] greater than about -1, SN Ia contribute iron but not oxygen. Magnesium seems a fair substitute because its abundance in SN II ejecta is insensitive to the mass cut and to the episode of explosive nucleosynthesis (King 2002). There is less controversy about the Mg abundance of metal-poor stars but this may be because we have not probed as closely into the abundance uncertainties!

To an observer, the theoretical relation between the Be and O (or Mg) abundances can take simple limiting forms. Suppose that the primary process of spallation is a collision between a cosmic ray proton (or α) and an interstellar O nucleus. Cosmic rays have a short life in the Galaxy and their flux must be rejuvenated regularly or continuously. One might reasonably identify the cosmic ray flux with the supernova rate, say $\phi_{CR} \propto dn_{SN}/dt$. The production rate of Be (or B) can be written as $d\epsilon(\text{Be})/dt \propto \phi_{CR}\epsilon(\text{O})_{ISM}$. Recognising that O is a product of Type II SN, we may write $dn_{SN}/dt \propto d\epsilon(\text{O})/dt$. On combining these simple relations, we find $d\epsilon(\text{Be})/dt \propto \epsilon(\text{O})d\epsilon(\text{O})/dt$ or $\epsilon(\text{Be}) \propto \text{O}^2$. This 'primary' relation for the direct spallation process is to be contrasted with the prediction for the inverse spallation process in which production of Be occurs through collisions between cosmic ray O nuclei and interstellar protons and αs. In this case, the production rate $d\epsilon(\text{Be})/dt \propto \phi_{CR}$ which with $\phi_{CR} \propto dn_{SN}/dt$ and $\phi_{CR} \propto dn_{SN}/dt$ gives $\epsilon(\text{Be}) \propto \epsilon(\text{O})$, a linear instead of the quadratic relation predicted for the direct spallation process. Gilmore et al. (1992) note that a linear relation is also obtained if spallation occurs in the ejecta of a supernova between fast protons and O nuclei. Of course, the linear and quadratic relations are modified when the modellers of Galactic chemical evolution enter the fray. For this topic I defer to Matteucci's chapter; a recent reference Fields et al. (2001) and astro-ph may serve to track the voluminous literature.

At present, the Be versus O and the Be versus Mg relations for metal-poor stars appear to be quasi-linear; King (2002) gives indices of about 1.1 for both Be and B when Mg is adopted. Recent determinations of the O abundance seem likely to confirm this near-linear trend. All interpretations assume that spallation processes are the sole source of the Be but not necessarily of the B (see below). Ingredients of the modelling other than the mode of synthesis are varied to fit the Be vs O (or Mg) relations.

A result not often commented upon is the lack of dispersion in the Be vs O (or Mg) relations. Given the different origins of Be and O, a dispersion in Be abundance at a fixed O abundance would not seem surprising. Yet, the Be vs O relation for metal-poor stars seems dispersionless beyond that ascribable to the errors of measurement. One might expect any 'cosmic' dispersion to increase at the lowest metallicities when few supernovae had contaminated the pre-galactic gas. This expectation was offered by Primas et al. (2000) as one way to account for their finding that the abundance in the most metal-poor star yet observed for Be indicated a flattening of the Be vs Fe trend at the low metallicity ($[Fe/H] \simeq -3$ and below) end. The competing scenario is that the Be abundances at low metallicity approach a plateau indicative of a pre-Galactic, if not a cosmological origin of Be. Since the Be abundance of the putative plateau is several orders of magnitude greater than predicted by the standard Big Bang model which accounts - almost - for the inferred primordial light element abundances, the prospect of a real detection of the Be plateau leads to the unleashing of speculations about non-standard Big Bang models (see Steigman's chapter). One may hope that observing time will be granted to find and observe suitable extremely metal-poor dwarf stars to map the Be abundances at the lowest metallicities to define the cosmic dispersion and search for the Be plateau. It is surely ironic that proposals to search for the Be plateau may lose in a competition with proposals about extragalactic objects - near and far - when discovery of the plateau seems certain to lead to major revisions to our understanding of the early Universe.

7.3.5. Galactic Evolution of Boron

Boron is likely not solely a product of spallation involving cosmic rays; some ^{11}B is probably produced by neutrino-induced spallation in Type II SN. The first indicator of the need for a supplement to cosmic ray spallation came from the isotopic boron ratio ($^{11}B/^{10}B = 4.05$) of meteorites. Spallation by high-energy (relativistic) cosmic-rays predicts a ratio of 2.5 and the meteoritic abundance ratios of ^6Li:^9Be:^{10}B. Although the prediction can be adjusted towards the observed isotopic boron ration by crafting of the cosmic ray spectrum at low energies, an additional source of ^{11}B would be welcomed in fitting observation to prediction. It appears that, if the ^{11}B deficit is made up by spallation run by low energy cosmic rays, the concomitant production of ^7Li necessarily raises the Li abundance in halo dwarfs above the observed level of the Spite plateau (Prantzos, Cassé & Vangioni-Flam 1993). Alternatively, one might suppose the meteoritic ratio is anomalous - fractionation? This latter supposition has been rendered implausible by the recent measurement of the isotopic ratio in two B-type stars from *HST-GHRS* observations of a B III resonance line at 2066 Å. Proffitt et al. (1999) observed and analysed two sharp-lined stars to find that the isotopic boron ratios are the same as the meteoritic value: the reported ratios were $4.7^{+1.1}_{-1.0}$ and $3.7^{+0.8}_{-0.6}$ for two stars with an elemental abundance similar to the meteoritic value. The elemental B abundances were 0.2 to 0.3 dex less than the meteoritic value given by Zhai & Shaw (1994). If these lower values are evidence of destruction of B by protons in subsurface layers, the isotopic B ratio will have been altered from its original value. The predicted sense of the alteration depends on which compilation of nuclear reaction rates is adopted. Given the listings in Table 1, it is reasonable to suppose that a ^{11}B supplement has been provided by Type II SN and their ν-process. Direct evidence for the ν-process' contribution to ^{11}B and other trace species (^7Li and ^{19}F) is presently lacking.

Timmes, Woosley & Weaver (1995) developed a chemical evolution model of the solar neighbourhood in an attempt to account for the observed abundances of elements from H to Zn in metal-rich and metal-poor stars. The ν-process contributions were included. With their predicted yields of ^{11}B and excluding ^{10}B and ^{11}B from cosmic ray driven spallation, they were able to reproduce the then fragmentary data on the run of the boron abundance with metallicity (see their Fig. 9) from $[Fe/H] \sim -2.5$ to $[Fe/H] \simeq 0$ and including a fit to the meteoritic abundance. Newer data on the B abundances is equally well fit. Obviously and as recognized by Timmes et al., the observed Be abundances imply that some of the observed B must come from spallation. A combination of the ν-process (with somewhat reduced yields) and spallation by relativistic cosmic rays would account for the rise of the boron abundance with metallicity and the ^{11}B/^{10}B ratio in meteorites and young stars. Precise measurements of the B/Be ratio across all metallicities could shed light on the relative roles of the two contributors to the synthesis of the boron isotopes.

Observations of boron in stars requires ultraviolet spectra. To trace the evolution of the B abundance with metallicity one must obtain *HST* high-resolution spectra of metal-poor dwarfs at 2497 Å, the wavelength region of the strongest of the B I resonance lines. Stars with undepleted lithium are the ideal candidates. At [Fe/H] less than about -3, the one unblended B I line is too weak to detect. Several attempts have been made to map the decrease of the B abundance with metallicity and to estimate the B/Be ratio. Duncan et al. (1997) by combining *HST*-based B abundances with Be abundances from ground-based spectra and including estimates of the (small) non-LTE effects on the B I and Be II lines obtained a B/Be ratio of "$\sim$ 15", and a near-linear relation between B (and Be) and metallicity. These authors concluded a ν-process contribution to B production must have been "at most a small part of total B production". García López et al. (1998) added a couple of new stars and reanalyzed archived *HST* spectra and found that the B/Be ratio is constant for metal-poor stars at a value of about 20, a value not significantly different from the earlier result of 15, and consistent with production by spallation. In an independent analysis, King (2002) claims B/Be $\sim 11 \pm 2 - 3$, which would further reduce the need for a contribution from Type II SN and the ν-process. Considerable light would be shed on the respective contributions of spallation and the ν-process by mapping the trend of the B isotopic ratio with metallicity. This will be exceedingly difficult given the small isotopic wavelength shifts and the crowded UV spectra in which the B I lines reside. But Duncan & Rebull (2000) have taken up the challenge.

8. Stellar spectroscopy and the s-process

8.1. Introduction

It is a consequence of the competition between the attractive strong but short range nuclear force and the repulsive weak but long range electrostatic force that the binding energy per nucleon reaches a minimum at iron. One consequence is that synthesis of elements beyond the iron-peak by reactions involving charged particles drains energy from the medium. This is in contrast to the synthesis of lighter elements by charged particle reactions in which energy is liberated. It is this energy that enables a massive star, for example, to maintain a high central pressure and so stave off gravitational collapse until the point that the core has been converted to iron. In principle, charged particle reactions in an explosion might push nucleosynthesis beyond the iron-peak. In practice, it is neutrons that dominate nucleosynthesis from the iron-peak to the heaviest stable and very heavy unstable nuclei. Synthesis occurs using lighter elements as seeds from which heavier elements are synthesized by a series of neutron captures and decays, most often β-decays, of neutron-rich unstable nuclides.

Neutron capture reactions can occur at low temperatures but stars can activate neutron sources only at high temperatures. It was apparent very early in the search for the origins of the chemical elements that two different neutron capture processes are at work in the Universe - see the classical papers by Cameron (1957) and Burbidge, Burbidge, Fowler & Hoyle (1957). The processes are distinguished by the neutron density achieved at the

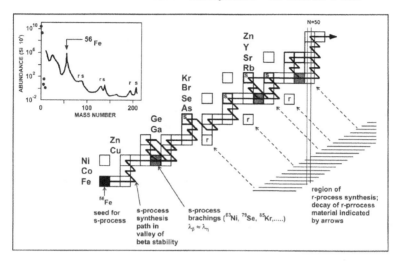

FIGURE 4. The initial steps of the s-process from ^{56}Fe through to the neutron magic number $N = 50$. s-only and r-only nuclei are labelled. Possible branches in the s-process path occur at nuclei identified by the shaded boxes. The branch at ^{85}Kr is discussed in the text. Several proton-rich nuclei (open boxes to left of the s-process path) are present; these are products of neither the s- not the r-process. The inset shows the standard abundance distribution with the iron-group peak, and r- and s-process peaks identified. After Käppeler et al. (1989).

site of synthesis: the s-process where s denotes 'slow' operates at low neutron density; the r-process where r denotes 'rapid' operates at high neutron density. Representative densities are 10^7 cm^{-3} for the s-process and 10^{22} cm^{-3} for the r-process. In the s-process, the chain of neutron captures and (mostly) β-decays remains close the valley of stability; if neutron capture forms an unstable isotope, that isotope will generally have sufficient time to decay before capturing an additional neutron - collisions with neutrons occur on a 'slow' timescale. In sharp contrast, a path for a r-process runs on the neutron-rich side far from the valley of stability; many neutron captures occur successively before β-decay occurs predominantly at neutron magic numbers. On cessation of the r-process, the unstable very neutron-rich nuclei β-decay to the valley of stability.

A further distinction may be made between these two neutron capture processes. The low neutron densities of the s-process are achievable in varieties of stable stars. This opens the possibility that the fruits of the s-processing may be seen at the surface of the star. The preeminent examples are the S-stars. In contrast, the very high neutron densities of the r-process are expected to be achievable almost exclusively in explosive situations. For the stellar spectroscopist this means that r-process products are not observable in the atmosphere of the host star. The r-process products seen in a star were synthesized by earlier generations of stars. The qualification 'almost exclusively' refers to theoretical ideas that a He-core flash in a low mass red giant may result in some r-processing leading to the possibility that r-process products might be dredged to the surface in the subsequent evolution of the red giant. The identities of effective r-process sites remain elusive. It is widely supposed that the deep interior of SN II is a principal site. Other suggestions include the merger of two neutron stars.

One may dissect the origins of the heavy nuclides in the valley of stability - see Fig. 4. Some may be synthesized only by the s-process: such a nucleus (Z, N) is shielded from manufacture by the r-process by the existence of a stable nuclide $(Z - 2, N + 2)$ which marks the end-point for the β-decays occurring on cessation of the r-process. Other nuclides are deemed to be products solely of an r-process; such a nuclide (Z, N) is preceded by one or more short-lived isotopes $(Z, N-1)$ at which the s-process path takes a turn to a $(Z + 1, N - 1)$. Many nuclides may be made by either the s- or r-process. There are a few proton-rich nuclei which cannot be made by neutron capture. One refers to these as p-nuclei or as products of a p-process. One presumes that the influence of the p-process extends to slightly more neutron-rich nuclides. However, the abundance of a p-nuclide is generally so much lower than that element's more neutron-rich isotopes that a p-process contribution to these other isotopes is neglected. In summary, one may designate nuclides as products of a p-process, s-process, r-process, or as a combination of s-process and r-process.

In principle, the abundance pattern of the s-process and r-process contributions are resolvable given sufficient abundance . In practice, the exercise is possible in detail only for the solar system for which abundances isotope by isotope are obtainable. Given the abundances of the s-only nuclides and a theory of the s-process one may predict the contribution of the s-process to the nuclides of mixed $(s + r)$ origin. Subtraction of this contribution gives the r-process contribution. The full pattern of the r-process abundances is obtained by combining the r-process abundance from the mixed nuclides with the abundances of those nuclides made exclusively by the r-process. A recent exercise in resolving s- from r-process using solar system abundances was reported by Arlandini et al. (1999). The abundance pattern of the s-process or r-process found in this way is not attributable to a single site but rather represents the cumulative effect of many sites whose ejecta were mixed into the interstellar gas prior to the formation of the solar system.

This latter qualification applies to all but a few stars. Exceptions include the S-type stars in which s-processing occurs internally. An extreme S star has an atmosphere enriched 10 to 100 times in s-process products - these products mask the s-process contribution of the original star. But for stars the information on heavy element abundances is very much less complete than for the solar system. Not only are many heavy elements not measureable in the stellar spectrum but information on the isotopic abundances is almost completely missing. The isotopic abundances are crucial to full resolution of the s- from the r-process in the solar system composition. All is not lost. Ratios of elemental abundances may serve as measures of the relative contributions of the s- and r-process. For example, the Ba/Eu ratio is high for the s-process and lower for the r-process.

In this section, I discuss some theoretical aspects of the s-process and a few observational investigations of the operation of the s-process in individual stars like the the S stars.

8.2. *Nuclear physics of the s-process*

The cross-section for neutron capture depends in part on the transit time of the neutron across the target nucleus: $\sigma \propto 1/v$ where σ is the cross-section and v is the relative velocity of the neutron and the nucleus. Neutrons of velocity v are captured at a rate proportional to σv which is independent of v. It follows immediately that the neutron capture rate in gas at temperature T is independent of T as long as the transit time is the dominant consideration.

The structure of the nucleus plays a part in determining the magnitude of the cross-section. This is especially the case at the neutron magic numbers where the cross-section experiences a marked decline. For example, the cross-section at 30 keV for ^{139}La at $N = 82$ is about 40 milli-barn but the typical cross-section for other odd Z nuclei between this magic number and the next at $N = 126$ is about 2000 milli-barn. A closed neutron shell at the neutron magic numbers is reluctant to accept an additional neutron. Much about the s-process is understood once the pattern of a small cross-section at neutron magic numbers and a quasi-constant cross-section between magic numbers is appreciated.

Measurements of neutron capture cross-sections have been made for nuclei along the s-process path. A useful compilation of results is provided by Bao et al. (2000). Much of the recent work has come from the Karlsruhe group (Käppeler et al. 1989) where targets are bombarded with a neutron beam with a 30 keV quasi-Maxwellian distribution of energies - a fair simulation of the neutron energies in the He-shell of an AGB star. These measurements are made on nuclei at room temperature, i.e., the targets are in their nuclear ground states.Under stellar conditions, excited nuclear states may be populated and may participate in the s-process. Excited states, especially isomeric states, can influence the s-process path. Experimental data on cross-sections, and lifetimes of excited states is understandably incomplete. Several unstable nuclei along the s-process path have been investigated. Note that relatively long-lived nuclei may serve as branch points for the s-process path: at low neutron density the unstable nucleus β-decays but at high neutron density, the nucleus may capture a neutron. These alternatives provide a fork or branch in the s-process path controlled by the ambient neutron density. Other branches involve a temperature sensitivity. See Käppeler et al. (1989) for a discussion of the branches examinable with the precise solar system abundances.

8.3. *Operation of the s-process*

The mathematics of the s-process is simple. Here, I suppose that there is a unique path, i.e., there are no branches. I also assume $\sigma \propto 1/v$. I introduce the time integrated neutron flux $\tau = \int \Phi dt$ where the instantaneous flux is $\Phi = n_n v_T$ and the thermal neutron velocity is $v_T = \sqrt{2kT/\mu}$ (μ = the reduced mass of the neutron - nucleus). The neutron flux τ has the dimensions of L^{-2} and is usually given in milli-barns. The s-process operates on the seed nuclei - generally, considered to be the Fe-peak nuclei - converting stable nuclide $m-1$ on the path to stable nuclide m which in turn is converted to stable nuclide $m+1$; the assumption of no branches means that unstable nuclides need not be introduced specifically into the differential equation describing the flow along the s-process path. The flow ends at ^{206}Pb in a closed cycle with ^{209}Bi. Three neutron captures convert ^{206}Pb to ^{209}Pb which β-decays to ^{209}Bi which on neutron capture is mainly decays by α-emission to return to ^{206}Pb.

In the limit that a quasi-equilibrium is reached, the product of neutron capture cross-section times the abundance (here, $\sigma_m n_m$) is constant along the path. In general, one expects this product to vary smoothly between neutron magic numbers. At these magic numbers where the cross-section goes through a minimum, $\sigma_m n_m$ will experience a discontinuity. Examples of solutions to the coupled equations were provided by Clayton et al. (1961). Revisions to the neutron capture cross-sections do not change their basic results.

For exposures equivalent to about 20 neutrons per Fe seed nucleus ($\tau \sim 0.5$), the seed nuclei are converted principally to elements of atomic mass A from 70 to 85 with very little synthesis of nuclei beyond the neutron magic number $N = 50$. At about 50 neutrons per Fe seed nucleus, the Fe seeds have been flushed to nuclei between the magic numbers $N = 50$ and $N = 82$. And at exposures of 130 neutrons per Fe seed nucleus, the greatest $\sigma_m n_m$ occur between $N = 82$ and Pb. Finally for higher exposures the Fe seeds are predominantly converted to Pb nuclei. A recent discovery of 'lead stars' shows that a high neutron to seed ratio is possible for metal-poor stars (Van Eck et al. 2001).

Examination of $\sigma_m n_m$ for the solar system abundance distribution shows that it cannot be fitted with a predicted distribution for a single τ – see Käppeler et al. (1989). Lighter s-nuclei between the iron-group and about Zr at $N = 50$ are attributed to the weak s-process involving mild exposure to neutrons. The site of this process is very likely the C and He shells of a hydrostatically burning massive star – see Langer's chapter. Overlapping the weak s-process at around $N = 50$ and extending to the termination at Pb-Bi, the s-process contributions are assigned to the main process. An exponential-like distribution of exposures τ is needed to fit all main process nuclei. The site of this process is now known to be the He-shell of an AGB star. Such a shell in an S star is the site of Merrill's technetium. The solar system material is, of course, a mix of contributions from many AGB stars of different masses and metallicities.

8.4. AGB stars and the s-process

An AGB star is a red giant with a C-O electron degenerate core surrounded by a thin He shell outside of which is the extensive H-rich convective envelope. The C-O core with the He shell is approximately the size of the Earth but contains several tenths of a solar mass of material. This core is a putative white dwarf. The radius of the AGB star is approximately the size of the Earth's orbit around the Sun. Stellar evolution in the nuclear sense is driven by H-burning in a shell just below the convective envelope and by He-burning in the thin He shell. Evolution is also driven by mass loss off the surface. At the highest luminosities, mass loss controls the evolution of the AGB star.

The H- and He-burning shells alternate as the primary source of nuclear energy. Two episodes of this alternation are sketched in Fig. 5. Shell H-burning proceeds with the He-shell inactive. The freshly produced He increases the mass of the He-shell which contracts slightly under gravity to higher densities and temperatures. When the ignition temperature of He is reached, there is a thermal runaway (thermal pulse) in the He shell which becomes almost completely convective with the result that products of He-burning - direct and associated - created in the shell are distributed across the convective region, which spans almost the entire thickness of the shell. At maximum extent, the convective shell almost reaches the boundary between the H-rich and He-rich layers. Convection ceases and the He burns quiescently at the base of the He-shell with energy transported outwards by radiation. Expansion of the shell pushes the H-He boundary out to lower temperatures. H-burning ceases at this point. The convective envelope may then descend to mix with regions previously below the H-He boundary and once part of the convective He-shell – the regions between A and B in the mass coordinate marked in Fig. 5 are swept into the convective envelope. Busso et al. (1999) refer to this as 'salting' the envelope with He-burning products. Others refer to salting as the third dredge-up. After a 'short' time, the supply of helium above its ignition temperature runs out, a collapse of the shell leads to reignition of H and the H-burning shell is reestablished. This sequence of a long H-burning phase, a thermal pulse in the He shell followed by a short period of steady He burning, a salting of the envelope, and resumption of H-burning is repeated many times. Mass loss eventually so slims the envelope mass that it is unable to sustain nuclear burning. The star evolves off the AGB to higher temperatures at almost a constant luminosity but may, as a hot post-AGB star experience a final thermal pulse. Mass loss via the slow (10 - 20 km s^{-1}) wind distributes the H- and He-burning products into the local interstellar medium.

That s-processing occurs in red giant AGB stars was clear from Merrill's discovery and Cameron's discussion of neutron production and capture. Full understanding of how the s-process operates in AGB stars may now be close – see Busso et al. (1999). Two neutron sources are recognized: $^{22}\mathrm{Ne}(\alpha, n)^{16}\mathrm{O}$ and $^{13}\mathrm{C}(\alpha, n)^{16}\mathrm{O}$ with ignition temperatures of

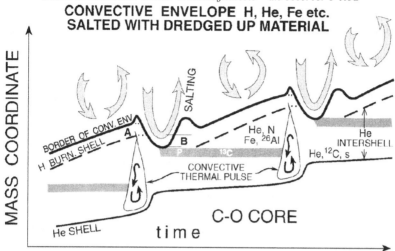

FIGURE 5. Evolution of the interior of a thermally pulsing AGB star. Locations of the lower boundary of the convective envelope, the H-burning shell, and lower edge to the He-shell are labelled. Two convective thermal pulses are shown. Key to understanding the *s*-process are the shaded bars of which the middle one is overwritten with a 'P' and '^{13}C' – see text. From Busso, Gallino & Wasserburg (1999).

approximately 300 and 100 million K, respectively. The ^{22}Ne source is expected to be the primary supplier of neutrons in intermediate-mass AGB stars. The strength of the source is fixed by a star's initial C, N, and O abundances: H-burning and α-capture prior to He-burning convert the initial C, N, and O to ^{14}N and then ^{22}Ne. Release of the neutrons occurs in the thermal pulse at high neutron density. The ^{13}C source is the principal one for low mass AGB stars but here there is presently no easy solution to the strength of the source. The ^{13}C residue from H-burning is too low for the source to be effective in running the *s*-process. It is assumed that the ^{13}C abundance in the He-shell is increased to satisfactory levels through the diffusion of protons across the H-He boundary following the thermal pulse and prior to reignition of the H-burning shell – see the gray bar in Fig. 5 and the label 'P'. Protons in the He-shell are captured by ^{12}C nuclei to form ^{13}C. Thanks to the high ratio of ^{12}C relative to protons, the vast majority of the protons are captured by ^{12}C rather than ^{13}C despite the latter's larger cross-section. Thus, after diffusion of protons, there is a thin layer in the He-shell containing ^{13}C, a potential neutron source – see the gray bar in Fig. 5 and the label '^{13}C. There is at present no *ab initio* way to predict the amount of ^{13}C created in the He-shell. The mass of ^{13}C remains the one key free parameter in predictions of *s*-processing by low-mass AGB stars. The existence of Pb-rich metal-poor stars suggests that the ^{13}C mass may not decrease sharply with decreasing initial metallicity. It is apparent from Fig. 5 that much of the material in the He-shell is re-exposed many times to the *s*-process, a repetition that simulated the exponetial-like distribution of exposures suggested by the solar system's main *s*-process abundances.

Here, I will discuss one aspect of *s*-process operation - what is the typical neutron density at which the *s*-process operates? The key to answering this question is found in branches along the *s*-process path (Fig. 4). Two branches may be exploited by the stellar spectroscopist: one at ^{85}Kr and one at ^{95}Zr. Here, I discuss the former; both were considered by Lambert et al. (1995). The ^{85}Kr branch is shown in Fig. 4. Details of the branch are not revealed by this figure. In particular, there is a low-lying short-lived isomeric state in ^{85}Kr which must be considered and provides a path to ^{85}Rb even at high neutron densities.

At low neutron density, unstable ^{85}Kr decays to ^{85}Rb before it can capture a neutron. The *s*-process path continues to ^{86}Sr and then to Y and Zr. At high neutron density, neutron capture occurs and stable ^{86}Kr is produced. From ^{86}Kr, the *s*-process path goes to stable ^{87}Rb and onto Sr and Y. In principle, the presence of ^{86}Kr in *s*-process material would be the signal that the high density path had been taken. Unfortunately, Kr is not observable in the stars for which Rb, Sr, Y, and Zr are measureable. Clues to the path taken at the ^{85}Kr branch come from the measurement of the Rb abundance relative to that of Sr, Y, or Zr. Fortunately, the neutron-capture cross-section of ^{87}Rb ($Z = 37$) is about an order of magnitude smaller than that of ^{85}Rb, a consequence of the neutron number increase from $N = 48$ to the magic number $N = 50$ for ^{87}Rb. As a result the Rb abundance (relative to Sr, Y, and Zr) increases by about an order of magnitude as the switch is made from the low to the high density path at ^{85}Kr.

Rubidium abundances in cool giants are measureable from the Rb I 7800Å line. Unfortunately, the Rb isotopic mix is not measureable because the hyperfine structure from ^{85}Rb and ^{87}Rb adds confusion. Our analyses of the 7800Å line in a sample of M, MS, and S giants showed Rb to increase with the *s*-process enrichment of the atmosphere. The Rb/Sr ratio suggested a low neutron density around 10^7 cm^{-3}.

At the outset of our investigation, it was thought that the *s*-process operated only during the thermal pulse in the hot convective shell. The inferred neutron density clearly excluded intermediate-mass AGB stars as the *s*-process site: ignition of the ^{22}Ne source at high temperatures leads to a very high neutron density. Ignition of the ^{13}C source in the thermal pulse of a low mass star also leads to predicted neutron densities greater than observed corresponding to a surface Rb/Sr ratio about a factor of 3 higher than measured. The resolution of this discrepancy came when it was realized that neutrons are released from ^{13}C in the He-shell prior to the thermal pulse (Straneiro et al. 1995). Consumption of ^{13}C at the lower temperatures of the radiative He-shell necessarily result in lower neutron densities. In fact, the predicted Rb/Sr is in good accord with the observed values for *s*-processed enriched stars. Additional *s*-processing with neutrons from any remnant of the ^{13}C supply and the available ^{22}Ne occurs during the thermal pulses but this has little affect on the Rb/Sr/Y/Zr ratios.

Recently, Abia et al. (2001 – see also Abia & Wallerstein, 1998) measured the Rb, Sr, Y, and Zr abundances in a sample of N-type carbon stars. Carbon enrichment is attributed to salting of the atmosphere with ^{12}C from the He-shell of an AGB star. The authors conclude that most of the carbon stars are "of low mass, experiencing *s*-process nucleosynthesis phenomena dominated by the neutron source provided by α-capture s on ^{13}C in radiative intershell layers." It is pleasing that the neutron density inferred from carbon stars is consistent with that from MS and S stars because the sequence of third dredge-ups is predicted to raise the low C/O ratio of MS and S stars to a value C/O in excess of unity which defines a cool carbon star. A similarity in mass is then expected.

8.5. *Weak s-process at low metallicities?*

If the *s*-process site is the He-shell of a low mass AGB star and the *r*-process site is deep within a SN II, the ratio of *s*- to *r*-process contamination of interstellar gas will have evolved over the Galaxy's lifetime. The *r*-process species will have appeared along with the other SN II products such as oxygen. Later, the much more slowly evolving low-mass stars will dump *s*-process products into the interstellar gas. This obvious piece

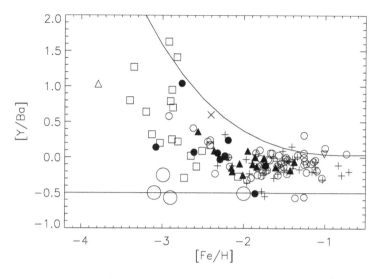

FIGURE 6. The abundance ratio [Y/Ba] versus [Fe/H] for metal-poor giants and dwarfs. Plotted data are tkaen from a variety of sources - please see Lambert & Allende Prieto (2002) for the key to the symbols. The solid lines sketch possible boundaries to the data.

of galactic chemical evolution leads to the idea that heavy elements in metal-poor main sequence and early giant stars should be dominated by r-process products. The s-process products are expected to make their first appearance in stars formed from gas with a moderate metallicity.

Abundance analyses of very metal-poor stars do show that r-process products are present without s-process contamination for [Fe/H] $\simeq$ -2 (Burris et al. 2000; Johnson & Bolte 2001). In particular, the Ba/Eu ratio approaches the value estimated for the r-process contributions to the solar system abundances. Several very metal-poor stars are known with severe overabundances of the heavy elements (Cowan et al. 1999; Westin et al. 2000; Hill et al. 2002; Cowan et al. 2002). For elements Ba and heavier, the relative abundances in the severely r-process overabundant stars are robustly equal to the solar r-process ratios. This solar-like mix also appears to be present in stars less enriched in the heavy elements. This robustness is a challenge to modellers of the r-process and its site of operation.

A quite different picture emerges for the elements lighter than barium. Fig. 6 shows [Y/Ba] versus [Fe/H] for metal-poor stars; [Sr/Ba] and [Zr/Ba] versus [Fe/H] give similar pictures. The lower bound to [Y/Ba] is very similar to that found by considering the solar r-process contributions to Y and Ba. Note that the stars very enriched in r-process products (large open symbols in the figure) define this lower bound. Other stars are spread between the lower bound and an upper envelope that rises with decreasing [Fe/H]. Is this spread in Sr, Y, and Zr abundances relative to Ba, which is in striking contrast to the lack of spread in the ratio of heavier elements to Ba among the same stars, an indicator of variations in r-process contaminations of the stars' natal clouds, or is it a signature of an early appearance of a weak s-process cotamination interstellar gas?

Wasserburg et al. (1996 - see also Wasserburg & Qian 2000; Qian & Wasserburg 2001; Sneden et al. 2000) suggest multiple sources of r-process products in the early

Galaxy. Qian & Wasserburg (2001) propose three sources: (i) a pre-Galactic component contributing the 'light' r-process products Sr, Y, and Zr; (ii)a common type of supernova providing Ba to U in solar r-process ratios and Sr, Y, and Zr in sub-solar proportions (relative to Ba), and (iii) a rare type of supernova contributing iron, Sr, Y, and Zr, and possibly some Ba. Stars with [Fe/H] $\sim$ -3 are formed from gas with the pre-Galactic composition after contaminated by ejecta from a few common supernovae. Given the difficulty of determining the abundances of elements between the iron-peak and Sr in very metal-poor stars, and the lack of quantitative predictions for r-process synthesis of these elements, it may prove difficult to definitively which of the two basic neutron capture schemes is responsible for the marked variation in [Y/Ba] at fixed [Fe/H] in Fig. 6. Surely, someone will assume this spectroscopic challenge!

9. Concluding Remarks

This ramble through the tales of stellar nucleosynthesis comes to an end. Much remains for young fresh minds to accomplish. Especially appealing, I dare to hope, is the blending of disciplines needed to address fully questions of nucleosynthesis: atomic/molecular physics and nuclear physics, physics of stellar atmospheres with stellar interiors.

One frontier for current exploration concerns the nucleosynthesis occurring in the Universe's dark ages between the time the cosmic microwave photons were set free and systems containing 'normal' stars had formed. In this connection, I close with an extensive quotation from Fred Hoyle's *Frontiers of Astronomy*:

'But how about stars that were formed very early in the history of the Galaxy, before very many supernovae had added their quota of elements to the interstellar gas? Where did such stars get their elements from? A partial answer is that the oldest stars contain only very low concentrations of elements other than hydrogen and helium. ..... But should not the very first stars have contained nothing but hydrogen, without even a trace of other elements. This is a deep question. .... We end the present discussion by noticing that while such stars are significantly deficient in heavy elements there seem to be no stars in which heavy elements are entirely absent. Whether or not this is in disagreement with our general evolutionary picture we must leave over for a subsequent settlement.'

Fifty years later, 'the general evolutionary picture' of nucleosynthesis and the observational evidence has greatly changed. The Big Bang inflationary Universe has replaced the steady state Universe. Basic concepts of stellar nucleosynthesis have been confirmed and developed. The metal abundance of the oldest stars has been extended from 'about one-twentieth of the concentration in the Sun' to about 10^{-5} that of the Sun. Yet, the origins of the first stars and the first metals remain questions for 'a subsequent settlement'.

I thank the winter school's organizers for their invitation and most especially for their patience with me over the long months it took to complete the manuscript. My research into cosmochemistry is supported by The Robert A. Welch Foundation of Houston, Texas.

REFERENCES

ABIA, C., BUSSO, M., GALLINO, R., DOMÍNGUEZ, STRANIERO, O. & ISERN, J. 2001 The ^{85}Kr s-process branching and the mass of carbon stars. *ApJ* 559, 1117.

ABIA, C. & WALLERSTEIN, G. 1998 Heavy-element abundances in seven SC stars and several related stars. *MNRAS* 293, 89.

ARLANDINI, C., KÄPPELER, F., WISSHAK, K., GALLINO, R., LUGARO, M., BUSSO, M. & STRANIERO, O. 1999 Neutron capture in low-mass asymptotic giant branch stars: Cross sections and abundance signatures. *ApJ* 525, 886.

BAO, Z.Y., BEER, H., KÄPPELER, F., WISSHAK, K. & RAUSCHER, T. 2000 Neutron cross sections for nucleosynthesis studies. *At. Nucl. Data Tables* 76, 70.

BOESGAARD, A.M., DELIYANNIS, C.P., KING, J.R., RYAN, S.G., VOGT, S.S. & BEERS, T.C. 1999 Beryllium abundances in halo stars from Keck/HIRES observations. *AJ* 117, 1549.

BÖHM-VITENSE, E. 1989 Introduction to stellar astrophysics - volume 2 - Stellar atmospheres. *Cambridge University Press*, Cambridge.

BURBIDGE, E.M., BURBIDGE, G.R., FOWLER, W.A. & HOYLE, F. 1957 Synthesis of elements in stars. *Rev. Mod. Phys.* 29, 547.

BURRIS, D.L., PILACHOWSKI, C.A., ARMANDROFF, T.E., SNEDEN, C., COWAN, J.J. & ROE, H. 2000 Neutron-capture elements in the early Galaxy: Insights from a large sample of metal-poor giants. *ApJ* 544, 302.

BUSSO, M., GALLINO, R. & WASSERBURG, G.J. 1999 Nucleosynthesis in asymptotic giant branch stars: Relevance for Galactic enrichment and solar system formation. *ARAA* 37, 239.

CAMERON, A.G.W. 1955 Origin of anomalous abundances of the elements in giant stars. *ApJ* 121, 144.

CAMERON, A.G.W. 1957 Nuclear reactions in stars and nucleogenesis. *PASP* 69, 201.

CAMERON, A.G.W. 1999 Adventures in cosmogony. *ARAA* 37, 1.

CAMERON, A.G.W. & FOWLER, W.A. 1971 Lithium and the *s*-process in red giant stars. *ApJ* 164, 111.

CHAMBERLAIN, J.W. & ALLER, L.H. 1951 The atmospheres of A-Type subdwarfs and 95 Leonis. *ApJ* 114, 52.

CHARBONNEL, C. & BALACHANDRAN, S.C. 2000 The nature of the lithium rich giants. Mixing episodes on the RGB and early-AGB. *A&A* 359, 563.

CLAYTON, D.D., FOWLER, W.A., HULL, T.E. & ZIMMERMAN, B.A. 1961 Neutron capture chains in heavy element synthesis. *Ann Phys* 12, 331.

COWAN, J.J., PFEIFFER, B., KRATZ, K.-L., THIELEMANN, F.-K., SNEDEN, C., BURLES, S., TYTLER, D., & BEERS, T.C. 1999 *r*-process abundances and chronometers in metal-poor stars. *ApJ* 521, 194.

COWAN, J.J., SNEDEN, C., BURLES, S., IVANS, I.I., BEERS, T.C., TRURAN, J.W., LAWLER, J.E., PRIMAS, F., FULLER, G.M., PFEIFFER, B. & KRATZ, K.-L. 2002 The chemical composition and age of the metal-poor halo star BD +17° 3248. *ApJ* 572, 861.

DELLA VALLE, M., PASQUINI, L., DAOU, D. & WILLIAMS, R.E. 2002 The evolution of Nova V382 Velorum 1999. *A&A* 390, 155.

DRAKE, N.A., DE LA REZA, R., DA SILVA, L. & LAMBERT, D.L. 2002 Rapidly rotating lithium-rich K giants: The new case of the giant PDS 365. *AJ* 123, 2703.

DUNCAN, D.K., PRIMAS, F., REBULL, L.M., BOESGAARD, A.M., DELIYANNIS, C.P., HOBBS, L.M., KING, J.R. & RYAN S.G. 1997 The evolution of Galactic boron and the production site of the light elements. *ApJ* 488, 338.

DUNCAN, D.K. & REBULL, L.M. 2000 Measurement of the B isotope ratio in old stars. *BAAS* 32, 730.

FIELDS, B.D., OLIVA, K.A., CASSÉ, M. & VANGIONI-FLAM, E. 2001 Standard cosmic ray energetics and light element production. *A&A* 370, 623.

FOWLER, W.A., BURBIDGE, G.R. & BURBIDGE, E.M. 1955 Nuclear reactions and element synthesis in the surfaces of stars. *ApJS* 2, 167.

GARCÍA LÓPEZ, R.G., LAMBERT, D.L., EDVARDSSON, B., GUSTAFSSON, B., KISELMAN, D. & REBOLO, R. 1998 Boron in very metal-poor stars. *ApJ* 500, 241.

GILMORE, G., GUSTAFSSON, B., EDVARDSSON, B. & NISSEN, P.E. 1992 Is beryllium in metal-poor stars of galactic or cosmological origin? *Nature* 357, 379.

GIRIDHAR, S., RAO, N.K. & LAMBERT, D.L. 1994 The chemical composition of the RV Tauri variable IW Carinae. *ApJ* 437, 476.

GIRIDHAR, S., LAMBERT, D.L. & GONZALEZ, G. 2000 Abundance analyses of field RV Tauri stars. V. DS Aquarii, UY Arae, TW Camelopardalis, BT Librae, U Monocerotis, TT Ophiuchi, R Scuti, and RV Tauri. *ApJ* 531, 521.

HARRIS, M.J., TEEGARDEN, B.J., WEIDENSPOINTNER, G., PALMER, D.M., CLINE, T.L., GEHRELS. N. & RAMATY, R. 2001 Transient Gamma-ray spectrometer observations of Gamma-ray lines from Novae. III. The 478 keV line from ^{7}Be decay. *ApJ* 563, 950.

HILL, V., PLEZ, B., CAYREL, R., BEERS, T.C., NORDSTRÖM, B., ANDERSEN, J., SPITE, M., SPITE, F., BARBUY, B., BONIFACIO, P., DEPAGNE, E., FRANÇOIS, P., PRIMAS, F. 2002 First stars. I. The extreme r-element rich, iron-poor halo giant CS 31082-001. Implications for the r-process site(s) and radioactive cosmochronology. *A&A* 387, 560.

HOYLE, F. 1994 Home is where the wind blows. *University Science Books*, Mill Valley, California.

JOHNSON, J.A. & BOLTE, M. 2001 The r-process in the early Galaxy. *ApJ* 579, 616.

JOSÉ, J. 2002 Nuclear ashes: Reviewing thirty years of nucleosynthesis in classical novae. *AIP Conference Proc.* 637, 104.

KÄPPELER, F., BEER, H. & WISSHAK, K. 1989 s-process nucleosynthesis - nuclear physics and the classical model. *Rept. Prog. Physics* 52, 945.

KING, J.R. 2002 The evolution of Galactic beryllium and boron traced by magnesium and calcium. 2002 *PASP* 114, 25.

LAMBERT, D.L., SMITH, V.V., BUSSO, M., GALLINO, R. & STRANIERO, O. 1995 The chemical composition of red giants. IV. The neutron density at the s-process site. *ApJ* 450, 302.

LAMBERT, D.L., HINKLE, K.H. & LUCK, R.E. 1988 The peculiar supergiant HR 4049. *ApJ* 333, 917.

LAMBERT, D.L. & ALLENDE PRIETO, C. 2002 The isotopic mixture of barium in the metal-poor subgiant HD 140283. *MNRAS* 335, 325.

LECKRONE, D.S., PROFFITT, C.R., WAHLGREN, G.M., JOHANSSON, S.G., & BRAGE, T. 1999 Very high resolution ultraviolet spectroscopy of a chemically peculiar star: results of the Chi Lupi pathfinder project. *AJ*, 117, 1454.

MCKELLAR, A. 1940 Intense λ6708 resonance doublet of Li I in the spectrum of WZ Cassiopeiae. *PASP* 52, 407.

MELÉNDEZ, J. & BARBUY, B. 2002 NIRSPEC infrared OH lines: Oxygen abundances in metal-poor stars down to [Fe/H] = -2.9. *ApJ*, 575, 474.

MERRILL, P.W. 1952 Spectroscopic observations of stars of class S. 1952 *ApJ* 116, 21.

NISSEN, P.E., PRIMAS, F., ASPLUND, M. & LAMBERT, D.L. 2002 O/Fe in metal-poor main sequence and subgiant stars. *A&A* 390, 235.

PAGEL, B.E.J. 1997 Nucleosynthesis and the chemical evolution of galaxies. *Cambridge University Press*.

PALACIOS, A., CHARBONNEL, C. & FORESTINI, M. 2001 The lithium flash. Thermal instabilities generated by lithium burning in RGB stars. 2001 *A&A* 375, L9.

PHILLIPS, A.C. 1993 The physics of stars 1993 *J. Wiley & Sons*.

PLEZ, B., SMITH, V.V. & LAMBERT, D.L. 1993 Lithium abundances and other clues to envelope burning in Small Magellanic Cloud asymptotic giant branch stars. *ApJ* 418, 812.

PRANTZOS, N., CASSÉ, M. & VANGIONI-FLAM, E. 1993 Production and Evolution of LiBeB isotopes in the Galaxy. *ApJ* 403, 630.

PRESTON, G.W. 1974 The chemically peculiar stars of the upper main sequence. *ARAA* 12, 257.

PRIMAS, F., ASPLUND, M., NISSEN, P.E. & HILL, V. 2000 The beryllium abundance in the very metal-poor halo star G64-12 from VLT/UVES observations. *A&A* 364, L42.

PROFFITT, C.R., JÖNSSON, P., LITZÉN, U., PICKERING J.C. & WAHLGREN, G.M. 1999 Goddard High-resolution Spectrograph observations of the B III resonance doublet in early B stars: Abundances and isotope ratios. *ApJ* 516, 342.

QIAN, Y.-Z., WASSERBURG, G.J. 2001 A model for abundances in metal-poor stars. *ApJ* 559, 925.

ROMANO, D., MATTEUCCI, F., VENTURA, P. & D'ANTONA, F. 2001 The stellar origin of ^{7}Li: Do AGB stars contribute a substantial fraction of the local Galactic lithium abundance. *A&A* 374, 646.

SACKMANN, I.-J. & BOOTHROYD, A.I. 1992 The creation of superrich lithium giants. *ApJ* 392, L71.

SAVAGE, B.D. & SEMBACH, K.R. 1996 Interstellar abundances from absorption-line observations with the Hubble Space Telescope. *ARAA* 34, 279.

SMITH, V.V. & LAMBERT, D.L. 1989 Synthesis of lithium and s-process elements in Small Magellanic Cloud asymptotic giant branch stars. *ApJ* 345, L75.

SMITH V.V. & LAMBERT, D.L. 1990 On the occurrence of enhanced lithium in Magellanic Cloud red giants. *ApJ* 361, L69.

SNEDEN, C., COWAN, J.J., IVANS, I.I., FULLER, G.M., BURLES, S., BEERS, T.C., LAWLER, J.E. 2000 Evidence of multiple r-process sites in the early Galaxy: New observations of CS 22892-052. *ApJ* 533, L139.

SPITE, F. & SPITE, M. 1982 Abundance of lithium in unevolved halo stars and old disk stars - Interpretation and consequences. *A&A* 115, 357.

STRANEIRO, O., GALLINO, R., BUSSO, M., CHIEFFI, C., RAITERI, C.M., LIMONGI, M. & SALARIS, M. 1995 Radiative ^{13}C burning in asymptotic giant branch stars and s-processing. *ApJ* 440, L85.

TAKADA-HIDAI, M. 1990 The sulfur abundance in HR 4049 (HD 89353). *PASP* 102, 139.

TAKEDA, Y., PARTHASARATHY, M., AOKI, W. ET AL. 2002 Detection of zinc in the very metal-poor post-AGB star HR 4049. *PASJ* 54, 765.

TIMMES, F.X., WOOSLEY, S.E. & WEAVER, T.A. 1995 Galactic chemical evolution: Hydrogen through zinc. 1995 *ApJS* 98 617

TRAVAGLIO, D., RANDICH, S., GALLI, D., LATTANZIO, J., ELLIOTT, L.M., FORESTINI, M. & FERRINI, F. 2001 Galactic chemical evolution of lithium: Interplay between stellar sources. *ApJ* 559, 909.

VAN ECK, S., GORIELY, S., JORISSEN, A. & PLEZ, B. 2001 Discovery of three lead-rich stars. *Nature* 412, 793.

VAN WINCKEL, H., WAELKENS, C., FERNIE, J.D. & WATERS, L.B.F.M. 1999 The RV Tauri phenomenon and binarity. *A&A* 343, 202.

VAN WINCKEL, H., WAELKENS, C. & WATERS, L.B.F.M. 1995 The extremely iron-deficient "Post-AGB" stars and binaries. *A&A* 293, L25.

WAELKENS, C., VAN WINCKEL, H., BOGAERT, E. & TRAMS, N.R. 1991 HD 52961 - an extremely metal-deficient, CNO- and S-rich, pulsating star embedded in a dust cloud. *A&A* 251, 495.

WASSERBURG, G.J., BUSSO, M. & GALLINO, R. 1996 Abundances of actinides and short-lived nonactinides in the interstellar medium: Diverse supernova sources for the r-processes. *ApJ* 466, L109.

WASSERBURG, G.J. & QIAN, Y.-Z. 2000 Prompt iron enrichment, two r-process components, and abundances in very metal-poor stars. *ApJ* 529, 21.

WESTIN, J., SNEDEN, C., GUSTAFSSON, B. & COWAN, J.J. 2000 The r-process enriched low metallicity giant HD 115444. *ApJ* 530, 783.

ZHAI, M. & SHAW, D.M. 1994 Boron cosmochemistry. Part I:Boron in meteorites. *Meteoritics*, 29, 607.

Abundance Determinations In H II Regions And Planetary Nebulae

By GRAŻYNA STASIŃSKA

Observatoire de Paris-Meudon, 5, place Jules Janssen, 92195 Meudon cedex, France

The methods of abundance determinations in H II regions and planetary nebulae are described, with emphasis on the underlying assumptions and inherent problems. Recent results on abundances in Galactic H II regions and in Galactic and extragalactic Planetary Nebulae are reviewed.

1. Introduction

H II regions are ionized clouds of gas associated with zones of recent star formation. They are powered by one, a few, or a cluster of massive stars (depending on the resolution at which one is working). The effective temperatures $T_\star$ of the ionizing stars lie in the range 35 000 – 50 000 K. The nebular geometries result from the structure of the parent molecular cloud. Stellar winds, at evolved stages, may produce ring-like structures, but the morphology of H II regions is generally rather complex on all scales. Typical hydrogen densities n are $10^3 - 10^4$ cm^{-3} for compact H II regions. The average densities in giant extragalactic H II regions are lower, typically 10^2 cm^{-3} since giant H II regions encompass also zones of diffuse material. The total supply of nebular gas is generally large, so that all (or at least a significant fraction) of the ionizing photons are absorbed.

Planetary nebulae (PNe) are evolutionary products of so-called intermediate mass stars (initial masses of $1 - 8$ M$_\odot$) as they progress from the asymptotic giant branch (AGB) to the white dwarf stage. It is the interaction of the slow AGB wind with the fast post-AGB wind which produces the nebula. Because the ionizing star is also the remnant of the PN progenitor, the morphology is much simpler that in the case of H II regions, although not all PNe are round! The temperature of the central star – or nucleus – can be much higher than that of main sequence massive stars, reaching values of the order 200 000 K for a remnant of about 0.6 M$_\odot$. The densities of the brightest (and therefore best studied) PNe are around $10^3 - 10^5$ cm^{-3}. PNe of lower densities, corresponding to more evolved stages, are fainter and therefore less observed. The amount of nebular gas is not always sufficient to trap all the stellar ionizing photons, and a significant part of these may leak out from the nebula.

This brief introduction points at two things. One is that the ionized plasmas in H II regions and PNe are similar from the physical point of view, and therefore can be analyzed with similar techniques (although the range of physical conditions is somewhat different). The other is that the astrophysical significance of the chemical composition in these two classes of objects is not the same. H II regions probe the state of the gas at the birth of massive stars (i.e. a few Myr ago). The status of the chemical composition of PN envelopes is more complex. Some constituents have not been changed and reflect the state of the gas out of which the progenitor of the PN was formed, 10^8 yr ago or more. Other elements, such as carbon and nitrogen, have had their abundances strongly affected by nucleosynthesis and mixing processes in the progenitor, and therefore probe the evolution of intermediate mass stars.

The text presented below is based on lectures given at the XIII Canary Islands Winterschool on Cosmochemistry, where I have been asked to review the status of abundances in planetary nebulae (both Galactic and extragalactic) and in Galactic H II regions. Abundances in extragalactic H II regions were treated by Don Garnett, and the determination of the primordial helium abundance using low metallicity H II galaxies was discussed by Gary Steigman. In my lectures, I have emphasized the methods for abundance determinations in ionized nebulae. In this respect, giant extragalactic H II regions provide interesting complementary information and methods used for giant H II regions were included for completeness.

The scope of this article is as follows. §2 summarizes the basic physics of photoionized nebulae, §3 presents the different families of methods for abundance determinations. §4 discusses the various sources of uncertainties. §5 outines some important recent results on abundances in the Milky Way H II regions, including ring nebulae. §6 presents a selection of recent results on abundances in planetary nebulae, that are relevant to our understanding of the chemical history of galaxies or of the nucleosynthesis in intermediate mass stars. Due to limited space (and limited knowledge!), §5 and §6 are not to be taken for extensive reviews. A large amount of interesting work could not be mentioned here. This text is rather to be understood as a guide line for the astronomer interested in nebular abundances, either to embark on his own abundance determinations or to be able to better understand the literature on this topic. The papers quoted below were preferably chosen among recent studies published in refereed journals. A few pioneering, older studies are occasionally mentioned.

2. Basic physics of photoionized nebulae

Excellent introductions are provided in textbooks such as those of Spitzer (1978), Aller (1984) or Osterbrock (1989). Here, we simply emphasize the properties to bear in mind when dealing with abundance determinations.

2.1. *Ionization and recombination*

2.1.1. *Global ionization budget*

Consider a source of photons surrounded by a cloud of nebular gas. The gas particles are ionized by those photons with energies above the ionization threshold. Once ionized, the particles tend to recombine with the free electrons, and an equilibrium stage is eventually established in which the rate of ionization equals the rate of recombination for each species.

Closer to the source, the density of ionizing photons is larger, therefore the resulting ionization state of the gas is higher. If there is enough nebular matter, all the ionizing photons can be absorbed, producing an ionization bounded nebula. If not, the nebula is called density bounded.

It is the most abundant species, (H and He in general, but this could be C, N, O in hydrogen-poor material) which absorb most of the Lyman continuum photons from the ionizing source, and thus define the size of the ionized region in the ionization bounded case.

In an ionization bounded nebula purely composed of hydrogen, the total number of recombinations per unit time balances the total number of photons with energies above 13.6 eV emitted per unit time either by the star, or during recombination to the ground level. One has:

$$Q(\mathrm{H}^0) + \int n(\mathrm{H}^+)n_e\epsilon\alpha_1(\mathrm{H}, T_e)dV = \int n(\mathrm{H}^+)n_e\epsilon\alpha_{tot}(\mathrm{H}, T_e)dV, \qquad (2.1)$$

where $Q(H^0)$ is the total number of photons with energies above 13.6 eV emitted by the star per second; $n(H^+)$ is the number density of H ions, n_e is the electron density, ϵ is the volume filling factor of the nebular gas; $\alpha_1(H, T_e)$ is the H recombination coefficient to the ground level while $\alpha_{tot}(H, T_e)$ is the total H recombination coefficient, which are both roughly inversely proportional to the electron temperature T_e. The integrations are performed over the nebular volume.

In the case of a constant density nebula with constant filling factor, the radius of the ionized region, or Strömgren radius is then:

$$R_S = \left(\frac{3Q(H^0)}{4\pi\epsilon n_e^2 \alpha_B(H, T_e)} \right)^{1/3}, \tag{2.2}$$

where $\alpha_B(H, T_e)$ is the H recombination coefficient to the excited states (in this equation, T_e represents an average electron temperature of the nebula). The thickness of the transition region between the fully ionized zone and the neutral zone is approximately one mean free path of an ionizing photon $d = 1/n(H^0)\alpha_\nu$, where α_ν is the hydrogen photoionization cross section at the typical frequency of the photons reaching the ionization front. This thickness is generally much smaller than the size of the nebula and justifies the concept of a Strömgren sphere. There are however cases when the transition region might be extended, such as in diffuse media or when the ionizing radiation field contains a large amount of X-ray photons (which are less efficiently absorbed by hydrogen).

During the recombination process captures to the excited levels decay to lower levels by radiative transitions. The total luminosity of the Hβ line is thus

$$L_{H\beta} = \int n(H^+) n_e \epsilon 4\pi j_{H\beta}(T_e) dV, \tag{2.3}$$

where $j_{H\beta}(T_e)$ is the emission coefficient of Hβ and is roughly proportional to $\alpha_B(H)$. Therefore the total luminosity in Hβ in an ionization bounded nebula is a direct measure of $Q(H^0)$. At $T_e = 10^4$ K, it is given by:

$$L_{H\beta} = 4.8 \ 10^{-13} Q(H^0) \text{erg s}^{-1}. \tag{2.4}$$

In the case of a density bounded nebula, though, some ionizing photons escape and $L_{(H\beta)}$ is then given by:

$$L_{H\beta} = 1.5 \ 10^{32} (T_e/10^4)^{-0.9} M_{neb} <n> \text{erg s}^{-1}, \tag{2.5}$$

where M_{neb} is the nebular mass in solar units and $<n>$ is defined as:

$$<n> = \int n^2 \epsilon dV / \int n\epsilon dV, \tag{2.6}$$

assuming that in the nebula $n(H^+) = n_e = n(H) \equiv n$.

Thus, in the density bounded case, the total Hβ luminosity does not say anything about $Q(H^0)$, except that $Q(H^0)$ has to be larger than the value required to obtain the observed luminosity in Hβ. For a given total nebular mass, $L_{H\beta}$ is larger for denser nebulae, since recombinations are then more frequent.

For nebulae composed of pure hydrogen, the maximum ionizable mass of gas for a given value of $Q(H^0)$ is, at $T_e = 10^4$ K:

$$M_{ion} = 3.2 10^{-45} Q(H^0) / <n> \text{ M}_\odot. \tag{2.7}$$

The following table gives the values of M_{ion} for a typical PN, an H II region ionized by an O7 star, and a giant H II region ionized by a cluster of stars representing a total mass of 10^4 M$_\odot$ (a Salpeter mass function is assumed and the star masses range between 1 and 100 M$_\odot$).

	$Q(\mathrm{H}^0)$	$M_\star$	M_{ion}	M_{ion}
			$n = 10^2 \ \mathrm{cm}^{-3}$	$n = 10^4 \ \mathrm{cm}^{-3}$
planetary nebula	$3\,10^{47}$ ph s^{-1}	$0.6\ \mathrm{M}_\odot$	$10\ \mathrm{M}_\odot$	$10^{-1}\ \mathrm{M}_\odot$
single star H II region	$3\,10^{48}$ ph s^{-1}	$30\ \mathrm{M}_\odot$	$10^2\ \mathrm{M}_\odot$	$1\ \mathrm{M}_\odot$
giant H II region	$3\,10^{50}$ ph s^{-1}	$10^4\ \mathrm{M}_\odot$	$10^4\ \mathrm{M}_\odot$	$10^2\ \mathrm{M}_\odot$

TABLE 1. Typical masses of the ionizing stars (or star clusters) and maximum nebular ionizable masses

The surface brightness of an object is an important parameter from the observational point of view. Indeed, for extended objects, it determines the detectability or the quality of the spectra. For illustrative purposes, let us consider here the simple case of an homogeneous sphere and define:

$$S_{\mathrm{H}\beta} = F_{\mathrm{H}\beta}/(\pi\theta^2) = L_{\mathrm{H}\beta}/(4\pi^2 R_{neb}^2), \tag{2.8}$$

where $F_{\mathrm{H}\beta}$ is the observed Hβ flux, θ is the angular radius of the nebula and R_{neb} its physical radius. With the help of the previous equations one obtains for the ionization bounded case:

$$S_{\mathrm{H}\beta} \propto (Q(\mathrm{H}^0)n^4\epsilon^2)^{1/3} \tag{2.9}$$

and for the density bounded case:

$$S_{\mathrm{H}\beta} \propto (M_{\mathrm{neb}}n^5\epsilon^2)^{1/3}. \tag{2.10}$$

Thus better data will be obtainable for objects of higher densities, and objects with higher M_{neb} or $Q(\mathrm{H}^0)$.

The number fractions of He and heavy elements (C, N, O...)† in real nebulae are about 10% and 0.1% respectively. Helium, although ten times less abundant than hydrogen, is the dominant source of absorption of photons at energies above 24.4 eV. For order of magnitudes estimates, however, the formulae given above can still be used, since each recombination of He roughly produces one photon that can subsequently be absorbed only by hydrogen. The same remark generally holds for photons above 54.4 eV in the spectra of PNe with hot nuclei (see however Stasińska & Tylenda 1986). Naturally, for detailed studies, a photoionization modelling is necessary that takes into account properly the transfer of the photons arising from the recombination to He^0 and He$^+$.

2.1.2. *The ionization structure*

At a distance r from the ionizing source, the number densities $n(\mathrm{X}_i^j)$ and $n(\mathrm{X}_i^{j+1})$ of the ions X_i^j and X_i^{j+1} are schematically related by the following expression:

$$n(\mathrm{X}_i^j)Q(\mathrm{H}^0)/r^2 K = n(\mathrm{X}_i^{j+1})n_e\alpha(\mathrm{X}^j), \tag{2.11}$$

where K is a factor taking into account the frequency distribution of the ionizing radiation field and the absorption cross section (note that, for simplicity, the charge exchange process is not included in this equation). Of course, ions X_i^{j+1} can exist only if the radiation field contains photons able to produce these ions, and the ratio $n(\mathrm{X}_i^{j+1})/n(\mathrm{X}_i^j)$ will be higher for higher effective temperatures of the ionizing source.

† It is a tradition in nebular studies, to refer to elements other than H and He as "heavy elements" or "metals".

Integrating Eq. (2.11) over the nebular volume and using Eq. (2.2), it can be shown that, for a spherical nebula of constant density and filling factor and with an ionizing radiation of given effective temperature, the average ionic ratios are proportional to $(Q(\mathrm{H}^0)n\epsilon^2)^{1/3}$. In other words, a nebula of density $n = 10^4$ cm^{-3} ionized by one star with $T_\star = 50\,000$ K will have the same ionization structure as a nebula of density $n = 10^2$ cm^{-3} ionized by one hundred such stars.

The ionization parameter is usually defined by

$$U = Q(\mathrm{H}^0)/(4\pi R^2 nc), \tag{2.12}$$

where R is either the Strömgren radius, or a typical distance from the gas cloud to the ionizing star, and c is the speed of light. U is thus directly proportional to $(Q(\mathrm{H}^0)n\epsilon^2)^{1/3}$ in the case of a constant density sphere and this parameter describes the ionization structure.

It is important to be aware that equation (1.12) shows that at a given distance from the source, ionization drops when the density is increased locally (like in the case of a density clump). On the other hand, of two nebulae with uniform density and ionized by the same star, the highest average ionization will occur for the densest one.

The presence of intense lines of low ionized species such as [N II] $\lambda 6584$, [S II] $\lambda 6716$, $\lambda 6731$, [O I] $\lambda 6300$, is often considered in the literature as a signature of the presence of shocks. Shock models indeed predict that these lines are strong, but it must be kept in mind that pure photoionization models can also produce strong low ionization lines. This is for example the case for nebulae containing regions of low ionization due to gas compression (e.g. Dopita 1997 Stasińska & Schaerer 1999). Another example is that of nebulae excited by very high energy photons, for which the absorption cross-section is small and which induce a warm, only partially ionized zone.

2.2. *Heating and cooling*

During the photoionization process, the absorption of a photon creates a free electron which rapidly shares its energy with the other electrons present in the gas by elastic collisions, and thus heats the gas. The energy gains are usually dominated by photoionization of hydrogen atoms, although photoionization of helium contributes significantly. Intuition might suggest that T_e will decrease away from the ionizing source, since the ionizing radiation field decreases because of geometrical dilution and absorption in the intervening layers. This is actually not the case. The total energy gains per unit volume and unit time at a distance r from the ionizing source are schematically given by:

$$G = n(\mathrm{H}^0) \int_{h\nu_0}^\infty (4\pi J_\nu(r)/h\nu), a_\nu(\mathrm{H}^0)(h\nu - h\nu_0)\mathrm{d}h\nu \tag{2.13}$$

where

$$4\pi J_\nu(r) = \pi F_\nu(r) = \pi F_\nu(0)(R_\star)^2/r^2 \mathrm{e}^{-\tau_\nu(r)}. \tag{2.14}$$

If ionization equilibrium is achieved in each point of the nebula, one has (in the "on-the-spot case")

$$n(\mathrm{H}^0) \int_{h\nu_0}^\infty (4\pi J_\nu(r)/h\nu)a_\nu(\mathrm{H}^0)\mathrm{d}h\nu = n(\mathrm{H}^+)n_e\alpha_B(\mathrm{H}). \tag{2.15}$$

Therefore, G can be written

$$G = n(\mathrm{H}^+)n_e\alpha_B(\mathrm{H}) <E>, \tag{2.16}$$

where

$$<E> = \int_{h\nu_0}^\infty (4\pi J_\nu(r)/h\nu)a_\nu(\mathrm{H}^0)(h\nu - h\nu_0)\mathrm{d}h\nu \Big/ \int_{h\nu_0}^\infty (4\pi J_\nu(r)/h\nu)a_\nu(\mathrm{H}^0)\mathrm{d}h\nu. \tag{2.17}$$

Thus $< E >$ can be seen as the average energy gained per photoionization, and is roughly independent of r. It can be shown (see e.g. Osterbrock 1989), that when the ionization source is a blackbody of temperature $T_\star$, one has $< E >\approx (3/2)kT_\star$. Therefore:

$$G \propto n^2 T_\star T_e^{-1}, \tag{2.18}$$

meaning that the energy gains are roughly proportional to the temperature of the ionizing stars.

Thermal losses in nebulae occur through recombination, free-free radiation and emission of collisionally excited lines. The dominant process is usually due to collisional excitation of ions from heavy elements (with O giving the largest contribution, followed by C, N, Ne and S). Indeed, these ions have low-lying energy levels which can easily be reached at nebular temperatures. The excitation potentials of hydrogen lines are much higher, so that collisional excitation of H^0 can become important only at high electron temperatures.

For the transition l of ion j of an element X^i, in a simple two-level approach and when each excitation is followed by a radiative deexcitation, the cooling rate can be schematically written as

$$L_{coll}^{ijl} = n_e n(X_i^j) q_{ijl} h\nu_{ijl} = 8.63\,10^{-6} n_e n(X_i^j) \Omega_{ijl}/\omega_{ijl} T_e^{-0.5} e^{(\chi_{ijl}/kT_e)} h\nu_{ijl}, \tag{2.19}$$

where Ω_{ijl} is the collision strength, ω_{ijl} is the statistical weight of the upper level, and χ_{ijl} is the excitation energy.

If the density is sufficiently high, some collisional deexcitation may occur and cooling is reduced. In the two-level approach one has:

$$L_{coll}^{ijl} = n_e n(X_i^j) n_e q_{ijl} h\nu_{ijl}(1/(1 + n_e(q_{12} + q_{21})/A_{21}). \tag{2.20}$$

So, in a first approximation, one can write that the electron temperature is determined by

$$G = L = \sum_{ijl} L_{coll}^{ijl}, \tag{2.21}$$

where G is given by Eq. (2.18) and L_{coll}^{ijl} by Eq. (2.20).

The following properties of the electron temperature are a consequence of the above equations:

– T_e is expected to be usually rather uniform in nebulae, its variations are mostly determined by the mean energy of the absorbed stellar photons, and by the populations of the main cooling ions. It is only at high metallicities (over solar) that large T_e gradients are expected: then cooling in the O^{++} zone is dominated by collisional excitation of fine structure lines in the ground level of O^{++}, while the absence of fine structure lines in the ground level of O^+ forces the temperature to rise in the outer zones (Stasińska 1980a, Garnett 1992).

– For a given $T_\star$, T_e is generally lower at higher metallicity.

– For a given metallicity, T_e is generally lower for lower $T_\star$.

– For a given $T_\star$ and given metallicity, T_e increases with density in regions where n is larger than a critical density for collisional deexcitation of the most important cooling lines (around $5\,10^2 - 10^3$ cm^{-3}).

2.3. *Line intensities*

In conditions prevailing in PNe and H II regions the observed emission lines are optically thin, except for resonance lines such as H Lyα, C IVλ1550, N Vλ1240, Mg IIλ2800, Si IVλ1400, and some helium lines. Also the fine structure IR lines could be optically thick in compact H II regions or giant H II regions (however, the velocity fields are generally such that this is not the case). The fact that most of the lines used for abundance determinations are optically thin makes their use robust and powerful.

The intensity ratios of recombination lines are almost independent of temperature. On the other hand, intensity ratios of optical and ultraviolet collisional lines are strongly dependent on electron temperature if the excitation levels differ.

Abundances of metals with respect to hydrogen are mostly derived using the intensity ratio of collisionally excited lines with Hβ. It is instructive to understand the dependence of such emission line ratios with metallicity. Let us consider the [O III] λ5007/Hβ line ratio and follow its behaviour as $n(O)/n(H)$ decreases (from now on the notation $n(O)/n(H)$ will be replaced by O/H). The temperature dependence of the [O III] λ5007 and Hβ lines implies that:

$$[\text{O III}]\,\lambda5007/\text{H}\beta \propto n(O)/n(H)\; T_e^{0.5}e^{-28764/T_e}. \tag{2.22}$$

– At high metallicity (O/H around 10^{-3} and above), cooling is efficient and T_e is low. Energy is mainly evacuated by the [O III] λ88 μm line, whose excitation potential is 164 K. The cooling rate is then approximately given by

$$L = n(O^{++})n_e T_e^{-0.5}e^{164/T_e}. \tag{2.23}$$

Eq. (2.21) implies that

$$[\text{O III}]\,\lambda5007/\text{H}\beta \propto T_\star e^{-(28764+164)/T_e}. \tag{2.24}$$

Since T_e increases with decreasing O/H, Eq. (2.24) shows that [O III] λ5007/Hβ increases. Note the value of [O III] λ5007/Hβ depends on $T_\star$, being larger for higher effective temperatures.

– At intermediate metallicities, (O/H of the order of $10^{-3} - 2\,10^{-4}$), cooling is still mainly due to the oxygen lines, but the abundance of O/H being only moderate, T_e is higher, allowing collisional excitation of the [O III] λ5007 line, which now becomes the dominant coolant. The cooling can then be roughly expressed by:

$$L = n(O^{++})n_e T_e^{-0.5}e^{-28764/T_e}. \tag{2.25}$$

Eqs. (2.21) and (2.22) imply:

$$[\text{O III}]\,\lambda5007/\text{H}\beta \propto T_\star, \tag{2.26}$$

i.e. [O III] λ5007/Hβ is proportional to $T_\star$ and independent of O/H.

– Finally, at low metallicity, when cooling is dominated by recombination and collisional excitation of hydrogen, T_e becomes independent of O/H. ¿From Eq. (2.22), it follows that [O III] λ5007/Hβ is proportional to O/H. It also depends on $T_\star$ and on the average population of neutral hydrogen inside the nebula.

3. Basics of abundance determinations in ionized nebulae

3.1. *Empirical methods*

These are methods in which no check is made for the consistency of the derived abundances with the observed properties of the nebulae. They can be schematically subdivided into direct methods and statistical methods.

3.1.1. *Direct methods*

The abundance ratio of two ions is obtained from the observed intensity ratio of lines emitted by these ions. For example, O^{++}/H^+ can be derived from

$$O^{++}/H^+ = \frac{[\text{O III}]\, \lambda 5007/H\beta}{j_{[\text{O III}](T_e, n)}/j_{H\beta(T_e)}}, \tag{3.27}$$

where $j_{[\text{O III}]}(T_e, n)$ is the emission coefficient of the [O III] $\lambda 5007$ line, which is dependent on T_e and n (assumed uniform in the nebula).

T_e can be derived using the ratio of the two lines [O III] $\lambda 4363$ and [O III] $\lambda 5007$, which have very different excitation potentials. Other line ratios can also be used as temperature indicators in nebulae, such as [N II] $\lambda 5755/6584$ and [S III] $\lambda 6312/9532$. The Balmer and Paschen jumps, the radio continuum and radio recombination lines also allow to estimate the electron temperature, but the measurements are more difficult.

The density is usually derived from intensity ratios of two lines of the same ion which have the same excitation energy but different collisional deexcitation rates. The most common such ratio is [S II] $\lambda 6731/6717$. Far infrared lines can also be used to determine densities. Each line pair is sensitive in a given density range (about 2 to 3 decades), which can be ranked as follows (Rubin *et al.* 1994): [N II] $\lambda 122\mu/205\mu$, [O III] $\lambda 52\mu/88\mu$, [S II] $\lambda 6731/6717$, [O II] $\lambda 3726/3729$, [S III] $\lambda 18.7\mu/33.6\mu$, [A IV] $\lambda 4740/4711$, [Ne III] $\lambda 15.5\mu/36.0\mu$, [A III] $\lambda 8.99\mu/21.8\mu$, C III] $\lambda 1909/1907$. The electron density can also be measured by the ratio of high order hydrogen recombination lines.

Plasma diagnostic diagrams combining all the information from temperature- and density-sensitive line ratios can also be constructed for a given nebula (e.g. Aller & Czyzak 1983), plotting for each pair of diagnostic lines the curve in the (T_e, n) plane that corresponds to the observed value. The curves usually do not intersect in one point, due to measurement errors and to the fact that the nebula is not homogeneous (and also to possible uncertainties in the atomic data) and provide a visual estimate of the uncertainty in the adopted values of T_e and n.

The total abundance of a given element relative to hydrogen is given by the sum of abundances of all its ions. In practise, not all the ions present in a nebula are generally observed. The only favourable case is that of oxygen which in H II regions is readily determined from:

$$O/H = O^+/H^+ + O^{++}/H^+. \tag{3.28}$$

Note that even if [O I] $\lambda 6300$ is observed, it should not be included in the determination of the oxygen abundance, since the reference hydrogen line is emitted by H^+, while O^0 is tied to H^0.

In almost all other cases (except in some cases when multiwavelength data are available), one must correct for unseen ions using ionization correction factors. A common way to do this in the 70' and 80' and even later was to rely on ionization potential considerations, which led to such simple expressions as:

$$N/O = N^+/O^+, \tag{3.29}$$

$$Ne/O = Ne^{++}/O^{++}, \tag{3.30}$$

$$C/O = C^{++}/O^{++}. \tag{3.31}$$

In high excitation PNe where He II lines are seen, oxygen can be present in ionization stages higher than O^{++}. A popular ionization correction scheme for oxygen (e.g. Torres-

Peimbert & Peimbert 1977) was:

$$\frac{O}{H} = \frac{(He^+ + He^{++})}{He^+} \frac{(O^+ + O^{++})}{H^+}.$$ (3.32)

Expressions (2.29 – 2.31) are based on the similarity the ionization potentials of C^+, N^+, O^+, Ne^+. Expression (2.32) is based on the fact that the ionization potentials of He^+ and O^{++} are identical.

However, photoionization models show that such simple relations do not necessarily hold. For example, the charge transfer reaction $O^{++} + H^0 \rightarrow O^+ + H^+$ being much more efficient than the $Ne^{++} + H^0 \rightarrow Ne^+ + H^+$ one, Ne^{++} is more recombined than O^{++} in the outer parts of nebulae and in zones of low ionization parameter.

Also, while it is true that no O^{+++} ions can be found outside the He^{++} Strömgren sphere, since the photons able to ionize O^{++} are absorbed by He^+, O^{++} ions can well be present inside the He^{++} zone.

Ionization correction factors based on grids of photoionization models of nebulae are therefore more reliable. Complete sets of ionization correction factors have been published by Mathis & Rosa (1991) for H II regions and Kingsburgh & Barlow (1994) for planetary nebulae, or can be computed from grid of photoionization models such as those of Stasińska (1990), Gruenwald & Viegas (1992) for single star H II regions, Stasińska et al. (2001) for giant H II regions, Stasińska et al. (1998) for PNe.

However, it must be kept in mind that ionization correction factors from model grids may be risky too, both because the atomic physics is not well known yet (see §4.1) and because the density structure of real nebulae is more complicated than that of idealized models. The most robust relation seems to be $N/O = N^+/O^+$ (but see Stasińska & Schaerer 1997). Such a circumstance is fortunate, given the importance of the N/O ratio both in H II regions (as a constraint for chemical evolution studies) and in PNe (as a clue on PNe progenitors).

In spite of uncertainties, ionization correction factors often provide more accurate abundances than summing up ionic abundances obtained combining different techniques in the optical, ultraviolet and infrared domains.

Note that there is no robust empirical way to correct for neutral helium to derive the total helium abundance. The reason is that the relative populations of helium and hydrogen ions mostly depend on the energy distribution of the ionizing radiation field, while those of ions from heavy elements are also a function of the gas density distribution.

In summary, direct methods for abundance determinations are simple, powerful, and provide reasonable results (provided one keeps in mind the uncertainties involved, which will be developed in the next sections). Until recently, abundances were mostly derived from collisionally excited optical lines. This is still the case, but the importance of infrared data is growing, especially since the ISO mission. IR line intensities have the advantage of being almost independent of temperature. They arise from a larger variety of ions than optical lines. They allow to probe regions highly obscured by dust. However, they suffer from beamsize and calibration problems which are far more difficult to overcome than in the case of optical spectra. Abundance determinations using recombination lines of heavy elements have regained interest these last years. They require high signal-to-noise spectroscopy since the strengths of recombination lines from heavy elements are typically 0.1% of those of hydrogen Balmer lines. They will be discussed more thoroughly in the next sections, since they unexpectedly pose one of the major problems in nebular astrophysics.

3.1.2. *Strong line or statistical methods*

When the electron temperature cannot be determined, for example because the observations do not cover the appropriate spectral range or because temperature sensitive lines such as [O III] λ4363 cannot be observed, one has to go for statistical methods or "strong line methods". These methods have first been introduced by Pagel et al. (1979) to derive metallicities in giant extragalactic H II regions. They have since then being reconsidered and recalibrated by many authors, among which Skillman (1989), McGaugh (1991, 1994), Pilyugin (2000, 2001).

Pagel et al. (1979) proposed to use the 4 strongest lines of O and H : Hα, Hβ, [O II] λ3727 and [O III] λ5007. ¿From §2, the main parameters governing the relative intensities of the emission lines in a nebula are : $<T_\star>$, the mean effective temperature of the ionization source, the gas density distribution (parametrized by U in the case of homogeneous spheres), and the metallicity, represented by O/H. Luckily oxygen is at the same time the main coolant in nebulae, and the element whose abundance is most straightforwardly related to the chemical evolution of galaxies. The spectra must be corrected for reddening, which is done by comparing the observed Hα/Hβ ratio with the case B recombination value at a typical T_e and assuming a reddening law (see §4.3). Therefore two independent line ratios, [O II] λ3727/Hβ and [O III] λ5007/Hβ, remain to determine three quantities. Statistical methods rely on the assumption that $<T_\star>$ (and possibly U) are closely linked to the metallicity, and that it is the metallicity which drives the observed line ratios. Basing on available photoionization model grids, Pagel et al. showed that ([O II] λ3727 + [O III] λ5007)/Hβ, later called O_{23}, could be used as an indicator of O/H at metallicities above half-solar. Skillman (1989) later argued that this ratio could also be used in the low metallicity regime, in cases when the observations did not have sufficient signal-to-noise to measure the [O III] λ4363 line intensity. McGaugh (1994) improved the method and proposed to use both [O III] λ5007/[O II] λ3727 and O_{23} to determine simultaneously O/H and U (his method should perhaps be called the O_{23+} method). For the reasons explained above, the same value of ([O II] λ3727 + [O III] λ5007)/Hβ can correspond to either a high or a low value of the metallicity. A useful discriminator is [N II] λ6584/[O II] λ3727, since it is an empirical fact that [N II] λ6584/[O II] λ3727 increases with O/H (McGaugh 1994).

The expected accuracy of statistical methods is typically 0.2 – 0.3 dex, the method being particularly insensitive in the turnover region at O/H around 3 10^{-4}.

On the low metallicity side, the method can easily be calibrated with data on metal-poor extragalactic H II regions where the [O III] λ4363 line can be measured. Recently, Pilyugin (2000) has done this using the large set of excellent quality observations of blue compact galaxies by Izotov and coworkers (actually, the strong line method proposed by Pilyugin differs somewhat from the O_{23} method, but it relies on similar principles). He showed that the method works extremely well at low metallicities (with an accuracy of about 0.04 dex). This is a priori surprising, since giant H II regions are powered by clusters of stars that were formed almost coevally. The most massive stars die gradually, inducing a softening of the ionizing radiation field on timescales of several Myr, which should affect the O_{23} ratio, as shown by McGaugh (1991) or Stasińska (1998). As discussed by Stasińska et al. (2001), data on H II regions in blue compact dwarf galaxies are probably biased towards the most recent starbursts, and the dispersion in $<T_\star>$ is not as large as could be expected a priori. Another possibility, advocated by Bresolin et al. (1999) in their study of giant H II regions in spiral galaxies is that some mechanism must disrupt the H II regions after a few Myr. Of course, the O_{23} method is expected of much lower accuracy when applied to H II regions ionized by only a few stars, since in that case the ionizing radiation field varies strongly from object to object.

On the high metallicity side (O/H larger than about $5\ 10^{-4}$), the situation is much more complex. In this regime, there is so far no direct determination of O/H to allow a calibration of the O_{23} method since the [O III] $\lambda4363$ line is too weak to be measured (at least with 4 m class telescopes). The calibrations rely purely on models but it is not known how well these models represent real H II regions. Besides, at these abundances, the [O II] $\lambda3727$ and [O III] $\lambda5007$ line intensities are extremely sensitive to any change in the nebular properties (Oey & Kennicutt 1993, Henry 1993, Shields & Kennicutt 1995). Note that the calibration proposed by Pilyugin (2001) of his related X_{23} method in the high metallicity regime actually refers to O/H ratios that are lower than $5\ 10^{-4}$.

Other methods have been proposed as substitutes to the O_{23} method. The S_{23} method, proposed by Vílchez & Esteban (1996) and Díaz & Pérez-Montero (2000) relies on the same principles as the O_{23} method, but uses ([S II] $\lambda6716$, $\lambda6731$ + [S III] $\lambda9069$, $\lambda9532$)/Hβ (S_{23}) instead of ([O II] $\lambda3727$ + [O III] $\lambda5007$)/Hβ. One advantage over the O_{23} method is that the relevant line ratios are less affected by reddening. Besides, the excitation levels of the [S II] $\lambda6716$, $\lambda6731$ and [S III] $\lambda9532$ lines are lower than those of the [O II] $\lambda3727$ and [O III] $\lambda5007$ lines, so that S_{23} increases with metallicity in a wider range of metallicities than O_{23} (the turnover region for S_{23} is expected at O/H around 10^{-3}). Unfortunately, [S III] $\lambda9532$ is more difficult to observe than [O III] $\lambda5007$. Oey & Shields (2000) argue that the S_{23} method is more sensitive to U than claimed by Díaz & Pérez-Montero (2000). This would require futher checks, but in any case, the S_{23} method could be refined into an S_{23+} method in the same way as the O_{23} was refined into the O_{23+} method.

Stevenson et al. (1993) proposed to use [Ar III] $\lambda7136$ / [S III] $\lambda9532$ as an indicator of the electron temperature in metal-rich H II regions, and therefore of their metallicity. This method relies on the idea that the Ar/S ratio is not expected to vary significantly from object to object, and that the Ar^{++} and S^{++} zones should be coextensive. However, photoionization models show that, because of the strong temperature gradients expected at high metallicity, this method could lead to spurious results.

Alloin et al. (1979) proposed to use [O III] $\lambda5007$/[N II] $\lambda6584$ as a statistical metallicity indicator. While this line ratio depends on an additional parameter, namely N/O, the accuracy of this method turns out to be similar to that of statistical methods mentioned above. More recently, Storchi-Bergmann et al. (1994), van Zee et al. (1998) and Denicoló et al. (2002) advocated for the use of the [N II] $\lambda6584$/Hβ ratio (N_2) as metallicity indicator. Similarly to [N II] $\lambda6584$/[O III] $\lambda5007$, this ratio shows to be correlated with O/H over the entire range of observed metallicities in giant H II regions. The reason why, contrary to the O_{23} ratio, it increases with O/H even at high metallicity is due to a conjunction of [N II] $\lambda6584$/Hβ being less dependent on T_e than O_{23}, N/O being observed to increase with O/H in giant H II regions (at high metallicity at least) and U tending to decrease with metallicity. The advantage of this ratio is that it is independent of reddening and of flux calibration, and is only weakly affected by underlying stellar absorption in the case of observations encompassing old stellar populations. This makes it extremely valuable for ranking metallicities of galaxies up to redshifts about 2.5.

As mentioned above, statistical methods for abundance determinations assume that the nebulae under study form a one parameter family. This is why they work reasonably well in giant H II regions. They are not expected to work in planetary nebulae, where the effective temperatures range between 20 000 K and 200 000 K. Still, it has been shown empirically that there is an upper envelope in the [O III] $\lambda5007$/Hβ vs. O/H relation (Richer 1993), probably corresponding to PNe with the hottest central stars.

The existence of such an envelope can be used to obtain lower limits of O/H in PNe located in distant galaxies.

3.2. *Model fitting*

3.2.1. *Philosophy of model fitting*

A widely spread opinion is that photoionization model fitting provides the most accurate abundances. This would be true if the constraints were sufficiently numerous (not only on emission line ratios, but also on the stellar content and on the nebular gas distribution) and if the model fit were perfect (with a photoionization code treating correctly all the relevant physical processes and using accurate atomic data). These conditions are never met in practise, and it is therefore worth thinking, before embarking on a detailed photoionization modelling, what is the aim one is pursueing.

Two opposite situations may arise when trying to fit observations with a model.

The first one occurs when the number of strong constraints is not sufficient, especially when no direct T_e indicator is available. Then various models may be equally well compatible with the observations. For example, from a photoionization model analysis Ratag et al. (1997) derive an O/H ratio of 2.2 10^{-4} for the PN M 2-5. However, if one explores the range of acceptable photoionization models one finds two families of solutions (see Stasińska 2002). The first has O/H $\simeq$ 2.4 10^{-4}, the second has O/H $\simeq$ 1.2 10^{-3}! The reason for such a double solution is simply the behaviour of [O III] $\lambda5007$/Hβ or [O II] $\lambda3727$/Hβ with metallicity, as explained in §2.3. Note that both families of models reproduce not only the observed line ratios (including upper limits on unobserved lines) but also the nebular size and total Hβ flux.

The other situation is when, on the contrary, one cannot find any solution that reproduces at the same time the [O III] $\lambda4363/5007$ line ratio and the constraints of the distribution of the gas and ionizing star(s) (e.g. Peña et al. 1998, Luridiana et al. 1999, Stasińska & Schaerer 1999). The model that best reproduces the strong oxygen lines has a different value of O/H than would be derived using an empirical electron-temperature based method. The difference between the two can amount to factors as large as 2 (Luridiana et al. 1999). It is difficult to say a priori which of the two values of O/H – if any – is the correct one.

The situation where the number of strong constraints is large and everything is satisfactorily fitted with a photoionization model is extremely rare. One such example is the case of the two PNe in the Sgr B2 galaxy, for which high signal-to-noise integrated spectra are available providing several electron temperature and density indicators with accuracy of a few %. Dudziak et al. (2000) reproduced the 33 (resp. 27) independent observables (including imagery and photometry) with two-density component models having 18 (resp. 14) free parameters for Wray 16-423 (resp. He 2-436). Still, the models are not really unique. The authors make the point that they can reproduce the present observations with a range of values for C/H and $T_\star$. Yet, the derived abundances are not significantly different from those obtained from the same observational data by Walsh et al. (1997) using the empirical method. The only notable difference is for sulfur whose abundance from the models is larger by 50%, and for nitrogen whose abundance from the models is larger by a factor of 2.8 in the case of He 2-436. This apparent discrepancy for the nitrogen abundance actually disappears if realistic error bars are considered for the direct abundance determinations (rather than the error bars quoted in the papers). Indeed, the fact that the nebular gas is rather dense, with different density indicators pointing at densities from 3 10^3 cm^{-3} up to over 10^5 cm^{-3} introduces important uncertainties in the temperature derived from [N II] $\lambda5755/6584$ due to collisional deexcitation. It must be noted that realistic error bars on abundances derived from model fitting are extremely difficult to obtain, since this would imply the construction of a tremendous number of models, all fitting the data within the observational errors.

To summarize, abundances are not necessarily better determined from model fitting. However, model fitting, if done with a sufficient number of constraints, provides ionization correction factors relevant for the object under study that should be more accurate than simple formulae derived from grids of photoionization models. This could be called a "hybrid method" to derive abundances. Such a method was for example used by Aller & Czyzak (1983) and Aller & Keyes (1987) to derive the abundances in a large sample of Galactic planetary nebulae, and is still being used by Aller and his coworkers. It must however be kept in mind that if photoionization models do not reproduce the temperature sensitive line ratios, this actually points to a problem that has to be solved before one can claim to have obtained reliable abundances.

Ab initio photoionization models are sometimes used to estimate uncertainties that can be expected in abundance determinations from empirical methods. For example Alexander & Balick (1997) and Gruenwald & Viegas (1998) explored the validity of traditional ionization correction factors in the case of spatially resolved observations. A complete discussion of uncertainties should also take into account uncertainties in the atomic data and the effect of a simplified representation of reality by photoionization models.

3.2.2. Photoionization codes

Photoionization codes are built to take into account all the major physical processes that govern the ionization and temperature structure of nebulae. In addition to photoionization, recombination, free-free radiation, collisional excitation they consider collisional ionization (this is important only in regions of coronal temperatures), charge exchange reactions, which are actually a non negligible cause of recombination for heavy elements, especially if the physical conditions are such that the population of residual hydrogen atoms in the ionized gas exceeds 10^{-3}. Some codes are designed to study nebulae that are not in equilibrium and they may include such processes as mechanical heating and expansion cooling.

Most nebular studies use static photoionization codes, which assume that the gas is in ionization and thermal equilibrium. The most popular one is CLOUDY developed by Ferland and co-workers, for which an extensive documentation is available and which is widely in use (see Ferland et al. 1998, and http://www.nublado.org/ for the latest release). Several dozens of independent photoionization codes suited for the study of PNe and H II regions have been constructed over the years. Some of them have been intercompared at several workshops (Péquignot 1986, Ferland et al. 1996 and Ferland & Savin 2001). The codes mainly differ in the numerical treatment of the transfer of the ionizing photons produced in the nebula: on the spot reabsorption, outward-only approximation (most codes presently), full treatment (either with classical techniques as in Rubin 1968 or Harrington 1968 or with Monte-Carlo techniques as in Och et al. 1998). They also differ in their capacity of handling different geometries. Most codes are built in plane parallel or spherical approximations, but a few are built in 3D (Gruenwald et al. 1997, Och et al. 1998). While 3D codes are better suited to represent the density distribution in real nebulae, their use is hampered by the fact that the number of free parameters is extremely large. Presently, simpler codes are usually sufficient to pinpoint difficulties in fitting observed nebulae within our present knowledge of the physical processes occuring in them and to settle error bars on abundance determinations.

When the timescale of stellar evolution becomes comparable to the timescale of re-combination processes, the assumption of ionization equilibrium is no more valid. This for example occurs in PNe with massive (>0.64 M$_\odot$) nuclei, whose temperature and luminosity drop in a few hundred years while they evolve towards the white dwarf stage. In that case, the real ionization state of the gas is higher than would be predicted by a static photoionization model, and a recombining halo can appear. To deal with such situations, one needs time dependent photoionization codes, such as those of Tylenda (1979), or Marten & Szczerba (1997).

The nebular gas is actually shaped by the dynamical effect of the stellar winds from the ionizing stars. This induces shocks that produce strong collisional heating at the ionization front or at the interface between the main nebular shell of swept-up gas and the hot stellar wind bubble. On the other hand, expansion contributes to the cooling of the nebular gas. Several codes have been designed to treat simultaneously the hydrodynamical equations and the microphysical processes either in 1D (e.g. Schmidt-Voigt & Köppen 1987a and 1987b , Marten & Schönberner 1991, Frank & Mellema 1994a, Rodríguez-Gaspar & Tenorio-Tagle 1998) or in 2D (Frank & Mellema 1994b, Mellema & Frank 1995, Mellema 1995). It may be that some of the problems found with static codes will find their solution with a proper dynamical description. However, so far, for computational reasons, the microphysics and transfer of radiation is introduced in a more simplified way in these codes. Also, it is much more difficult to investigate a given problem with such codes, since the present state of an object is the result of its entire history, which has to be modelled ab initio.

4. Main problems and uncertainties in abundance determinations

The validity of derived abundances depends on the quality of the data and on the method of analysis. Typical quoted values for the uncertainties are 0.1 – 0.25 dex for ratios such as O/H, N/H, Ne/H, a little more for S/H, A/H, C/H, a little less for N/O, Ne/O and a few % for He. The optimism of the investigator is an important factor in the evaluation of the accuracy. This section comments on the various sources of uncertainties in abundance determinations.

4.1. *Atomic data*

Reviews on atomic data for abundance analysis have been given by Mendoza (1983), Butler (1993), Storey (1997), Nahar (2002). On-line atomic data bases are available from different sites. For example http://plasma-gate.weizmann.ac.il/DBfAPP.html provides links to many sites of interest, including the site of CLOUDY. The XSTAR atomic data base, constructed by Bautista & Kallman (2001) and used in the photoionization code XSTAR can be found at http://heasarc.gsfc.nasa.gov/docs/software/xstar/xstar.html.

The OPACITY and IRON projects (Seaton 1987, Hummer et al. 1993) have considerably increased the reliability of atomic data used for nebular analysis in the recent years. In the following, we simply raise a few important points.

4.1.1. *Ionization, recombination and charge exchange*

Until recently, photoionization cross sections and recombination (radiative and dielectronic) coefficient sets used in photoionization computations were not obtained self-consistently. Photoionization and recombination calculations are presently being carried out using the same set of eigenfunctions as in the IRON project (Nahar & Pradhan 1997, Nahar et al. 2000). The expected overall uncertainty is 10 – 20%. Experimental checks on a few species (see e.g. Savin 1999) can provide benchmarks for confrontation with numerical computations.

Concerning charge exchange, only a few detailed computations are available (see references in the compilation by Kingdon & Ferland 1996). Coefficients computed with the Landau-Zener approximation are available for most ions of interest. They are unfortunately rather uncertain. Differences with coefficients from quantal computations, which are available for a few species only, can be as large as a factor 3.

Due to the uncertainties in atomic parameters, the ionization structure predicted by photoionization models is so far expected to be accurate only for elements from the first and second row of the Mendeleev table.

4.1.2. *Transition probabilities, collision strengths and effective recombination coefficients*

The atomic data to compute the emissivities of optical forbidden lines have been recently recomputed in the frame of the IRON project (Hummer et al. 1993). The expected accuracies are typically of 10% for second row elements, however, the uncertainty is difficult to determine internally. Comparison with laboratory data is scarce, and actually, PNe are sometimes used as laboratories to test atomic physics calculations. For example, van Hoof et al. (2000) studied 3 PNe in detail and concluded that the [Ne V] collision strenghts computed by Lennon & Burke (1994) should be correct within 30 %, contrary to previous suggestions by Oliva et al. (1996) and Clegg et al. (1987). Another example is the density derived from [O III] $\lambda 52 \,\mu$m/[O III] $\lambda 88 \,\mu$m, which is significantly lower than derived from [S II] $\lambda 6731/6717$ and [Ar IV] $\lambda 4711+4740$ for a large sample of PNe observed by ISO (Liu et al. 2001b). These authors argue that [O III] IR lines can be emitted from rather low density components but it could just be that the atomic data are in error.

Concerning recombination lines, the effective recombination rates for lines from hydrogenic ions have been recomputed by Storey & Hummer (1995) and by Smits (1996) for He I $\lambda 5876$. For C, N, O, estimates for all important optical and UV transitions are given by Péquignot et al. (1991). Detailed computations of effective recombination coefficients are now available for lines from several ions of C, N, O and Ne (see e.g. a compilation in Liu et al. 2000). Note however that these do not include dielectronic recombination for states with high quantum number, which may have important consequences for the interpretation of recombination line data (see §4.6)

4.2. *Stellar atmospheres*

The ionization structure of nebulae obviously depends on the spectral distribution of the stellar radiation field. The theory of stellar atmospheres has made enormous progress these last years, due to advanced computing facilities. Several sets of models for massive O stars and for PNe nuclei are now available. The most detailed stellar atmosphere computations now include non-LTE effects and blanketing for numerous elements (e.g. Dreizler & Werner 1993, Hubeny & Lanz 1995, Rauch et al. 2000) and supersede previous works. The effect of winds, which is especially important for evolved stars such as Wolf-Rayet stars, is included in several codes, although with different assumptions (Schaerer & de Koter 1997, Hillier & Miller 1998, Koesterke et al. 2000, Pauldrach et al. 2001).

The resulting model atmospheres differ considerably between each other in the extreme UV. This has a strong impact on the predicted nebular ionization structure (see e.g. Stasińska & Schaerer 1997 for the Ne and the N^+/O^+ problems). Actually, the confrontation of photoionization models with observations of nebulae is expected to provide tests of the ionizing fluxes from model atmospheres (see Oey et al. 2000, Schaerer 2000, Giveon et al. 2002, Morisset et al. 2002). This is especially rewarding with the ISO data which provide accurate measurements for many fine-structure lines of adjacent ions.

For exploration purposes, it is sometimes sufficient to assume that the ionizing stars radiate as blackbodies, e.g. when interested in a general description of the temporal evolution of PNe spectra as their nuclei evolve from the AGB to the white dwarf stage (e.g. Schmidt-Voigt & Köppen 1987a, b, Stasińska et al. 1998). On the other hand, for a detailed model analysis of specific objects, the black body approximation is generally not well suited. For example, the emission of [Ne V] lines in PNe cannot be understood when using blackbodies of reasonable temperatures.

4.3. *Reddening correction*

The usual dereddening procedure is to derive the logarithmic extinction at Hβ, C, from the observed Hα/Hβ ratio, assuming that the intrinsic one has the value (Hα/Hβ)$_B$ predicted by case B recombination:

$$C = [\log(\mathrm{H}\alpha/\mathrm{H}\beta)_B - \log(\mathrm{H}\alpha/\mathrm{H}\beta)_{obs}]/(f_\alpha - f_\beta), \qquad (4.33)$$

where f_α and f_β represent the values of the reddening law at the wavelengths of the Hα and Hβ lines respectively.

Then, for any observed line ratio $(F_{\lambda 1}/F_{\lambda 2})_{obs}$ one can obtain the reddening corrected value $(F_{\lambda 1}/F_{\lambda 2})_{corr}$ from:

$$\log(F_{\lambda 1}/F_{\lambda 2})_{corr} = \log(F_{\lambda 1}/F_{\lambda 2})_{obs} + C(f_{\lambda 1} - f_{\lambda 2}). \qquad (4.34)$$

Ideally, one can iterate after having determined the electron temperature of the plasma, to use a value of (Hα/Hβ)$_B$ at the appropriate temperature.

There are nevertheless several problems. One is that the extinction law is not universal. As shown by Cardelli et al. (1989), it depends on the parameter $R_V = A_V/E(B-V)$, where A_V is the absolute extinction in V and $E(B-V)$ is the color excess. While the canonical value of R_V is $3-3.2$, the actual values range from 2.5 to 5 (Cardelli et al. 1989, Barbaro et al. 2001, Patriarchi et al. 2001). Objects located in the Galactic bulge suffer from an extinction characterized by a low value of R_V (e.g. Stasińska et al. 1992, Liu et al. 2001a). Cardelli et. el. (1989) attribute these differences in extinction laws between small and large values of R_V to the presence of systematically larger particles in dense regions. These variations in R_V have a significant effect on line ratios when dealing with ultraviolet spectra. It is therefore convenient to link the optical and ultraviolet spectra by using line ratios with known intrinsic value, such as He II λ1640 / He II λ4686.

Another difficulty is that dust is not necessarily entirely located between the object and the observer as in the case of stars. Some extinction may be due to dust mixed with the emitting gas. In that case, the wavelength dependence of the extinction is different and strongly geometry dependent (Mathis 1983). One way to proceed, which alleviates this problem, is to use the entire set of observed hydrogen lines and fit their ratios to the theoretical value, which then gives an empirical reddening law to deredden the other emission lines. This, however, is still not perfect, since the extinction suffered by lines emitted only at the periphery of the nebula, or, on the contrary, only in the central parts, is different from the extinction suffered by hydrogen lines which are emitted in the entire nebular body. The problem is further complicated by scattering effects (see e. g. Henney 1998).

In the case of giant H II regions, where the observing slit encompasses stellar light, one must first correct for the stellar absorption in the hydrogen lines. This can be done in an iterative procedure, as outlined for example by Izotov et al. (1994).

A further problem is that the intrinsic hydrogen line ratios may deviate from case B theory. This occurs for example in nebulae with high electron temperature ($\sim 20\,000$ K), where collisional contribution to the emissivity of the lowest order Balmer lines may become significant. In that case, a line ratio corrected assuming case B for the hydrogen lines, $(F_{\lambda 1}/F_{\lambda 2})_{\rm B}$ is related to the true line ratio $(F_{\lambda 1}/F_{\lambda 2})_{true}$ by:

$$\log(F_{\lambda 1}/F_{\lambda 2})_{\rm B} - \log(F_{\lambda 1}/F_{\lambda 2})_{true} = [\log({\rm H}\alpha/{\rm H}\beta)_{\rm B} - \log({\rm H}\alpha/{\rm H}\beta)_{true}](f_{\lambda 1} - f_{\lambda 2})/(f_\alpha - f_\beta).$$
$$(4.35)$$

The error is *independent of the real extinction* and can be large for $\lambda 1$ very different from $\lambda 2$. For example, it can easily reach a factor 1.5 – 2 for C III] $\lambda 1909/$[O III] $\lambda 5007$ (see Stasińska 2002).

Whatever dereddening procedure is adopted, it is good practise to check whether the Hγ /Hβ value has the expected value. If not, the [O III] $\lambda 4363/5007$ ratio will be in error by a similar amount.

4.4. *Aperture correction, nebular geometry and density inhomogeneities*

Observations are made with apertures or slits that often have a smaller projected size on the sky than the objects under study. When combining data obtained with different instruments, one needs to correct for aperture effects. To merge spectra obtained by IUE with optical spectra, one can use pairs of lines of the same ion such as He II $\lambda 1640$ and He II $\lambda 4686$. However, ionization stratification and reddening make the problem difficult to solve. One can also use C III] $\lambda 1909$ and C II $\lambda 4267$, but this involves additional difficulties (see §4.6). Summarizing, aperture corrections can be wrong by a factor as large as 2 (Kwitter & Henry 1998, van Hoof et al. 2000).

Interpretation of emission line ratios should care whether the observing slit covers the entire nebula, at least in the estimation of error bars on derived quantities. This is especially important when the observations cover only a small fraction of the total volume. Gruenwald & Viegas (1992) have published line of sight results for grids of H II region models, that can be used to estimate the ionization correction factors relevant to H II region spectra observed with small apertures. Alexander & Balick (1997) and Gruenwald & Viegas (1998) have considered the case of PNe, and shown that traditional ionization correction factors may strongly overestimate (or underestimate) the N/H ratio in the case when the slit size is much smaller than the apparent size of the nebula. The ratio N/O is less affected by line of sight effects. The problem is of course even worse in real nebulae than in those idealized models, due to the presence of small scale density variations. Integrated spectra have the merit on being less dependent on local conditions and of being more easily comparable to models. For extended nebulae, they can be obtained by scanning the slit across the face of the nebula (van Hoof et al. 2000, Liu et al. 2000), or by using specially designed nebular spectrophotometers (Caplan et al. 2000).

Tailored modelling taking explicitly into account departure from spherical symmetry is still in its infancy. One may mention the work of Monteiro et al. (2000) who constructed a 3D photoionization model to reproduce the narrow band HST images and velocity profiles of the PN NGC 3132 and concluded that this nebula has a diabolo shape despite its elliptical appearance. For the abundance determination however, which is the topic of this review, their finding has actually no real incidence.

More relevant for abundance determinations are the works of Sankrit & Hester (2000) and Moore et al. (2000), who modelled individual filaments in large nebulae, trying to reproduce the emission line profiles in several lines. Such a method uses many more constraints than classical T_e-based methods to derive abundances, but would need additional line ratios, and especially the T_e indicators, to be validated.

If large density contrasts occur in ionized nebulae, the use of forbidden lines for abundance determinations may induce some bias if collisional deexcitation is important. These biases have been explored by Rubin (1989) and his "maximum bias table" can be used to confine errors in abundances due to these effects.

4.5. *Spatial temperature variations*

4.5.1. *Temperature gradients*

At high metallicities, as explained above, large temperature gradients are expected in ionized nebulae. Therefore, empirical methods based on [O iii] λ4363/5007 will underestimate the abundances of heavy elements, since the [O iii] λ4363 line will be essentially emitted in the high temperature zones, inducing a strong overestimate of the average T_e. Therefore, although with very large telescopes it will now be possible to measure [O iii] λ4363 even in high metallicity giant H ii regions, one should refrain from exploiting this line in the usual way. Doing this, one would necessarily find sub-solar oxygen abundances, even for giant H ii regions with metallicities well above solar (Fig. 3 of Stasińska 2002). High metallicity *luminous* PNe offer a much safer way to probe the metallicity in central parts of galaxies (see §6.3 for the relevance of PNe as metallicity indicators of their environment). Indeed the higher effective temperatures and the higher densities in *luminous* PNe induce higher values of T_e in the O^{++} zone and a shallower temperature gradient, leading to a negligible bias in the derived abundances (see Fig. 4 of Stasińska 2002).

While T_e-based empirical methods are biased for metal rich giant H ii regions, tailored photoionization modeling to reproduce the *distribution* of the emission in the Hα, Hβ, He i λ5876, [O ii] λ3727 and [O iii] λ5007 lines are worth trying. As suggested by Stasińska (1980a), at high metallicity, regions emitting strongly [O iii] λ5007 will be decoupled from the regions emitting strongly in the recombination lines, and would be almost cospatial with regions emitting most of [O ii] λ3727. While the models of Stasińska (1980a) were made under spherical symmetry, the statement is more general, because it relies of the principles of ionization and thermal balance outlined in Sects. 2.1 and 2.2.

4.5.2. *Small scale temperature variations*

If the temperature in a nebula is not uniform, T_e-based empirical abundances are biased. Peimbert (1967) developed a mathematical formulation to evaluate the bias. It is based on the Taylor expansion of the average temperature

$$T_o(N_i) = \frac{\int T_e N_i n_e \mathrm{d}V}{\int N_i n_e \mathrm{d}V} \qquad (4.36)$$

defined for each ion N_i, using the *r.m.s.* temperature fluctuation

$$t^2(N_i) = \frac{\int (T_e - T_o(N_i))^2 N_i n_e \mathrm{d}V}{T_o(N_i)^2 \int N_i n_e \mathrm{d}V}. \qquad (4.37)$$

From comparison of temperatures measured by different methods, this temperature fluctuation scheme led to conclude that temperature fluctuations are common in nebulae, with typical values of $t^2 = 0.03 - 0.05$ (see references in Peimbert 1996, Stasińska 1998, Mathis et al. 1998, Esteban 2002). The case is not always easy to make: the determination of the continuum in the vicinity of the Balmer jump is difficult, the combination of data from different instruments for the comparison of far infrared data with optical ones involves many potential sources of errors, lines of O^{++} and of H are not emitted in coextensive zones etc ... Nevertheless, the observational results seem overwhelming. And, as noted by Peimbert (2002), the value of t^2 found in such a way is never negative! Note that in the PNe NGC 6153, NGC 7009, M1-42 and M2-36, (Liu et al. 1995b, 2000, 2001b, Luo et al. 2001) much larger values of t^2, of the order of 0.1, would be derived from the comparison of optical recombination lines (ORL) to collisionally excited lines (CEL). But this may be another problem (see §4.6).

A value of $t^2 \sim 0.04$, in the scheme of Peimbert (1967), typically leads to an underestimation of O/H by about 0.3 dex †. It is thus extremely important to determine whether temperature fluctuations exist or whether they are an artefact of the techniques employed. And, if they really exist, to understand their nature and possibly derive some systematics to account for them in abundance derivations. Note that, so far, the evidences are always *indirect*, based on the comparison of different methods to estimate T_e. Only mapping the nebulae with appropriate sensitivity and spatial resolution in the temperature diagnostic lines could give *direct* evidence of small scale fluctuations. In the planetary nebula NGC 6543, HST mapping of [O III] $\lambda 4363/5007$ shows much smaller spatial temperature variations than expected for this object from indirect measurements (Lame et al. 1997). In NGC 4361, long slit spectroscopy gives a *surface* temperature fluctuation $t_s^2 \sim 0.002$ (Liu 1998). In Orion, long slit mapping of the Balmer decrement gives $t_s^2 \sim 0.001$ (Liu et al. 1995a). All these observed t_s^2 translate into *volume* temperature fluctuations $t^2 \leq 0.01$.

Actually, the value of t^2 defined by Eq. (4.37) is not strictly speaking equal to the value of t^2 derived observationally, for example from the comparison of temperatures derived from [O III] $\lambda 4363/5007$ and from the Balmer jump. Kingdon & Ferland (1995) introduced the notation t_{str}^2 for the former (*str* meaning "structural") and t_{obs}^2 for the latter. Photoionization models of planetary nebulae and H II regions generally fail to produce such large values of t_{obs}^2 as observed in real nebulae (e.g. Kingdon & Ferland 1995, Pérez 1997), except in the case of high metallicities, i.e. equal to the canonical "solar" value or larger. Note that in this case, what produces t_{obs}^2 in the model is actually the temperature gradient discussed above. Density fluctuations could be a source of temperature fluctuations, due to increased collisional deexcitation in zones of higher density, but photoionization models including such density fluctuations also fail to return large enough values of t^2 (Kingdon & Ferland 1995). Note that introducing a density condensation shifts the dominant oxygen ion to a less charged one, and consequently the increase in $t^2(O^{++})$ is not as important as might have been thought a priori. Viegas & Clegg (1994) argued that very high density clumps ($n > 10^5$ cm^{-3}) could *mimic* the effects of temperature fluctuations by collisionally deexciting the [O III] $\lambda 5007$ line. The existence of such clumps is however not confirmed by the densities derived from [Ar IV] $\lambda 4740/4711$, from the ratio of fine structure [O III] lines and from high order Balmer decrement lines (Liu et al. 2000, 2001b). Note that, if they existed, such clumps should be located very close to the star in order to emit significantly in [O III] $\lambda 5007$ with respect to the rest of the nebula.

As will be discussed in §4.6, abundance inhomogeneities have been proposed to solve the ORL / CEL problem (Torres-Peimbert et al. 1990, Péquignot et al. 2002). Carbon and/or oxygen rich pockets would produce zones of lower temperature (due to increased cooling). In PNe the existence of carbon rich pockets is attested from direct observations in at least a few objects (e.g. Abell 30 and Abell 78, Jacoby & Ford 1983) and these carbon rich inclusions are thought to be material coming from the third dredge up in the progenitor star (Iben et al. 1983). But the existence of oxygen rich pockets in PNe is more difficult to understand from the present day evolution models for intermediate mass stars (see §6). On the other hand, in H II regions, especially in giant H II regions,

---

† N/O and Ne/O ratios are less affected by temperature fluctuations than N/H, since N/O and Ne/O abundance determinations rely on lines with similar temperature dependences and emitted in roughly the same zones.

oxygen rich pockets could be made of material ejected by Type II supernovae and not yet mixed with the gas (Elmegreen 1998).

Other origins of temperature fluctuations have also been proposed, involving additional heating processes. The fact that several detailed photoionization studies of planetary nebulae (Peña et al. 1998) or giant H II regions (García-Vargas et al. 1997, Stasińska & Schaerer 1999, Luridiana, et al. 1999, Luridiana & Peimbert 2001, Relaño et al. 2002) predict significantly lower [O III] $\lambda4363/5007$ ratios than observed indeed argues for additional heating. Shock heating or conductive heating are among the possibilities to investigate. Heat conduction from hot bubbles has been examined by Maciejewski et al. (1996) and shown to be insufficient to explain the t^2 derived from observations. The energy requirements to produce the observed values of t^2 have been evaluated by Binette et al. (2001) in the hypothesis of hot spots caused by an unknown heating process. In H II regions, the mechanical energy associated with the ionizing sources (stellar winds, supernova explosions) does not seem sufficient to produce the required value of t^2 (Binette et al. 2001, Luridiana et al. 2001). In planetary nebulae, a considerable amount of kinetic energy is available from the central star winds. The radiative hydrodynamical models of Perinotto et al. (1998) present a temperature spike at the external shock front. This temperature increase, located in a relatively narrow external zone, is not expected to produce a higher [O III] $\lambda4363/5007$ temperature than derived from the Balmer jump. In the radiative hydrodynamical models computed by Mellema & Frank (1995) for aspherical nebulae, there are areas of lower density in which cooling is inefficient and the temperature is higher due to shock heating. Mellema & Frank suggest that this may explain the differences in temperatures derived from different methods. However, a quantitative analysis remains to be done in order to check whether the predicted effect indeed reproduces what is observed. Simulations taking into account the evolution of the velocity and mass-loss rate of the fast central star wind (Dwarkadas & Balick 1998) lead to considerably more structure on smaller scales, which could be even more favorable to solve the temperature fluctuation problem. In a slightly different context, Hyung et al. (2001) have tried to explain the high temperature observed in the inner halo of NGC 6543, (15 000 K as opposed to 8 500 K for the bright core) by means of a simulation using a hydrodynamic code coupled to a photoionization calculation. These authors showed that mass loss and velocity variations in the AGB wind can simultaneously explain the existence of shells in the halo and the higher O^{++} temperature.

Recently, Stasińska & Szczerba (2001) proposed a completely different origin for temperature fluctuations, related to photoelectric heating by dust grains. This hypothesis is also very promising and can be checked observationally (see Sect 4.7.5).

Although the t^2 scheme has proved very useful to uncover the possibile existence of temperature inhomogeneities, it may not be appropriate to describe the real situation. In the case where abundance inhomogeneities are the source of the temperature variations the Peimbert (1967) description is obviously inadequate. But it can also be inappropriate for nebulae of homogeneous chemical composition, as shown on a simple two-component toy model. Consider two homogeneous zones of volumes V_1 and V_2 with temperatures T_1 and T_2, electron densities n_1 and n_2, and densities of the emitting ions (e.g. O^{++}) N_1 and N_2. Calling f the ratio $(N_2 n_2 V_2)/(N_1 n_1 V_1)$ of the weigths of the emitting regions, the mean electron temperature defined by Peimbert (1967) can be expressed as:

$$T_0 = \frac{T_1 + fT_2}{1 + f} \qquad (4.38)$$

and t^2 as:

$$t^2 = \frac{(T_1 - T_0)^2 + f(T_2 - T_0)^2}{(1+f)T_0^2}. \tag{4.39}$$

For $T_0=10\,000$ K, the case $f = 1$ (i.e. regions of equal weight) corresponds to $T_1=12\,000$ K and $T_2 =8\,000$ K. It must be realized that this $4\,000$ K difference in temperatures requires a difference of a factor 3 in the heating or cooling rates between the two zones. When $f \gg 1$, there is a high weight zone at $T_2 \leq T_0$ and a low weight zone at $T_1 \gg T_0$. Such a situation could correspond to a photoionized nebula with small volumes being heated by shocks or conduction. When $f \ll 1$, there is a high weight zone at $T_1 \geq T_0$ and a low weight zone at $T_2 \ll T_0$ which could correspond to high metallicity clumps. With such a toy model, one can explore the biases in abundance obtained for O^{++} using the [O III] $\lambda4363/5007$ temperature and different lines emitted by this ion. Examples are shown in Figs. 8 and 9 of Stasińska (2002). Following expectations, O^{++} derived from [O III] $\lambda5007$ is generally underestimated, but it is interesting to note that the magnitude of the effect depends both on f and on T_0. The bias is very small when $T_0 \gtrsim 15\,000$ K. It is small in any case if $f > 3 - 4$, because [O III] $\lambda4363$ saturates above $\sim 50\,000$K. At $T_0 \sim 8\,000$ K, O^{++} is underestimated by up to a factor of 2–3 in the regime where [O III] $\lambda4363$ is significantly emitted in both zones. As expected, O^{++} derived from infrared fine structure lines and from the optical recombination line O II $\lambda4651$ is correct. Such a toy model demonstrates that the classical temperature fluctuation scheme can be misleading. Even in a simple two zone model, the situation needs at least three parameteres to be described, not two. In our representation, these parameters would be T_1, T_2 and f, but other definitions can be used.

4.6. *The optical recombination lines mystery*

It has been known for several decades that optical recombination lines in PNe and H II regions indicate higher abundances than collisionally excited lines (see Liu 2002 for a review). Most of the former studies concerned the carbon abundance as derived from C II $\lambda4267$ and from C III] $\lambda1909$, but more recent studies show that the same problem occurs with lines from O^{++}, N^{++} and Ne^{++} (Liu et al. 1995b, 2000, 2001a, Luo et al. 2001). The ORL abundances are higher than CEL abundances by factors of about 2 for most PNe, discrepancies over a factor 5 are found in about 5 % of the PNe and can reach factors as large as 20 (Liu 2002). For a given nebula, the discrepancies for the individual elements C, N, O, Ne are found to be approximately of the same magnitude.

The explanations most often invoked are: i) temperature fluctuations, ii) incorrect atomic data , iii) fluorescent excitation, iv) upward bias in the measurement of weak line intensities, v) blending with other lines, vi) abundance inhomogeneities. None of them is completely satisfactory, some are now definitely abandoned.

The completion of the OPACITY project has allowed accurate computation of effective recombination coefficients needed to analyze ORL data. The advent of high quantum efficiency, large dynamic range and large format CCDs now allows to obtain high quality measurements of many faint recombination lines for bright PNe, thus hypothesis iv) cannot be invoked anymore. In addition, Mathis & Liu (1999) have measured the weak [O III] $\lambda4931$, whose intensity ratio with [O III] $\lambda5007$ depends only on the ratio of transition probabilities from the upper levels. They found $(4.15 \pm 0.11)\,10^{-4}$ compared to the theoretical values $4.09\,10^{-4}$ (Nussbaumer & Storey 1981), $4.15\,10^{-4}$ (Froese Fisher & Saha 1985), $2.5\,10^{-4}$ (Galavís et al. 1997). If, as expected, the latter computations give the more accurate results, the bias in the measurement of extremely weak lines could amount to 60%. This is far below what is needed to explain the ORL/CEL discrepancy. A large number of faint recombination lines have now been measured, and the observed relative intensities of permitted transitions from C^{++}, N^{++}, O^{++} and Ne^{++} are in agreement with the predictions of recombination theory, which goes against ii), iii), iv) and v). As mentioned in the previous subsection, the values of t^2 derived from the comparison of temperatures from [O III] $\lambda4363/5007$ and from the Balmer jump are too small to account for the large abundances derived from the ORL, therefore i) does not seem to be the good explanation. This is true even adopting a two-zone toy model instead of Peimbert's fluctuation scheme.

On the basis of detailed studies of several PNe, Liu (2002) notes that for a given nebula, the discrepancies for the individual elements C, N, O, Ne are found to be approximately of the same magnitude. The ORL/CEL abundance ratios correlate with the difference between the temperatures from [O III] $\lambda4363/5007$ and from the Balmer jump. Liu et al. (2000) favour the hypothesis of an inhomogeneous composition, with clumps having He/H = 0.4 and C, N, O, Ne abundances around 400 times that in the diffuse gas in the case of NGC 6153. It is indeed possible to construct a photoionization model with components of different chemical composition that reproduces the observed integrated line ratios satisfactorily (Péquignot et al. 2002). However, such a model is difficult to reconcile with the present theories of element production in intermediate mass stars (e.g. Forestini & Charbonnel 1997). Also, such super metal rich knots are not in pressure equilibrium with the surroundings and should be short lived, unless they are very dense.

Spatial analyses of NGC 6153 (Liu et al. 2000) and of NGC 6720 (Garnett & Dinerstein 2001) show that the ORL/CEL discrepancy decreases with distance to the central star. A possible explanation for the large intensities of the recombination lines of C, N, O, Ne, mentioned by Liu et al. (2000), is high temperature dielectronic recombination for states with high quantum numbers, a process so far not included in the computations of the effective recombination coefficients. Then, the ORL would be preferentially emitted in regions of temperatures of $(2-5)\,10^4$ K. There remains to find a way to obtain such high temperature material in planetary nebulae. Apart from conduction and shock fronts, there is also the possibility of heating by dust grains (see §4.7.7).

4.7. *The role of internal dust*

Until now we have omitted the solid component of nebulae, which, although not important by mass (usually of the order of 10^{-3}, see Hoare et al. 1991, Natta & Panagia 1981, Stasińska & Szczerba 1999) importantly affects the properties of PNe and H II regions. The discussion below deals only with aspects that are explicitly linked with the derivation of the chemical composition of nebulae.

4.7.1. *Evidence for the presence of dust in the ionized regions*

Numerous mid- and far- IR spectral observations of PNe and H II regions have shown the presence of a strong continuous emission at a temperature around $100-200$ K, attributed to dust grains heated by the ionizing stars (see the discussion in Pottasch 1984). Near- and mid-IR array observations have shown that the distribution of this emission is comparable to the distribution of [Ne II] $\lambda12.8\,\mu m$ and [S IV] $\lambda10.5\,\mu m$ radiation, implying that dust is not only found in the neutral outskirts, but also inside the ionized regions (see review by Barlow 1993). This does not necessarily imply, however, that grains are intimately mixed with ionized gas. A priori, they could be located exclusively in tiny, dusty neutral clumps, such as observed in the Helix nebula NGC 7293 (e. g. O'Dell & Handron 1996) or in the Ring Nebula NGC 6720 (Garnett & Dinerstein 2001). A crucial piece of evidence is provided by the following argument. It has been demonstrated by Kingdon et al. (1995) and Kingdon & Ferland (1997) that, in nebulae of normal chemical composition, numerous lines of elements such as Mg, Al, Ca, Cr, Fe, should be detectable in ultraviolet or optical spectra. What observations show is that these elements are depleted in PNe by factors around 10 – 100 (Shields 1978, Shields et al. 1981, Shields 1983, Harrington & Marionni 1981, Volk et al. 1997, Perinotto et al. 1999, Casassus et al. 2000). The same holds for H II regions (Osterbrock et al. 1992, Esteban et al. 1998).

4.7.2. *Heavy element depletion*

One important consequence of the above mentioned observational fact is that analyses of ionized nebulae do not provide the real abundance of such elements as Mg, Al, Ca, Cr, Fe, which can be incorporated in grains. Carbon can also be significantly depleted in carbon-rich grains – graphite or PAHs. The measurement of carbon abundances from nebular lines therefore gives only a lower limit to the total carbon content. This is also true for oxygen, although to a much smaller extent. In H II regions it is possible to estimate the amount of oxygen trapped in dust grains from the observation of the Mg, Si and Fe depletions (see Esteban et al. 1998). Also, the consideration of the Ne/O ratio can be useful, since Ne, being a noble gas, cannot enter in the composition of grains.

4.7.3. *The effect of dust on the ionization structure*

Dust internal to H II regions and PNe competes with the gas in absorbing Lyman continuum photons, therefore lowering the Hβ luminosity. The nebular ionization structure is affected by two competing processes. The ionization parameter drops due to the fact that part of the Lyman continuum photons are absorbed by dust and not by gas. This alone would tend to lower the general ionization level. The ionizing radiation field seen by the atomic species depends on the wavelength dependence of the dust absorption cross section. For conventional dust properties, the absorption cross section per H nucleon smoothly decreases for energies above 13.6 eV (see e.g. Fig. 1 from Aanestad 1989), favouring the ionization of He with respect to H. In the model of the Orion nebula presented by Baldwin et al. (1991), the net effect of absorption by dust is to bring the H$^+$ and He$^+$ zones into closer agreement.

4.7.4. *The effect of dust obscuration on the emission line spectrum*

The presence of dust inside the ionized regions affects the emission line spectrum by selectively absorbing (and scattering) the emitted photons. Since the emission lines from various ions are formed in different zones, their relative fluxes as measured by the observer do not only depend on a general extinction law, but also on the differences in the geometrical paths of the photons in the different lines. This, in principle, can be modelled using a photoionization code including dust but the problem is complex and the solution extremely geometry-dependent. For practical purposes, as explained in §4.3, it is more convenient to deredden an observed spectrum by adjusting the observed Balmer decrement to a theoretical one. If comparisons need to be made with a photoionization model, then they should be made with the theoretical emitted spectrum without dust attenuation. Of course, such a procedure is only approximate.

Resonance lines have an increased path length with respect to other lines, and are therefore subject to stronger absorption by dust. This is the case of H Lyα, which may be entirely trapped by grains in the case of very dusty nebulae (Lyα absorption is actually one of the main heating agents of dust particles in planetary nebulae, see e.g. Pottasch 1984). Other resonance lines, such as C IV λ1550, N V λ1240 or Si IV λ1400, are also affected by this selective absorption process. Usually, an escape probability formalism is used to account for it (Cohen et al. 1984). The observed intensity of the resonance lines depends on the amount of dust, on the ionization structure and on the velocity fields both in the nebula and in the surrounding halo and intervening interstellar medium (see e.g. Middlemass 1988). The inclusion of dust attenuation in a tailored photoionization model of the PN NGC 7662 results in a derived gas phase C abundance twice as large as would be deduced using classical methods (Harrington et al. 1982).

Another consequence of selective dust absorption is that it prevents the 100 % conversion of high-n Lyman lines into Lyα and Balmer lines (the case B). For dusty environments such as the Orion Nebula, the Hβ emissivity can be reduced by 15 % (Cota & Ferland 1988).

4.7.5. *The effects of grains on heating and cooling of the gas*

An obvious effect of the presence of grains on the thermal balance of ionized nebulae, is due to the depletion of strong coolants such as Si, Mg, Fe, which enhances the electron temperature with respect to a dust-free situation. This aspect is important not only for detailed model fitting of nebulae, but also when using grids of photoionization models to calibrate strong line methods for abundance determinations (Henry 1993, Shields & Kennicutt 1995).

Grains have also a *direct* influence on the energy balance. The photoelectric effect on dust grains has been shown by Spitzer (1948) to be a potential heating source in the interstellar matter. Baldwin et al. (1991) have introduced the physical effects of dust in the photoionization code CLOUDY. They constructed a detailed model of the Orion nebula and found that in this object, heating by photoelectric effect can amount to a significant proportion of the total heating while collisions of the gas particles with the grains contribute somewhat to the cooling.

The effect of dust heating is dramatic in the H-poor and extremely dusty planetary nebula IRAS 18333-2357 in which m_d/m_H is estimated around 0.4 (Borkowski & Harrington 1991). In this object, heating is almost entirely due to photoelectric effect.

In nebulae in which dust-to-gas mass ratio, dust properties and grain size distributions have the canonical values, the relative importance of dust heating is generally not very large. If, however, there is a large proportion of *small* dust grains, then the contribution of dust heating to the total energy gains may become important, as demonstrated by Dopita & Sutherland (2000) on a grid of dusty photoionization models of planetary nebulae. The effect is more pronounced in the central parts of the nebulae, being proportional to the mean intensity of the ultraviolet radiation field, and gives rise to a strong temperature gradient.

If such small grains do exist (and there is now growing evidence for that, Weingartner & Draine, 2001), their presence in planetary nebulae would solve a number of problems that have found no satisfactory solution so far (see Stasińska & Szczerba 2001): i) the thermal energy deficit inferred in some objects from tailored photoionization modelling; ii) the large negative temperature gradients inferred directly from spatially resolved observations and indirectly from integrated spectra in some PNe; iii) the Balmer jump temperatures being systematically smaller than temperatures derived from forbidden lines; iv) the intensities of [O I] λ6300 often observed to be larger than predicted by photoionization models: indeed, near the ionization front, Lyman continuum photons are exhausted and the only photons still present are photons below the Lyman limit. Those are not absorbed by hydrogen but can heat the gas via photoelectric effect on dust grains. One should however remember that dust is not the only way to enhance [O I] emission, other mechanisms have been mentioned in §2.1.

The energy gains per unit volume of gas due to photoelectric effect, G_d, are proportional to the number density of dust grains and to the intensity of the stellar radiation field. Combining with Eq. (2.16) which expresses the gains due to photoionization of hydrogen, G_H, it is easy to show that G_d/G_H is proportional to $(m_d/m_H)U$, where U is the ionization parameter. This has important consequences in the case of filamentary structures. If small grains are present, the photoelectric effect will boost the electron temperature in the low density components. This will result in important small-scale temperature variations in the nebula. The models of Stasińska & Szczerba (2001) show that *moderate* density inhomogeneities (such as inferred from high resolution images of PNe) give rise to values of t^2 similar to the ones obtained from observations. Note that, contrary to the dust-free case, the tenuous component has a higher T_e than filaments or clumps, therefore the clumps are better confined.

Stasińska & Szczerba (2001) also point out that if, as expected, dielectronic recombinations for high level states strongly enhance the emissivities of recombination lines, the presence of small grains in filamentary planetary nebulae would boost the emission of recombination lines from the diffuse component, principally in the inner zone. Therefore, small grains could solve in a natural way both the temperature fluctuation problem and the ORL/CEL discrepancy.

The presence of small grains in planetary nebulae can be tested observationally by measuring the temperature *in* and *between* filaments.

4.8. *The specific case of the helium abundance determination*

The determination of the helium abundance follows the same principles as that of other elements. But one is much more demanding about the accuracy. To follow the production of helium in stars, and the evolution of the helium content in galaxies, 10% accuracy is a goal that one would like to achieve. Helium abundances compiled from the literature over the years must be considered with caution, because of the different treatments adopted by various authors. On the other hand, the required accuracy should be reachable with consistent observations and modern data treatments. To determine the primordial helium abundance, Y_p, one needs a much better accuracy, since quite different cosmologies are predicted for values Y_p differing by a only few percent. ¿From low metallicity giant extragalactic H II regions, Olive et al. (1997) find $Y_p = 0.234 \pm 0.002$ while Izotov & Thuan (1998) find $Y_p = 0.244 \pm 0.002$. These two estimates are mutually exclusive. Is it possible to say which of the two – if any – is correct?

The first step is to obtain the intrinsic values of the intensities of the helium and hydrogen lines in an observed spectrum. If the spectrum contains stellar light, as in the case of giant H II regions, one must correct the observed intensities for underlying stellar absorption. The recent evolutionary synthesis models of González Delgado et al. (1999) provide a theoretical framework for that. One also has to correct the intensity ratios for reddening, assuming a given reddening "law" and a given intrinsic value of the ratios of the hydrogen line intensities. The latter mainly depends on the electron temperature, which can be estimated from the [O III] $\lambda 4363/5007$ ratio, with a correction due the fact that the O^{++} region is only a part of the H^+ region. Using an appropriate number of lines, one can estimate iteratively the reddening and the correction for underlying absorption (e.g. Izotov & Thuan 1998). However, as commented by Davidson & Kinman (1985) and Sasselov & Goldwirth (1995), and as mentioned in §4.3, collisional excitation of H Balmer lines may become important, especially in H II regions of high T_e. So far, this effect has always been omitted in the determination of the abundance of primordial He. It may induce an overestimation of the reddening, and therefore an underestimation of the He^+ abundance derived from He I $\lambda 5876$ by up to 5 % (Stasińska & Izotov 2001). The importance of this effect depends on the abundance of residual H^0.

Then, from the corrected ratios of emission lines one has to determine the value of He^+/H^+, or, to be more exact, of $\int n(He^+)dV/ \int n(H^+)dV$. This assumes that the line emissivities do not vary strongly over the nebular volume. The emissivities depend on T_e, and also on n_e in the case of some helium lines, due to enhancement by collisional excitation from the metastable 2^3S level. If one knows n_e from plasma diagnostics, the contribution of collisional excitation can be obtained. The recent tables of Benjamin et al. (1999), based on a resolution of the statistical equilibrium of the He atom using the best available atomic data, can be used for this purpose. Note that these authors also provide analytical fits, with the warning that some of them lead to values that may differ by 1% or more from the actually computed values of the emissivities. Some He line emissivities are also affected by self absorption of the pseudo resonance lines from the 2^3S level. Using a sufficient number of helium lines, one can in principle determine iteratively and self-consistently the characteristic temperature and density of the helium line emission, and the relative abundance of He^+. The treatment of radiation transfer in the lines remains to be improved and is announced as a next step by Benjamin et al. (1999). However, this is a complex problem: it depends on the velocity field and on the amount of internal dust which selectively absorbs resonant photons. Therefore, one does not expect models to be easily applicable to real objects. However, since this is a second order effect, this is perhaps not too problematic, if one discards the lines likely to be most affected by this process. One must not forget that the emissivities of the H Balmer lines too may be in question, both because of the contribution of collisional excitation mentioned above and because the presence of dust deviates the hydrogen spectrum from case B (see Hummer & Storey 1992). Another problem is to take into account the non uniformity of T_e. Sauer & Jedamczik (2002) have computed a grid of photoionization models for this purpose, and introduce the concept of a "temperature correction factor" which they compute in their models. Note, however, that the real temperature structure of nebulae is not obtained from "first principles", as the preceding sections made clear. Therefore, the distribution of T_e in real objects has most probably a larger impact than predicted by the models of Sauer & Jedamczik (2002). Peimbert et al. (2002) have adopted a semi-empirical approach, based on the Peimbert's (1967) temperature fluctuation scheme. But the temperature fluctuation scheme may give spurious results in the hypothesis of zones of highly different temperatures, as argued in §4.5.2.

If He II lines are present in the spectra, they have to be accounted for, to determine $\int n(He^{++})dV / \int n(H^+)dV$. The major uncertainty in that case comes probably from the lack of knowledge of the temperature characterising the emission of He II lines. An additional difficulty is due to the fact that part of the He II emission may be of stellar origin.

The He/H abundance is obtained after considering ionization structure effects. For low values of the mean effective temperature of the radiation field, a zone of neutral helium is present. Unfortunately, no ionization correction formula can be safely applied, since the ionization structure of helium with respect to hydrogen mainly depends on the hardness of the radiation field, while the ionization structure of the heavy elements also strongly depends on the gas distribution (e.g. Stasińska 1980b). In the case of an H II region ionized by very hot stars, photoionization models show that the He^+ region may on the contrary extend further than the H^+ region (see for example Stasińska 1980b or Sauer & Jedamczik 2002). Whether this is the case for an object under study should be tested by models.

Olive & Skillman (2001) stress the importance of having a sufficient number of observational constraints and of using them in a self consistent manner with a Monte-Carlo treatment of all sources of errors. Unfortunately, the errors on the temperature structure and on the ionization structure of real nebulae are very difficult to evaluate, and this, combined with uncertainties in atomic parameters and deviations from case B theory implies that the uncertainty in derived helium abundances is certainly larger than claimed.

5. Observational results on abundances in H II regions of the Milky Way

5.1. *The Orion nebula: a benchmark*

The Orion nebula is the brightest and most observed H II region in the galaxy. Therefore it is a benchmark in many respects. O'Dell (2001) and Ferland (2001) have summarized our present knowledge on this object. In the following, we only discuss aspects related to the chemical composition in the ionized gas.

It is of interest, beforehand, to mention that it is with the Orion nebula that the concept of filling factor started. Using a spherical representation, Osterbrock & Flather (1959) showed that the optical surface brightness data could be reconciled with the observed [O II] $\lambda3726/3729$ intensity ratios only when assuming extreme density fluctuations. They proposed a schematic model in which these fluctuations are represented as condensations immersed in a vacuum, with the relative volume of the condensations being only 1/30 of the total volume of the nebula. But a more realistic model of the Orion nebula (Zuckerman 1973, Balick et al. 1974, see also discussion in O'Dell 2001) is to represent the Orion nebula as an ionized blister on a background molecular cloud. ¿From a detailed comparison of the Hβ surface brightness map and of the [S II] $\lambda6731/6717$ map, Wen & O'Dell (1995) derived a 3D representation of the nebula. The ionized skin is very thin with respect to the overall size of the nebula, which justifies the plane parallel approximation for photoionization modelling.

The extinction in the Orion nebula is well known to differ from the standard reddening law, and has been studied in detail (see Baldwin et al. 1991, Bautista et al. 1995, Henney 1998 for recent references)

Abundances have been derived both from T_e-based empirical methods and from photoionization models, using optical data with increased signal to noise and spectral resolution, with the addition of ultraviolet data obtained with IUE (and more recently with HST) and infrared data from ground-based telescopes and from KAO, ISO, and MSX. Table 2 summarizes the abundances derived during the last decade. All the abundances are given in ppM units ($10^6 \times$ the number of particles of a given species with respect to hydrogen)

There is rather good agreement for the oxygen abundances, the value of Esteban et al. (1998) with $t^2 = 0$ being the lowest and the one with $t^2 = 0.024$ being the highest. Note that the preferred abundances of Esteban et al. (1998) are those obtained with $t^2 = 0.024$, which is the value indicated by the ORL/CEL comparison. However, the comparison of the [O III] $\lambda4363/5007$ and Balmer jump temperatures is consistent with $t^2 = 0$. One must be aware that abundances from models are not always the most reliable, since the models do not reproduce the ionization structure perfectly. The values of Simpson et al. (1998) for Ne, S, and Ar are obtained from simultaneous observations of the most abundant ionic stages.

The Mg, Si, Fe and Ni abundances are heavily depleted with respect to the Sun (indicating the presence of grains intimately mixed with the gas phase in the ionized region). There is actually a controversy with respect to the interpretation of Fe lines (Bautista et al. 1994, Baldwin et al. 1996, Bautista & Pradhan 1998). Esteban et al. (1999a) recommend to derive Fe abundances from Fe^{++} lines as done in the works quoted in Table 2.

	He	C	N	O	Ne	Mg	Si	S	Ar	Fe	Ni
a	100000	280	68	400	81:		4.5	8.5	4.5		
b	90000	210	87:	380	390:	3.2:		13.3	2.1	4.2:	
c	101000		52:	310	40			9.4	2.6	2.7:	0.14:
d	97700	250	60	440	78			14.8	6.3	1.3	
e	100000	250	42	300	50			9.3	3.3	2.2	
f					99			8.6	2.6		
g			326								

a Rubin et al. (1991, 1993), optical + IR spectroscopy, model
b Baldwin et al. (1991), long slit optical + IR + UV spectroscopy, model
c Osterbrock et al. (1992), optical spectroscopy, empirical
d Esteban et al. (1998) ($t^2 = 0.024$), optical spectroscopy, empirical
e Esteban et al. (1998) ($t^2 = 0.0$), optical spectroscopy, empirical
f Simpson et al. (1998), IR spectroscopy, empirical
g Deharveng et al. (2000), optical integrated spectroscopy, empirical

TABLE 2. recent measurements of the Orion nebula abundances (ppM units)

	He	C	N	O	Ne	Mg	Si	S	Ar	Fe	Ni
Sun											
a	98000:	363	112	851	123:	38	35	16	3.6:	47	1.8
b	85000:	331	83	676	120:	38	35	21	2.5:	32	1.8
c		391	85	544		34	34			28	
d				490							
Gas phase local interstellar medium											
e		141	75	319		22	19.5	16.6		7.4	0.26
Be stars											
f		224	44.5	407							
g		190	64.7	350		23.0	18.8			28	

a Anders & Grevesse (1989)
b Grevesse & Sauval (1998)
c Holweger (2001)
d Allende Prieto et al. (2001)
e Meyer et al. (1998) (O), Meyer et al. (1997) (N), Sofia et al. (1997) (C), Cardelli et al. (1996), Sembach & Savage (1996) (Si, S, Fe, Ni) Sofia & (Meyer 2001) (Mg)
f Cunha & Lambert 1994
g Compilation from Sofia & Meyer (2001)

TABLE 3. Solar vicinity abundances (ppM units)

5.2. *Abundance patterns in the solar vicinity and the solar abundance discrepancy*

Stars and nebulae provide a different perspective of the solar vicinity chemical composition. The methods for abundance determinations differ (and might be in error in different ways) and the astrophysical significance of the abundances is not necessarily the same. One expects a priori the surface composition of the Sun to be identical with that of other objects in the solar vicinity. It turns out that the abundances from nearby H II regions (Orion being the best example) are significantly smaller than the solar abundances from the works of Anders & Grevesse (1989) or Grevesse & Sauval (1998). It is to be noted that, despite of this fact, the reference abundance is often chosen to be that of the Sun from Anders & Grevesse (1989). Table 3 summarizes the abundances in the Sun, in the local interstellar medium (ISM) and in local B stars from recent references.

Peimbert et al. (2001) notes that a decade ago, the oxygen abundance in the Sun was 0.44 dex higher than in Orion but when using the value from Esteban et al. (1998) with $t^2 = 0.024$ and the solar value of Grevesse & Sauval (1998), the difference is only 0.19 dex. When accounting for the fraction of oxygen that is contained in dust grains (which can be done assuming a standard chemical composition of the dust grains, and the constraints provided by the Mg, Si and Fe abundances), the oxygen abundance is multiplied by a factor 1.2 and the difference between the Solar value and Orion is only 0.11 dex.

The oxygen abundance in Orion obtained with $t^2 = 0$ is actually similar to the one in the local interstellar medium (obtained from high resolution and high signal to noise absorption measurements, Meyer et al. 1998) and in local B stars (e.g. Cunha & Lambert 1994). Several possible explanations have been invoked. The ones listed by Meyer et al. 1998 are: i) an early enrichment of the Solar system by a local supernova (not really tenable if the abundances of *all* the elements in the local ISM are 2/3 solar); ii) a recent infall of metal poor gas in the local Milky Way; iii) an outward diffusion of the Sun from a smaller Galactocentric distance. A more recent discussion Sofia & Meyer (2001) definitely rejects the hypothesis of the local ISM standard being 2/3 of the Sun. Indeed, new determinations give much smaller values for O/H: 544 ppM (Holweger 2001), 490 ppM (Allende Prieto et al. 2001). The support for the 2/3 solar value is also invalidated from carbon (see their discussion). Note that Sofia & Meyer (2001) also argue that B stars have metal abundances that are *too low* to be considered valid representations of the ISM. According to Meyer et al. (1998), the local standard oxygen abundance should be 540 ppM (gas + dust).

In conclusion, the "solar abundance discrepancy" has gradually disappeared, mostly because modern derivations of the solar oxygen abundance give much lower values than earlier ones. This reinforces confidence in H II regions as probes of the ISM abundances and in the methods used to analyze them. This is good news, since H II regions are practically the only way to derive oxygen abundance in external galaxies, if one excepts the abundance analysis in giant stars of local galaxies which require very large telescopes. Giant H II regions can be used as abundance indicators up to large redshifts (see Pettini in the same volume).

5.3. *Abundance gradients in the Galaxy from H II regions*

Abundance gradients in disk galaxies constitute one of the more important observational constraints for models of galaxy chemical evolution. As a matter of fact, abundance gradients were first recognized to exist in external galaxies, where radial trends of emission line ratios were noted as far back as in the fourties (Aller 1942) and were attributed to abundance gradients in the early seventies (Searle 1971, Shields 1974).

In our own galaxy, gradients are more difficult to determine, due to distance uncertainties and because many H II regions are highly obscured by dust lying close to the galactic plane. The first determination of an abundance gradient in our galaxy from H II regions was made by Peimbert, Rayo & Torres-Peimbert (1978). It is worth the effort to derive abundance gradients in the Milky Way because it is a benchline for chemical evolution of galaxies. Only in the Milky Way can one have direct access to abundance measurements from so many sources as H II regions, planetary nebulae, individual B, F, G stars etc..., which all probe different epochs in the Milky Way history. Esteban & Peimbert (1995) and Hou et al. (2000) provide excellent reviews on this topic. Table 4 presents a compilation of Galactic abundance gradients from H II regions in units of d log(X/H) / dR in kpc^{-1}. Column 9 indicates the spanned range of galactocentric distances in kpc. Column 10 lists the total number of objects used to derive the gradients. Note that the errors quoted for the gradients include only the scatter in the nominal values of the derived abundances about the best fit line. They do not take into account the uncertainties in the abundances and the possible errors on the galactocentric distances. Most abundances were obtained using empirical methods.

It must be noted that, even in the case of similar methods, some details in the procedures employed may lead to significantly different results. For example, the much larger oxygen gradient found by Peimbert et al. (1978) probably results from their using the temperature fluctuation scheme (with $t^2 = .035$).

A possible flattening of abundance gradients in the outer disk has been mentioned by Fich & Silkey (1991) and Vílchez & Esteban (1996) but Rudolph et al. (1997) and Deharveng et al. (2000) find no clear evidence for that.

The situation with the N/O ratio is not clear. N/O ratios determined from N^{++}/O^{++} using far infrared (FIR) lines (Simpson et al. 1995, Afflerbach et al. 1997, see also Lester et al. 1987 and Rubin et al. 1988) are up to twice the values derived from N^+/O^+ using optical data. Actually, what is found is that N^{++}/O^{++} is larger than N^+/O^+, so it cannot be an ionization correction factor problem. Rubin et al. (1988) suggest that the discrepancy may be due to the neglect of the recombination component of the [O II] $\lambda 3727$ emission. Such an explanation can indeed hold at low T_e (say below 6 000 K) but is not expected to work at high T_e. Another possibility suggested by Rubin et al. (1988) is that the [O III] $\lambda 52\,\mu$m and [O III] $\lambda 88\,\mu$m lines are optically thick, thus increasing the derived N^{++}/O^{++}. FIR lines from N^{++} and O^{++} have now been observed by ISO (Peeters et al. 2002), but in their analysis Martín-Hernández et al. (2002) do not use them to derive abundance gradients. It is not clear why, since they have constructed photoionization model grids to correct for unseen ions.

The only data on a possible carbon abundance gradient comes from optical recombination lines measures in 3 objects! Obviously more work is needed in this respect.

5.4. *The Galactic center*

The central parsec of the galaxy, identified with the Sagittarius A nebula, contains ionized gas powered by about 10^{40} ionizing photons sec^{-1} (Lacy el al. 1980). A cluster of He I emission line stars has been observed and spectroscopically analyzed (Tamblyn et al. 1996, Najarro et al. 1997). The complete spectrum of infrared fine structure lines that has been observed, combined with the H Brα and Brγ lines (see Shields & Ferland 1994 for a compilation) should in principle allow to perform an abundance analysis. From a two-component photoionization model Shields & Ferland (1994) estimate that the abundance of Ar should be about twice solar, but Ne seems rather to have the solar value. The evidence for over solar metallicity is thus mixed. The N/O ratio is estimated to about 3 – 4 times solar. However, the derived abundances may be clouded by errors in the reddening corrections (the extinction is as high as $A_V=31$, so, even at far infrared wavelengths, reddening become important) and uncertainties in the atomic parameters (mainly those determining the ionization structure). As a consistency check, Shields & Ferland (1994) compared the electron temperature measured from recombination lines with their model predictions. For that, they included heating by dust, and assumed the same grain content as in the model of Baldwin et al. (1991) for Orion. They found the

	He	C	N	O	Ne	S	Ar	range	nb
a	0.02 ± 0.01		-0.23 ± 0.06	-0.13 ± 0.04				8–14	5
b	-0.001 ± 0.008		-0.090 ± 0.015	-0.070 ± 0.015		-0.010 ± 0.020	-0.060 ± 0.015	4–14	35
c					-0.086 ± 0.013	-0.051 ± 0.013		0-12	95
d			-0.100 ± 0.020		-0.080 ± 0.020	0.070 ± 0.020		0-10	23
e			+0.002 0.020		-0.051	-0.013 0.020		12-18	15
f				-0.047 ± 0.009				0–17	28
g			-0.072 ± 0.006	-0.064 ± 0.009		-0.063 ± 0.006		0–12	34
h			-0.111 ± 0.012			-0.079 ± 0.009		0–17	28
i	-0.004 ± 0.005	-0.086 ± 0.002	-0.048 ± 0.017	-0.049 ± 0.017	-0.045 ± 0.017	-0.055 ± 0.017	-0.044 ± 0.030	6-9	3
j				-0.040 ± 0.005				5–15	34
k					-0.039 ± 0.007		-0.045 ± 0.011	0–15	34

a Peimbert et al. (1978), optical spectroscopy, $t^2=.035$
b Shaver et al. (1983), optical spectroscopy for 30 objects, radio data for 67 objects, $t^2=0$
c Simpson & Rubin (1990), FIR data from IRAS, no icfs
d Simpson et al. (1995), FIR data from KAO, models
e Vílchez & Esteban (1996), long slit optical spectroscopy, $t^2=0$
f Afflerbach et al. (1996), models to reproduce the T_e measured from radio recombination lines in 28 ultracompact H II regions
g Afflerbach et al. (1997), FIR data from KAO: 15 objects + sources from Simpson, models
h Rudolph et al. (1997), FIR data from KAO of 5 H II regions in the outer Galaxy + results from Simpson models
i Esteban et al. (1999b), optical echelle spectroscopy, $t^2 > 0$
j Deharveng et al. (2000), absolute integrated optical fluxes, $t^2=0$, rediscussion of distances
k Martín-Hernández et al. (2002), FIR data from ISO, model grids, rediscussion of distances

TABLE 4. Galactic abundance gradients from H II regions d log(X/H) / dR in kpc^{-1}

measured temperatures to be consistent with a metallicity 1 – 2 times solar, while 3 times solar would be only marginally consistent. However, with a population of small grains, photoelectric heating would be more important, and larger metal abundances could be acceptable.

The Galactic center has since then been reobserved by ISO (Lutz et al. 1996), but a detailed discussion of the new results remains to be done.

5.5. *Nebulae around evolved massive stars*

Evolved massive stars are associated with nebulae which result from the interaction of stellar winds and stellar ejecta with the ambient interstellar medium. By studying the chemical composition of these nebulae, together with their morphology, kinematics and total gas content, one can get insight into the previous evolutionary stages of the stars and unveil some of the nucleosynthesis and mixing processes occuring in their interiors.

Schematically, during main sequence evolution, the fast wind creates a cavity in the interstellar medium and sweeps out a shell of compressed gas. After departure from the main sequence, the nature of the mass loss changes and the star loses chemically enriched material. When the star reaches the Wolf-Rayet phase, its outer layers are almost hydrogen free. This material is lost at high velocity and catches up with material lost in previous stages (see Chu 1991 or Marston 1999 for a review).

Imaging surveys of the environments of WR stars have found that in 50% of cases a ring like nebula is seen (Marston 1999). Ring nebulae have been classified by Chu (1981) into R type – radiatively excited H II regions and subsonic expansion velocities, E type – nebulae formed out of stellar ejecta (chaotic internal motion, large velocities) and W type – wind-blown bubbles showing thin sheets or filaments. Atlases are published by Chu, Treffers & Kwitter (1983), Miller & Chu (1993) and Marston (1997). Known examples of R types are RCW 78 (amorphous, containing a WN 8 star) and RCW 118 (shell, surrounding a WN 6 star). Known cases of nebulae containing ejecta are M 1-67 (WN 8 star), RCW 58 (WN 8 star). Known W types are NGC 6888 (WN 6 star), S 308 (WN5 star), RCW 104 (WN4 star), although Esteban et al. (1992) consider NGC 6888 as an Bubble/Ejecta type in their classification.

Luminous Blue Variable stars are regarded as precursors of WR stars with the most massive progenitors. They are usually found to be associated with small ejecta type nebulae like η Car, AG Car (Nota et al. 1995).

The first spatially resolved and comprehensive study of abundances in Wolf-Rayet ring nebulae is that of Esteban and coworkers (Esteban et al. 1990, 1991, 1992, 1993, Esteban & Vílchez 1992), in which 11 objects have been analyzed with similar procedures. In a plot relating the N/O and O/H differential abundances (i.e. abundances with respect to interstellar medium ones) Esteban et al. (1992) find that most objects lie close to the $(O/H + N/H) = (O/H + N/H)_{Orion}$ line, indicating that oxygen has been converted into nitrogen. This is indeed what is predicted by the Maeder (1990) stellar evolution models at the beginning of the WN phase. Dividing their objects into 3 categories from their chemical composition (H II for objects with abundances close to those of the environing ISM, DN for diluted nebulae in which stellar ejecta are mixed with ambient gas and SE for pure stellar ejecta), Esteban et al. (1992) show that there is a rather good correspondence between the chemical classes and the morpho-kinematical classes. They also note that the masses of SE nebulae are small and compatible with the hypothesis of pure stellar ejecta, while the H II nebulae have larger dynamic ages, consistent with the idea of being composed of large quantities of swept up gas. Esteban et al. (1992) find that the SE nebulae surrounding WR stars are associated with stars showing variability and thus probably having unstable atmospheres. This is also true for the nebulae associated with

LBVs. In plots relating the N/O mass fraction to the He mass fraction, Esteban et al. (1992) find that SE nebulae lie close to the stellar evolution tracks of Maeder (1990) for initial masses 25 – 40 $M_\odot$, which become WN stars after a red supergiant (RSG) phase. This is consistent with the initial masses estimated from the star luminosities (Esteban et al. 1993).

Since this pioneering study, detailed computations have been performed to simulate the evolution of the circumstellar gas around massive stars (García-Segura et al. 1996a and 1996b), coupling hydrodynamics with stellar evolution. The fate of the circumstellar gas results from interactions between the fast wind from the star while on the main sequence, the slow wind from the red supergiant or luminous blue variable stage and the fast wind from the WR stage. The resulting masses, morphologies and chemical composition of the circumstellar envelopes strongly depend on the initial stellar masses, both because of different nucleosynthesis and different time dependence of the winds. Stars with initial masses around 35 $M_\odot$ are predicted to go through a RSG stage, and produce massive nebular envelopes (~ 10 $M_\odot$) with composition only slightly enriched in He and CNO processed material. Stars with initial masses around 60 $M_\odot$ are predicted to go through a LBV stage, and produce less massive nebular envelopes (~ 4 $M_\odot$) with helium representing about 70% of the total mass fraction, and CNO equilibrium abundances (C depleted by a factor 23, N enriched by a factor 13, and O depleted by a factor 18). The composition and morphology of NGC 6888 and Sh 308 well agree with the theoretical prediction of a RSG progenitor. On the other hand, Smith (1996) notes that recent abundance determinations in nebulae associated with LBV stars do not agree with the composition predicted by the García-Segura et al. (1996b) model of evolution of a 60 $M_\odot$ star through the LBV stage. The abundance paterns of these nebulae are rather similar to those of SE nebulae surrounding RSG stars, with mild enrichments in He and N and mild depletion in O, suggesting that the star went through a RSG phase. It must be noted that abundance determinations in such objects are often difficult, because few diagnostic lines are available, so that ratios like N/H or O/H may be rather uncertain, but N/O is more reliable. In a rediscussion of nebulae around LBV stars, Lamers et al. (2001) conclude that the stars have not gone through a RSG phase. The chemical enhancements are due to rotation-induced mixing, and the ejection is possibly triggered by near-critical rotation.

6. Observational results on abundances in planetary nebulae

Until recently, a little less than 20 elements were available for abundance studies in planetary nebulae. These were: H, He, C, N, O, F, Ne, Na, Mg, Si, P, S, Cl, Ar, K, Ca, Mn, Fe, although routine abundance determinations are available for only about 10 elements. As already mentioned and will be made clearer in the next sections, these elements can serve either as probes of the ISM abundances or as probes of the nuclear and mixing processes in the progenitor stars. It has also been mentioned that some elements are heavily depleted in dust grains, so that the abundances of these elements in PNe (Mg, Si, P, K, Ca, Mn, Fe) rather give information on the chemistry of dust grains. This is of great interest since it is now believed that a large portion of grains found in the ISM were actually seeded in the atmospheres of evolved, intermediate mass stars (Dwek 1998).

Recently, ultra deep spectroscopy of bright PNe allowed to detect and measure lines from elements of the fourth, fifth and even sixth row of the Mendeleev table (Péquignot & Baluteau 1994, Baluteau et al. 1995, Dinerstein 2001, Dinerstein & Geballe 2001): V, Cr, Co, Ni, Cu, Zn, Se, Br, Kr, Rb, Sr, Y, Te, I, Xe, Cs, Ba, Pb. When the atomic

	He	C	N	O	Ne	Na	Mg	Si	S	Cl	A	K	Ca	Fe
a	10600	600	160	410	100	1.2	22	6.2	9.4	0.11	2.3	.05		
b	11000	955	162	508	137									
c	11100	600	150	300	95	2	50	5	6.9	0.18	2.0	.16	.4	
d	11000	1000	182	436	129		25		10		2.5			1
e		1300	330	420										
f	10000	3000	200	730	220		35		17					
g	10800	2500	275	700	154						18.2			1
h	9120		331	910	275				148	22.4	15.1	.14	.14	

a Bernard Salas et al. (2001), FIR data from ISO + optical + UV, empirical
b Kwitter & Henry (1996), optical + UV data, model
c Keyes et al. (1990), optical + UV data, model
d Middlemass (1990), optical + UV + FIR data, model
e Perinotto et al. (1980), optical + UV data, empirical
f Péquignot et al. (1978), optical + UV + FIR data, model
g Shields (1978), optical + UV + FIR data, model
h Aller 1954, optical data, empirical

TABLE 5. Abundances in NGC 7027

data for a quantitative analysis of these lines become available (and some atomic physics work has already been fostered by these discoveries, see e.g. Schöning & Butler 1998), this will open a new possibility to test PNe progenitors as production sites of r- and s-process elements.

The determination of isotopic abundance ratios in PNe would allow serious constraints on the nucleosynthesis in post-AGB stars (see e.g. Forestini & Charbonnel 1997, Marigo 2001). They strongly depend on stellar mass, metallicity and mixing length. Unfortunately, from the observational point of view, this field is still in its infancy. The ^{12}C/^{13}C ratio has been measured in only a couple of nebulae in either hyperfine UV transitions (Clegg et al. 1997, Brage et al. 1998) or in millimetric lines of CO (Bachiller et al. 1997, Palla et al. 2000). The ^{3}He abundance has been determined in a few nebulae from the hyperfine transition at 8.665 GHz (Balser et al. 1997, see also Galli et al. 1997) .

6.1. NGC 7027 and IC 418: two test cases

It is instructive to compare the abundances determined by various authors for two bright and well studied PNe.

NGC 7027 is the PN with the highest optical surface brightness despite of 3.5 mag absorption by dust and is a benchmark for PN spectroscopists. It is a very high excitation nebula, with lines of [Ne VI] now measured (Bernard Salas et al. 2001). The central star temperature is estimated to be 140 000 –180 000 K, the gas density is around 5 10^4 cm^{-3}. The nebula is surrounded by a dusty neutral shell. Table 5 lists the abundances derived for this object. Substantial differences are seen among the results obtained by various authors. It is interesting to recall that the concommittent models of Shields (1978) and of Péquignot et al. (1978) produced [O II] and [N II] intensities about one order of magnitude smaller than observed. Multidensity geometries and modifications of the stellar continuum failed to resolve this difficulty. Péquignot et al. (1978) postulated the existence of efficient charge transfer processes, and obtained an excellent fit to the observations by adjusting the charge transfer rates. These charge transfer rates were

	He	C	N	O	Ne	Mg	S	Cl	Ar
a	90000	219	86	153	9.2				
b	70000	300	70	180	3	6.9	2.5	.1	0.5
c	110000			288	52				2.7
d	72000		66	275	13		3:		0.8
e	93000	616:	74	436	74		4.2	.09	2.3
f		710				25			
g			45	398	19				
h	> 76000	794	100	760	40				

a Henry et al. (2000), optical + UV data, model
b Hyung et al. (1994), opt echelle + UV + a few IR data, model
c de Freitas Pacheco et al. (1992), optical data, empirical
d Gutiérrez-Moreno & Moreno (1988), optical data, empirical
e Aller & Czyzak (1983), optical data, hybrid method
f Harrington et al. (1980), optical + UV data, empirical
g Barker (1978), optical data, empirical
h Torres-Peimbert & Peimbert (1977), optical data, empirical with $t^2 = 0.035$

TABLE 6. Abundances in IC 418

later confirmed by atomic physics computations. In spite of the different approaches adopted by Shields (1978) and Péquignot et al. (1978), the resulting abundances are rather similar. On the other hand, they are significantly different from the abundances obtained later for this object. This is not only due to the use of different atomic data or to the number and quality of observational constraints (e.g. ISO spectroscopy provided high quality measurements on a large number of IR lines): when models are not entirely satisfactory, the abundances finally adopted are a matter of the author's personal choice.

IC 418 is also a bright and relatively dense ($n \sim 5\ 10^4$ cm^{-3}) PN, but with a central star of low effective temperature ($T_\star \sim 38\,000$ K), so that fewer ions are observed. The nebula is surrounded by an extended neutral shell. Here again, there are substantial differences among the published abundances. In this case, the differences in O/H cannot be attributed to ionization correction, since O is observed in all its ionization stages. It is actually the observational data which strongly differ from one author to another! Besides, results from empirical methods depend, as we know, on the assumptions made for the temperature structure. As for models, they do not give satisfactory fits for this object and therefore do not provide reliable abundances.

These two examples may serve as a warning that abundances are not necessarily as well determined as might be thought from error bars quoted in the literature.

6.2. What do PN abundances tell us?

The chemical composition of PNe envelopes results from a mixing of elements produced by the central star and dredged up to the surface with the original material out of which the star was made. Basically, the evolution of the central star can be described as follows (Blöcker 1999, Lattanzio & Forestini 1999). After completion of central hydrogen burning through the CNO bicycle, hydrogen burns in a shell around the He core. Due to core contraction the envelope expands. The star evolves towards larger radii and lower effective temperatures and ascends the red giant branch (RGB). During evolution on the RGB, the envelope convection moves downward reaching layers which have previously experienced H-burning (first dredge up), and brings up processed material to the surface. This material is mainly ^{14}N, ^{13}C, ^{12}C, and ^{4}He (Renzini & Voli 1981).

The ascent on the giant branch is terminated by ignition of the central helium. The subsequent evolution is characterized by helium burning in a convective core and a steadily advancing hydrogen shell. The fusion of helium produces ^{12}C by the triple α process, and this carbon is in turn subject to α capture to form ^{16}O. Eventually the helium supply is totally consumed, leaving a core of carbon and oxygen. The star begins to ascend the giant branch again, now called the asymptotic giant branch (AGB). When a star reaches the AGB, it has the following structure: a CO core, a He burning shell, a He intershell, a H burning shell, and a convective envelope. In stars more massive than 4 $M_\odot$, the envelope penetrates the region where H burning has occured, dredging up some of its material to the stellar surface (second dredge up). During this episode, ^{14}N and ^{4}He increase, while ^{12}C and ^{13}C decrease with ^{12}C/^{13}C staying around one, and ^{16}O slightly decreases.

While on the AGB, the star experiences further nucleosynthesis. Thermal pulses of the He shell induce a flash-driven convection zone, which extends from the helium shell almost to the H shell and deposits there some ^{12}C made in the He shell. As the helium flash dies away, the energy deposited causes expansion and cooling, and the external convective region reaches down the carbon-rich region left after the flash, bringing ^{12}C and ^{4}He to the star surface (third dredge up). During thermal pulses, elements beyond iron are produced by slow neutron capture (s-process). This requires partial mixing of hydrogen into the carbon rich intershell (Lattanzio & Forestini 1999): these protons are captured by ^{12}C to produce ^{13}C which later releases neutrons via the ^{13}C(α,n)^{16}O reaction. For stars above 5 $M_\odot$(at solar metallicity) a second important phenomenon is hot bottom burning. The convective envelope penetrates into the top of the H-burning shell. Temperatures can reach as high as 10^8 K. This results in the activation of the CN cycle within the envelope, and the consequent processing of ^{12}C into ^{13}C and ^{14}N, with the result that ^{12}C/^{16}O is smaller than one.

In summary, nucleosynthesis in PNe progenitors mainly affects the abundances of He, N and C in the envelope. The He abundance increases during the first, second and third dredge up. The ^{14}N abundance increases during the first, second and third dredge up. In the case of hot bottom burning, primary N is produced out of C synthesized in the He shell and brought to the H shell after the flash. The ^{12}C abundance decreases during first and second dredge up but increases during third dredge up, and decreases during hot bottom burning. From the synthetic evolutionary models of Marigo (2001), the resulting enrichment in PNe envelopes with respect to the ISM may be as large as a factor of 10 or more for ^{12}C and ^{14}N.

The abundance of ^{16}O is slightly reduced as a consequence of hot bottom burning while, as pointed out by Marigo (2001), low mass stars may produce positive yields of ^{16}O, which is brought to the surface by third dredge up. Globally, the oxygen abundance is expected to be little affected by nucleosynthesis in PN progenitors (Renzini & Voli 1981, Forestini & Charbonnel 1997, van den Hoek & Groenewegen 1997, Marigo 2001). From the synthetic evolutionary models of Marigo (2001), the PN progenitors modify the PN oxygen abundance by at most a factor of 2, the effect being strongest at low metallicities (1/4 solar). At solar and half solar metallicity, the effect is practically negligible. As a consequence, the abundance of oxygen should be representative of the chemical composition of the matter out of which the progenitor star was made. The same holds for the abundances of elements such as Ne, Ar, S. On the other hand, the abundances of He, C, N and the s-process elements tell about the nuclear and mixing processes in the PN progenitors.

progenitor mass	central star mass	progenitor's birth	PN type[a]
2.4 – 8 $M_\odot$	> 0.64 $M_\odot$	1 Gyr ago	Type I
1.2 – 2.4 $M_\odot$	0.58 – 0.64 $M_\odot$	3 Gyr ago	Type II
1.0 – 1.2 $M_\odot$	~ 0.56 $M_\odot$	6 Gyr ago	Type III
0.8 – 1.0 $M_\odot$	~ 0.555 $M_\odot$	10 Gyr ago	Type IV

[a] PN types according to Peimbert (1978, 1990)

TABLE 7. Schematical classification of PNe and their progenitors

When using PNe as indicators of the chemical evolution of galaxies, one should be aware that PNe with different central star masses probe different epochs and are subject to different selection effects. The mere existence of the PN phenomenon requires that the star must have reached a temperature sufficient to ionize the surrounding gas before the ejected envelope has vanished into the interstellar space. Now, the evolution of the central star is more rapid for higher masses. PNe ionized by more massive nuclei reach higher luminosities, and they will be the ones for which abundances will be preferentially measured in distant galaxies. In nearby galaxies and in the Milky Way, observations are feasible for lower luminosity PNe. The observability of a PN depends on the detection threshold, but if it is low enough, PNe with less massive nuclei will be visible for a considerably longer time than PNe with massive nuclei. This results from the post-AGB evolution time being a strongly decreasing function of core mass (see e. g. the models of Blöcker 1995). Another point is that, because of the existence of an initial-final mass relation (e.g. Weidemann 1987), PNe with less massive nuclei correspond to stars with lower initial masses, which are far more numerous according to the Salpeter initial mass function. For these two reasons, samples of nearby PNe will not contain a large proportion of objects with high mass progenitors. They will not contain many PNe with central star masses below 1–1.5 $M_\odot$ either, because such stars are believed to turn into very slowly evolving post-AGB stars and the ejected envelope will have dispersed into the interstellar medium before being ionized. This is why the distribution of central star masses is so strongly peaked around 0.6 $M_\odot$ (Stasińska et al. 1997). PNe of different central star masses probe different epochs of galaxy history. Schematically, they can be classified as shown in Table 7 (which however must be taken only as a rough guideline). The subdivision of PNe into four types by Peimbert (1978) was motivated by this kind of considerations (but several revisions to his initial scheme were proposed later, as will be discussed in §6.4.1). All the above considerations need confirmation from observational data on PNe samples.

6.3. *PNe as probes of the chemical evolution of galaxies*

6.3.1. *The universal Ne/H versus O/H relation*

From a compilation of PNe abundances in the Galaxy and in the Magellanic Clouds, Henry (1989) found that the Ne/H versus O/H relation for PNe is very narrow and linear in logarithm. It is also identical to the one found for H II regions (Vigroux et al. 1987). This implies that Ne and O abundances in intermediate mass stars are not significantly altered by dredge up, and therefore that oxygen and neon abundances in PNe can indeed be used to probe the interstellar abundances of oxygen over a large portion of the history of galaxies.

	He	C	N	O	Ne	S	Ar	range	nb
a			-.084 ± .034	-.054 ± .019	-.069 ± .034	-.064 ± .035		5-12	43
b				-.058 ± .007	-.036 ± .010	-.077 ± .011	-.051 ± .010	4-14	128
c				-.069 ± .006	-.056 ± .007	-.067 ± .006	-.051 ± .006	4-13	91
d	-.009 ± .01		-.072 ± .024	-.03 ± .01	-.05 ± .02			1-14	277
e	-.004 ± .003	.023 ± .026	-.030 ± .014	-.030 ± .010	-.030 ± .012	-.016 ± .019	-.042 ± .013		74
f	-.023 ± .0033		-.086 ± .045	-.031 ± .0199		-.082 ± .027	-.072 ± .021	1-11	15
g	-.011 ± 0.003		-.073 ± .026	-.014 ± .016	.102 ± .064	-.079 ± .047	-.049 ± .021	1-9	21
h	-.019 ± .003	-.069 ± .023	-.072 ± .028	.072 ± .012		-.098 ± .022		7-14	42

a Martins & Viegas (2000), Type II, homogeneous rederivation of abundances from compiled intensities
b Maciel & Quireza (1999), Type II, abundances compiled from the literature
c Maciel & Köppen (1994), Type II, abundances compiled from the literature
d Pasquali & Perinotto (1993), Type I + II , abundances compiled from the literature
e Amnuel (1993), Type In (according to his classification), abondances compiled from the literature
f Samland et al. (1992), Type II, homogeneous observational material an automated photoionization model fitting
g Köppen & al. (1991), Type II, homogeneous observational material and empirical abundance derivations
h Faúndez-Abans & Maciel (1986), Type II, abundances compiled from the literature.

TABLE 8. Galactic abundance gradients from PNe d log(X/H) / dR in kpc^{-1}

6.3.2. Abundance gradients from PNe in the Milky Way

Table 8 presents a compilation of Galactic abundance gradients from PNe in units of d log(X/H) / dR in kpc^{-1}. Column 9 shows the spanned range of galactocentric distances in kpc. Column 10 gives the total number of objects used to derive the gradients. Note that, as in the case of H II regions, the quoted uncertainties in the published abundance gradients include only the scatter in the nominal values of the derived abundances. In the case of PNe, distances are not known with good accuracy, they are usually derived from statistical methods, typically within a factor of 2 or more. However, if a gradient is found with erroneous distances, this means that a gradient is indeed most likely present, since one does not expect a conspiration of errors in distances to produce a spurious gradient. On the other hand, the values of the computed gradient strongly depend on the adopted PNe distance scale, as noted by Amnuel (1993). Only PNe arising from disk population stars are suitable to determine abundance gradients in the Galactic disk. Therefore, high velocity PNe (Type III according to the classification by Peimbert 1978) and a fortiori PNe belonging to the halo (Type IV PNe) are not suitable for this purpose.

It is to be noted that, while the existence of gradients seems established, there are significant differences in the magnitudes of these gradients as found by different authors. At present, it is not possible to say how accurate these gradients are. Note that accounting for possible "temperature fluctuations" would probably steepen the derived gradients (Martins & Viegas 2000).

From the most recent results, galactic gradients found for O, Ne and S using PNe appear to be similar to the ones found from H II regions and young stars (Maciel & Quireza 1999). This suggests that abundance gradients in the Galaxy have not changed during the last 3 Gyr. N and C gradients are different between PN and H II regions, which is expected of course. Their values have been reported in Table 8 only to be complete, but the existence of N or C gradients in PNe populations would rather tell something on the stellar populations from which the PN arise. As for the C gradients, they are highly unreliable anyway.

We can compare the average O/H in PNe and in H II regions of the solar vicinity, using the gradients given in Tables 4 and 8 and adopting for simplicity that the galactocentric distance of the Sun is 8.5 kpc. We find that $12+\log$ O/H $= 8.81 \pm 0.04$ using the H II regions data from Shaver et al. (1983), 8.606 ± 0.06 using those from Afflerbach et al. (1997), and 8.63 ± 0.05 using Type II PNe from the compilation of Maciel & Quireza (1999). There is therefore no compelling evidence that O/H differs between Type II PNe and H II regions in the solar vicinity. This is a further indication that ISM abundances have remained constant during the last few Gyr and that there is no significant modification of O/H in PNe due to mixing in the progenitors.

Maciel & Köppen (1994) have examined whether abundance gradients in the Galaxy steepen with time, by comparing the gradients from Type I, Type II and Type III PNe. The evidence is marginal.

The question of possible vertical abundance gradients has been investigated by Faúndez-Abans & Maciel (1988), Cuisinier et al. (1996) and Köppen & Cuisinier (1997), the latter study being the most detailed. Adopting careful selection criteria on the quality of the spectra and the location of the PNe in the Galaxy in a sample of 94 PNe, the latter authors find a systematic decrease of the abundances of He, N, O, S and Ar with height above the plane. The N/O ratio also exhibits a clear decrease with height. These findings are compatible with a simple empirical model that the authors work out for the kinematical and chemical evolution of the solar neighbourhood in which the progenitor stars are supposed to be born in the galactic plane and reach greater heights due to the velocity dispersion that increases with age.

6.3.3. *PNe in the Galactic bulge*

Planetary nebulae offer one the best means to investigate the oxygen abundance in the Galactic bulge. Table 9 presents the mean oxygen abundances for PNe thought to be physically located in the Galactic bulge, using recent data. In all cases, the abundance derivations were made with T_e-based methods. The abundances derived by Ratag et al. (1997) have not been included, since many of them rely on modelling of objects with no observed constraint of T_e, and are therefore highly suspect (see discussion in §3.2.1).

ref	He	N	O	Ne	S	Ar	nb
a	.101 ± .028	8.13 ± 0.42	8.48 ± 0.43	7.96 ± .36			85
b	.107 ± .019	8.12 ± 0.37	8.74 ± 0.15		6.86 ± 0.20	6.28 ± 0.37	30
c	.126 ± .027	7.64 ± 0.55	8.22 ± 0.43	7.25 ± 0.46	6.48 ± 0.93	5.95 ± 0.55	45

a Stasińska et al. (1998), compiled intensities, homogeneous abundance derivations
b Cuisinier et al. (2000), homogeneous data, results kindly provided by A. Escudero
c Escudero & Costa (2001), homogeneous data

TABLE 9. Mean abundances of PNe in the Galactic bulge

	He	N	O	Ne	S	Ar
a		8.86 ± 0.29	9.16 ± 0.16	8.51 ± 0.30	7.83 ± 0.30	
b			9.13 ± .05	8.29 ± .08	7.52 ± .08	6.79 ± .08
c			9.25 ± 0.05	8.46 ± 0.06	7.46 ± 0.05	7.07 ± 0.07
f	10.91 ± 0.014	7.74 ± 0.22	8.73 ± 0.09		6.83 ± 0.13	6.22 ± 0.10
g			8.81 ± .08		6.89 ± 0.12	6.39 ± 0.05

a Martins & Viegas (2000)
b Maciel & Quireza (1999)
c Maciel & Köppen (1994)
f Samland & al. (1992)
g Köppen & al. (1991)

TABLE 10. Abundances at the galactic center extrapolated from disk PNe

However, the observations from Ratag et al. (1997) were used in the compilation of line intensities by Stasińska et al. (1998), and the abundances rederived in a consistent way with T_e-based methods (for objects for which this was possible). The objects of Escudero & Costa (2001) are newly discovered PNe from the list of Beaulieu et al. (1999). We see that the abundances of PNe in the Galactic bulge have clearly higher mean values and dispersions in O/H than the extrapolation from disk PNe towards the Galactic center, which are shown in Table 10. The effect may be even stronger than suggested from these tables, since samples of PNe considered as belonging to the Galactic bulge may actually contain PNe of the disk population that are found physically in the same region as the bulge.

Combining data on about 100 PNe in the Galactic bulge from the works of Cuisinier et al. (2000), Webster (1988), Aller & Keyes (1987) with their own data, Escudero & Costa (2001) suggest the existence of a vertical abundance gradient in the bulge, with lower O/H at high latitudes.

6.3.4. PNe in the Galactic halo

Only a small number of PNe in the halo are known so far, less than 20, for an expected total of several thousands (see e.g. Tovmassian et al. 2001). This number is however rapidly growing, thanks to systematic sky surveys at high Galactic latitudes for the search of emission line galaxies, and in which PNe are discovered serendipitously. Halo PNe belong to an old metal poor stellar population, and therefore serve as probes of the halo chemical composition at the time of the formation of their progenitors. They also give the opportunity to study mixing processes in metal poor intermediate mass stars.

Using published spectral line data, Howard et al. (1997) rederived the chemical composition of 9 halo PNe in a consistent way. They found that all had subsolar O/H, the most oxygen poor being K648, with log O/H + 12=7.61 (i.e. about 1/20 of the Anders & Grevesse 1989 solar value). They also found that the spread in Ne/O, S/O and Ar/O is much larger than can be accounted for by uncertainties alone. This scatter in PNe abundances is similar to the scatter observed in halo stars (Krishnaswamy-Gilroy et al. 1988), and suggests that accretion of extragalactic material occured during formation of the halo. It must be noted however that, among PNe considered to be in the halo, some actually probably belong to an old disk population (Torres-Peimbert et al. 1990).

After the study of Howard et al. (1997), a few other PNe were discovered in the halo and their chemical composition analyzed (Jacoby et al. 1997, Napiwotzki et al. 1994, Tovmassian et al. 2001). The most spectacular one is SBS 1150+599A (renamed PN G 135.9+55.9), which has an oxygen abundance less than 1/100 solar (Tovmassian et al. 2001). This makes it by far the most oxygen poor PN known (and perhaps the most oxygen poor *star* known). One may ask whether the oxygen abundance in this object really reflects that of the initial star. Indeed, bright giants in metal poor globular clusters seem to present star to star oxygen abundance variations (see e.g. Ivans et al. 1999), and mixing processes have been invoked to explain these abundance patterns (see Charbonnel & Palacios 2001 for a review). One could invoke that a similar process affects the oxygen abundance in PN G 135.9+55.9. However, Ne is found to be also strongly underabundant in this object (Ne/O ∼ 0.3, paper in preparation), indicating that this object is indeed extremely metal poor. In that case, the progenitor must have formed very early in the Galaxy but given rise to a PN only recently. Alternatively, it could have formed out of infalling metal poor material at a more recent epoch.

6.3.5. PNe probe the histories of nearby galaxies

A wealth of data exist for large samples of PNe in the Magellanic Clouds, both in the optical and in the UV (Monk et al. 1988, Boroson & Liebert 1989, Meatheringham & Dopita 1991a, 1991b, Vassiliadis et al. 1992, Leisy & Dennefeld 1996, Vassiliadis et al. 1996, 1998). PNe in the Magellanic Clouds represent a statistically significant sample at a common distance, suffering little extinction along the line of sight, and sufficiently bright to allow the measurement of diagnostic lines from various ions.

The oxygen abundances of PNe in the Magellanic Clouds span a relatively small range: log O/H + 12 = 8.10 ± 0.25 from a compilation of 125 objects for the LMC, log O/H + 12

= 7.74 ± 0.39 from a compilation of 48 objects for the SMC reanalyzed in a homogeneous way by Stasińska et al. (1998). If one considers only the high luminosity sample ($L_{[O\ III]}$ > 100 $L_\odot$), the spread is smaller and the mean abundance is significantly larger: 8.28 ± 0.13 (40 objects) for the LMC, 8.09 ± 0.11 (11 objects) for the SMC. This has been interpreted as due to the fact that, as a class, high luminosity PNe have progenitors of higher masses, therefore younger and made of more chemically enriched gas. The mean oxygen abundance in the high luminosity class compares well with that from H II regions in the Magellanic Clouds: 8.35 ± 0.06 for LMC, 8.03 ± 0.10 for SMC (Russell & Dopita 1992). This indicates that the oxygen abundance in luminous PNe is a very good proxy of the present day ISM oxygen abundances.

Dopita et al. (1997) have produced self consistent photoionization models to fit the observed line fluxes between 1200 and 1800 Å for 8 PNe in the LMC. With these models they obtain not only the elemental abundances, but also the temperatures and luminosities of the central stars. This allows them to place the objects in the HR diagram and derive the central star masses and post-AGB evolution times by comparison with theoretical tracks for post-AGB stars of various masses (the choice of H-burning or He-burning track for each object is made by the requirement of consistency with the observed expansion age of the nebula). Assuming the initial-final mass relation of Marigo et al. (1996), Dopita et al. (1997) are able to estimate the masses of the progenitors. This allows them to trace the age-metallicity relationship in the LMC. As a proxy of metallicity, they use the sum of the abundances from the α-process elements Ne, S, Ar (in order to alleviate any doubts that might come from the use of O whose abundance can be slightly affected by mixing processes). They find that the LMC experienced a long period of quiescence, followed by a short period activity within the past 3 Gyr which multiplied its metallicity by a factor 2. A further study is under way by the same autors to include 20 additional PNe in the LMC and 10 PNe in the SMC.

PN spectroscopy is now possible with relatively high signal-to-noise even in more distant galaxies. For example, observations of 28 PNe in the bulge of M31 and 9 PNe in the companion dwarf galaxy M32 allowed to obtain T_e-based abundances for these objects (Richer et al. 1999). The oxygen abundances of the PNe observed in the bulge of M31 are found to be very similar to those of the luminous PNe in the Galactic bulge (the comparison, in order to be meaningful, must be done on nebulae with similar luminosities, since the oxygen abundances has been shown to depend on luminosity in the Magellanic Clouds and the Galactic bulge). One finds log O/H + 12 = 8.64 ± 0.23 for the M31 bulge sample and 8.67 ± 0.21 for the high luminosity PNe in the Galactic bulge (Stasińska et al. 1998). Jacoby & Ciardullo (1999) obtained spectroscopic data on 12 PNe in the bulge and 3 in the disk of M31. They span a larger luminosity range than Richer et al. (1999) who were mainly interested in bright PNe. For the three objects in common with Richer et al. (1999), the oxygen abundances are in excellent agreement. Yet, for their entire sample, Jacoby & Ciardullo (1999) find log O/H + 12 = 8.50 ± 0.23 which is significantly lower than the value found by Stasińska et al. (1998), possibly because of the larger range of PNe luminosities in their sample.

The data on M32 by Richer et al. (1999) confirm the suggestion by Ford (1983) that the PNe in M32 are nitrogen rich. It seems unlikely that all the luminous PNe have high enough central star masses to undergo second dredge up, and this finding suggests that in M32 nitrogen was already enhanced in the precursor stars.

Other local group galaxies have smaller masses and therefore contain only a few PNe. Abundance data exist for PNe in NGC 6822, NGC 205, NGC 185, Sgr B2, Fornax (see references in Richer & Mc Call 1995 and Richer et al. 1998).

Richer & Mc Call (1995) compared the oxygen abundances from PNe in diffuse ellipticals and dwarf irregulars. They found that diffuse ellipticals have higher abundances than similarly luminous dwarf irregulars. This seems consistent with the idea that diffuse ellipticals would be the faded remnants of dwarf ellipticals. However, when considering also the O/Fe ratios, obtained by combining stellar abundance measurements, they conclude that diffuse ellipticals and dwarf ellipticals have had in fact fundamentally different star formation histories.

Combining the data on PNe in these dwarf spheroidals galaxies with those on PNe in M32 and in the bulge M31 and of the Milky Way, Richer et al. (1998) have shown that the mean oxygen abundance correlates very well with the mean velocity dispersion. Since the oxygen abundance of luminous PNe is a good proxy of the oxygen abundance in the ISM at the time when star formation stopped, this implies that there is a correlation between the energy input from supernovae and the gravitational potential energy. Such a correlation arises naturally if chemical evolution in these systems is stopped by Galactic winds.

The oxygen abundances found in the elliptical galaxy NGC 5128 (Centaurus-A) by Walsh et al. (1999) show a mean value of about 8.4, i.e. smaller than the mean value determined for the bright PNe in M31. This result is somewhat difficult to understand for such a massive galaxy, unless the most metal rich stars do not produce observable PNe. This possibility is known as the the AGB manqué phenomenon (see e.g. Greggio & Renzini 1990), by which intermediate mass stars do not reach the top of the AGB due to intense stellar winds.

6.4. *PNe probe the nucleosynthesis in their progenitor stars*

6.4.1. *Global abundance ratios*

It is clear from the diagrams presented by Henry et al. (2000) that PNe show significantly higher values of He/H, N/O, C/O than H II regions of the same O/H. This indicates that He, N and C have been synthesized in PNe progenitors, as theory predicts. More quantitative comparison with theory is difficult because of the number of determining parameters (stellar mass, parametrization of the mixing processes) and of complex selection effects. In the following we draw a few examples of more detailed interpretations of abundance ratios that have been proposed.

The nature of Type I PNe is a good example of the difficulty in the interpretation. Peimbert (1978) had defined type I PNe as objects having He/H > 0.125 and N/O > 0.5. Kaler et al. (1978) interpreted the high N/O and He/H together with the (He/H , N/O) correlation observed in such objects as due to second dredge up, implying initial stellar masses larger than 3 $M_\odot$. Later, the He/H criterion to define Type I PNe was abandoned (it must be noted that old determinations of He/H did not include proper correction for collisional excitation of He lines). Henry (1990) found that Type I PNe showed an (N/O, O/H) anticorrelation and concluded that in these objects N is produced at the expense of O (due to ON cycling). Kingsburgh & Barlow (1994) contested the existence of such an anticorrelation and propose a new definition of Type I PNe, as being PNe that underwent envelope burning conversion to N of dredged up primary C. Thus they are objects in which the present N/H is larger than the initial (C+N)/H (equal to 0.8 in the solar vicinity). Costa et al. (2000) on the contrary define Type I PNe using only the criterion He/H > 0.11. They find a (N/O, O/H) anticorrelation when PNe are segregated by types. They interpret this by saying that the oxygen abundance is not modified by the PN progenitor but reflects the metallicity of the site where the progenitor was born, and that dredge-up is more efficient at low metallicity. It must be noted that, whatever the definition, there is actually no clear dichotomy between Type I and other

PNe, the distribution of the N/O ratios is rather continuous (and this is also what is predicted at least at solar and half solar metallicities from the models of Marigo 2001).

Concerning carbon, (C+N+O)/H is found to increase with C/H and becomes dominated by C/H for the most carbon rich objects. This is seen both in Galactic samples (Kingsburgh & Barlow 1994) and in Magellanic Clouds samples (Leisy & Dennefeld 1996). This is in agreement with a scenario where carbon is produced by 3-α from He and brought to the surface by third dredge up. From the number of PNe with observed C enhancement, one concludes that third dredge up is common in PNe progenitors. Among the PNe in which the carbon abundance could be determined, about 40% (in the Galactic sample) and 70% (in the Magellanic Clouds sample) have C/O > 1. This is well in line with theoretical predictions that third dredge up is more efficient at low metallicity. Note that PNe with C/O > 1, the so-called carbon-rich PNe, are likely to contain carbon rich dust, since their progenitors must have developed a carbon chemistry to form grains in their atmospheres.

More detailed comparisons of PNe results with the predictions of post AGB models have been attempted by Henry et al. (2000) and Marigo (2001). Interpretations are difficult, due to the number of parameters involved and to the difficulty to derive accurate central star masses and to relate them to initial masses.

Péquignot et al. (2000) discuss two PNe in the Sgr B2 galaxy, He 2-436 and Wray 16-423, whose nuclei are interpreted as belonging to the same evolutionary track. The authors perform a differential analysis of these two PNe, based on tailored photoionization modelling, and argue that while systematic errors may substantially shift the derived abundances, the conclusions based on *differences* between the two models should not be influenced. The main conclusion is that third dredge up O enrichment is observed in He 2-436, at the 10 % level.

6.4.2. *Abundance inhomogeneities*

Many studies have suggested that structures seen in planetary nebulae (extended haloes, condensations) have different composition from the main nebular body, indicating that they are formed of material arising in distinct mass loss episodes characterized by different chemical compositions of the stellar winds. However, these differences in chemical composition may be spurious, due inadequacies of the adopted abundance determination scheme. For example, the knots and other small scale structures seen in PNe are possibly the result of instabilities or magnetic field shaping, and their spectroscopic signature could be due to a difference in the excitation conditions and not in the chemical composition. In the following, some examples of such studies are presented, adopting the view of their authors.

a) Extended haloes

NGC 6720, the "ring nebula" is surrounded by two haloes: an inner one, with petal-like morphology, and an outer one, perfectly circular, as seen in the pictures of Balick et al. (1992). Guerrero et al. (1997) have studied the chemical composition of these haloes, and found that the inner and outer halo seem to have same composition, suggesting a common origin: the red giant wind. On the other hand, the N/O ratio is larger in the main nebula by a factor of 2, indicating that the main nebula consists of superwind and the haloes of remnants of red giant wind.

NGC 6543, the "cat eye nebula" also shows two halo structures: an inner one, consisting of perfectly circular rings, and an outer one with flocculi attributed to instabilities (Balick et al. 1992). Unlike what is advocated for NGC 6720, the rings and the core in NGC 6543 seem to have same chemical composition (Balick et al. 2001). It must be noted however, that the abundances may not be reliable, since a photoionization model for the core of NGC 6543 predicts a far too high [O III] $\lambda4363/5007$ (Hyung et al. 2000). Another puzzle is the information provided by *Chandra*. Chu et al. (2001) estimated that the abundances in the X-ray emitting gas are similar to those of the fast stellar wind and larger than the nebular ones. On the other hand, the temperature of the X-ray gas ($\sim 1.7\ 10^6$ K) is lower by two orders of magnitudes than the expected post shock temperature of the fast stellar wind. This would suggest that the X-ray emitting gas is dominated by nebular material. These findings are however based on a crude analysis and more detailed model fitting is necessary.

b) FLIERs and other microstructures

A large number of studies have been devoted to microstructures in PNe, and their nature is still debated. Fast Low Ionization Emission Regions (FLIERs) have first been considered to show an enhancement of N and were interpreted as being recently expelled from the star (Balick et al. 1994). However, Alexander & Balick (1997) realized that the use of traditional ionization correction factors may lead to specious abundances. Dopita (1997) made the point that enhancement of [N II] $\lambda6584/H\alpha$ can be produced by shock compression and does not necessarily involve an increase of the nitrogen abundance. Gonçalves et al. (2001) have summarized data on the 50 PNe known to have low ionization structures (which they call LIS) and presented a detailed comparison of model predictions with the observational properties. They conclude that not all cases can be satisfactorily explained by existing models.

c) Cometary knots

The famous cometary knots of the Helix nebula NGC 7293 have been recently studied by O'Dell et al. (2000) using spectra and images obtained with the HST. The [N II] $\lambda6584/H\alpha$ and [O III] $\lambda5007/H\alpha$ ratios were shown to decrease with distance to the star. Two possible interpretations were offered. Either this could be the consequence of a larger electron temperature close to the star due to harder radiation field. Or the knots close to the star would be more metal-rich, in which case they could be relics of blobs ejected during the AGB stage rather than formed during PN evolution. Obviously, a more thorough discussion is needed, including a detailed modelling to reproduce the observations before any conclusion can be drawn.

d) Planetary nebulae with Wolf-Rayet central stars

About 8% of PNe possess a central star having Wolf-Rayet charateristics, with H-poor and C-rich atmospheres. The evolutionary status of these objects is still in question. A late helium flash giving rise to a "born-again" planetary nebula, following a scenario proposed by Iben et al. (1983), can explain only a small fraction of them. The majority appear to form an evolutionary sequence from late to early Wolf-Rayet types, starting from the AGB (Górny & Tylenda 2000, Peña et al. 2001). This seemed in contradiction with theory which predicted that departure from the AGB during a late thermal pulse does not produce H-deficient stars. Recently however, it has been shown that convective overshooting can produce a very efficient dredge up, and models including this process are now able to produce H-deficient post-ABG stars following a thermal pulse on the AGB (Herwig 2000, 2001, see also Blöcker et al. 2001). It still remains to explain why late type Wolf-Rayet central stars seem to have atmospheres richer in carbon than early type ones (Leuenhagen & Hamann 1998, Koesterke 2001). Also, one would expect the chemical composition of PNe with Wolf-Rayet central stars to be different from that of the rest of PNe. This does not seem to be the case, as found by Górny & Stasińska (1995), basing on a compilation of published abundances: PNe with Wolf-Rayet central stars are indistinguishable from other PNe in all respects except for their larger expansion velocities. Peña et al. (2001) obtained a homogeneous set of high spectral resolution optical spectra of about 30 PNe with Wolf-Rayet central stars and reached a similar conclusion, as far as He and N abundances are concerned. Their data did not allow to draw any conclusion as regards the C abundances.

e) H-poor PNe

There are only a few PNe which show the presence of material processed in the stellar interior. They are referred to as H-poor PNe, although the H-poor material is actually embedded in an H-rich tenuous envelope. The two best known cases are A 30 and A 78, whose knots are bright in [O III] λ5007 and He II λ4686 but in which Jacoby (1979) could not detect the presence of H Balmer lines. With deeper spectra, Jacoby & Ford (1983) estimated the He/H ratio to be $\sim$ 8 in these two objects. Harrington & Feibelman (1984) obtained IUE spectra of a knot in A 30, and found that the high C/He abundance implied by C II λ4267 is not apparent in the UV spectra, suggesting that the knot contains a cool C-rich core. Guerrero & Manchado (1996) obtained spectra of the diffuse nebular body of A30, showing it to be H-rich. A similar conclusion was obtained by Manchado et al. (1988) and Medina & Peña (2000) for the outer shell of A 78. However, quantitatively, the results obtained by these two sets of authors are quite different and a deeper analysis is called for.

Three other objects belong to this group: A 58, IRAS1514-5258 and IRAS 18333-2357, the PN in the globular cluster M22, already mentioned in §4.7.5.

One common characteristic of this class of objects is their extremely high dust to gas ratio, and the fact that the photoelectric effect on the grains provides an important (and sometimes dominant) contribution to the heating of the nebular gas (see Harrington 1996). This may lead to large point-to-point temperature variations (see §4.7.5) and strongly affect abundance determinations.

Harrington (1996) concludes his review on H-poor PNe by noting that the H-poor ejecta cannot be explained by merely taking material with typical nebular abundances and converting all H to He. There is additional enrichment of C, N, perhaps O, and most interestingly, of Ne. However, more work on these objects is needed – and under way (e.g. Harrington et al. 1997) – before the abundances can be considered reliable. Stellar atmosphere analysis of H-deficient central stars (e.g. Werner 2001) is providing complementary clues to the nature and evolution of these objects.

In conclusion, we have shown how nebulae can provide powerful tools to investigate the evolution of stars and to probe the chemical evolution of galaxies. Nevertheless, is necessary to keep in mind the uncertainties and biases involved in the process of nebular abundance derivation. These are not always easy to make out, especially for the non specialist. One of the aims of this review was to help in maintaining a critical eye on the numerous and outstanding achievements of nebular Astronomy.

ACKNOWLEDGMENTS: It is a pleasure to thank the organizers of the XIII Canary Islands Winter School of Astrophysics, and especially César Esteban, for having given me the opportunity to share my experience on abundance determinations in nebulae. I also wish to thank the participants, for their attention and friendship. I am grateful to Miriam Peña, Luc Jamet and Yuri Izotov for a detailed reading of this manuscript, to Daniel Schaerer for having provided useful information on stellar atmospheres and to André Escudero for having kindly computed a few quantities related to planetary nebulae in the galactic bulge. I would like to thank my collaborators and friends, especially Rosa González Delgado, Slawomir Górny, Claus Leitherer, Miriam Peña, Michael Richer, Daniel Schaerer, Laerte Sodré, Ryszard Szczerba and Romuald Tylenda for numerous and lively discussions in various parts of the World.

Finally, I would like acknowledge the possibility of a systematic use of the NASA ADS Astronomy Abstract Service during the preparation of these lectures.

REFERENCES

AANESTAD, P. A., 1989, ApJ, 338, 162

AFFLERBACH, A., CHURCHWELL, E., ACORD, J. M., HOFNER, P., KURTZ, S., DEPREE, C. G., 1996, ApJS, 106, 423

AFFLERBACH, A., CHURCHWELL, E., WERNER, M. W., 1997, ApJ, 478, 190

ALEXANDER, J., BALICK, B., 1997, AJ, 114, 713

ALLENDE PRIETO, C., LAMBERT, D. L., ASPLUND, M., 2001, ApJ, 556, L63

ALLER, L. H., 1942, ApJ, 95, 52

ALLER, L. H., 1954, ApJ, 120, 401

ALLER, L. H., 1984, Physics of thermal gaseous nebulae, Reidel, Dordrecht

ALLER, L. H., CZYZAK, S. J., 1983, ApJS, 51, 211

ALLER, L. H., KEYES, C. D., 1987, ApJS, 65, 405

ALLOIN, D., COLLIN-SOUFFRIN, S., JOLY, M., VIGROUX, L., 1979, A&A, .78, 200

AMNUEL, P. R., 1993, MNRAS.261, 263

ANDERS, E., GREVESSE, N., 1989, GeCoA, 53, 197

BACHILLER, R., FORVEILLE, T., HUGGINS, P. J., COX, P., 1997, A&A, 324.1123

BALDWIN, J. A., CROTTS, A., DUFOUR, R. J., FERLAND, G. J., HEATHCOTE, S., HESTER, J. J., KORISTA, K. T., MARTIN, P. G., O'DELL, C. R., RUBIN, R. H., TIELENS, A. G. G. M., VERNER, D. A., VERNER, E. M., 1996, ApJ, 468, L115

BALDWIN, J. A., FERLAND, G. J., MARTIN, P. G., CORBIN, M. R., COTA, S. A., PETERSON, B. M., SLETTEBAK, A., 1991, ApJ, 374, 580

BALICK, B., GAMMON, R. H., HJELLMING, R. M., 1974, PASP, 86, 616

BALICK, B., GONZÁLEZ, G., FRANK, A., JACOBY, G., 1992, ApJ, 392, 582

BALICK, B., PERINOTTO, M., MACCIONI, A., TERZIAN, Y., HAJIAN, A., 1994, ApJ, 424, 800

BALICK, B., WILSON, J., HAJIAN, A. R., 2001, ApJ, 121, 354

BALSER, D. S., BANIA, T. M., ROOD, R. T., WILSON, T. L., 1997, ApJ, 483, 320

BALUTEAU, J.-P., ZAVAGNO, A., MORISSET, C., PÉQUIGNOT, D., 1995, A&A, 303, 175

BARBARO, G., MAZZEI, P., MORBIDELLI, L., PATRIARCHI, P., PERINOTTO, M., 2001, A&A, 365, 157

BARKER, T., 1978, ApJ, 220, 193

BARLOW, M. J., 1993, in Planetary nebulae, IAU Symposium no. 155, Eds. R. Weinberger and A. Acker, Kluwer Academic Publishers, Dordrecht, p.163

BAUTISTA, M. A., KALLMAN, T. R., 2001, ApJS 134, 139

BAUTISTA, M. A., POGGE, R. W., DEPOY, D. L., 1995, ApJ, 452, 685

BAUTISTA, M. A., PRADHAN, A. K., 1998, ApJ, 492, 650

BAUTISTA, M. A., PRADHAN, A. K., OSTERBROCK, DONALD E., 1994, ApJ, 432, L135

BEAULIEU, S. F., DOPITA, M. A., FREEMAN, K. C., 1999, ApJ, 515, 610

BENJAMIN, R. A., SKILLMAN, E. D., SMITS, D. P., 1999, ApJ, 514, 307

BERNARD SALAS, J., POTTASCH, S. R., BEINTEMA, D. A., WESSELIUS, P. R., 2001, A&A, 367, 949

BINETTE, L., LURIDIANA, V., HENNEY, W. J., 2001, RMxAC, 10, 19

BLÖCKER, T., 1995, A&A, 299, 755

BLÖCKER, T., 1999, in Asymptotic Giant Branch Stars, IAU Symposium no. 191, Ed. T. Le Bertre, A. Lèbre, & C. Waelkens, p. 21

BLÖCKER, T., T., OSTERBART, R., WEIGELT, G., BALEGA, Y., MEN'SHCHIKOV, A., 2001, in Post-AGB Objects as a Phase of Stellar Evolution, Eds. R. Szczerba & S. K. Górny, Kluwer, p. 241

BORKOWSKI, K. J., HARRINGTON, J. P., 1991, ApJ, 379, 168

162   G. Stasińska: *Abundances In H II Regions And Planetary Nebulae*

BOROSON, T. A., LIEBERT, J., 1989, ApJ, 339, 844

BRAGE, T., JUDGE, P. G., ABOUSSAID, A., GODEFROID, M. R., JOENSSON, P., YNNERMAN, A., FROESE FISCHER, C., LECKRONE, D. S., 1998, ApJ, 500, 507

BRESOLIN, F., KENNICUTT. R. C. JR., GARNETT, D. R., 1999, ApJ, 510, 104

BUTLER, K., 1993, in Planetary nebulae, IAU Symposium no. 155, Eds. R. Weinberger and A. Acker, Kluwer Academic Publishers, Dordrecht, p.73

CAPLAN, J., DEHARVENG, L., PEÑA, M., COSTERO, R., BLONDEL, C., 2000, MNRAS, 311, 317

CARDELLI, J. A., CLAYTON, G. C., MATHIS, J. S., 1989, ApJ, 345, 245

CARDELLI, J. A., MEYER, D. M., JURA, M., SAVAGE, B. D., 1996, ApJ, 467, 334

CASASSUS, S., ROCHE, P. F., BARLOW, M. J., 2000, MNRAS, 314, 657

CHARBONNEL, C., PALACIOS, A., 2001, Ap&SS, 277, 157

CHU, Y.-H., 1981, ApJ, 249, 195

CHU, Y.- H., 1991, in Wolf-Rayet Stars and Interrelations with Other Massive Stars in Galaxies, IAU Symposium no. 143, Ed. K. A. van der Hucht and B. Hidayat, Kluwer Academic Publishers, Dordrecht, p. 349

CHU, Y.-H., GUERRERO, M. A., GRUENDL, R. A., WILLIAMS, R. M., KALER, J. B., 2001, ApJ, 553, L69

CHU, Y.-H., TREFFERS, R. R., KWITTER, K. B., 1983, ApJS, 53, 937

CLEGG, R. E. S., HARRINGTON, J. P., BARLOW, M. J., WALSH, J. R., 1987, ApJ, 314, 551

CLEGG, R. E. S., STOREY, P. J., WALSH, J. R., NEALE, L., 1997, MNRAS, 284, 348

COHEN, M., HARRINGTON, J. P., HESS, R., 1984, ApJ, 283, 687

COSTA, R. D. D., DE FREITAS PACHECO, J. A., IDIART, T.P., 2000, A&A, 145, 467

COTA, S. A., FERLAND, G. J., 1988, ApJ, 326, 889

CUISINIER, F., ACKER, A., KÖPPEN, J., 1996, A&A, 307, 215

CUISINIER, F., MACIEL, W. J., KÖPPEN, J., ACKER, A., STENHOLM, B., 2000, A&A, 353, 543

CUNHA, K., LAMBERT, D. L., 1994, ApJ, 426, 170

DAVIDSON, K., KINMAN, T. D., 1985, ApJS, 58, 321

DE FREITAS PACHECO, J. A., MACIEL, W. J., COSTA, R. D. D., 1992, A&A, 261, 579

DEHARVENG, L., PEÑA, M., CAPLAN, J., COSTERO, R., 2000, MNRAS, 311, 329

DENICOLÓ, G., TERLEVICH, R. J., TERLEVICH, E., 2002, MNRAS, 330, 69

DÍAZ, A. I., PÉREZ-MONTERO, E., 2000, MNRAS, 312, 130

DINERSTEIN, H. L., 2001, ApJ, 550, L223

DINERSTEIN, H. L., GEBALLE, T. R., 2001, ApJ, 562, 515

DOPITA, M. A., 1997, ApJ, 485, L41

DOPITA, M. A., SUTHERLAND, R. S., 2000, ApJ, 539, 742

DOPITA, M. A., VASSILIADIS, E., WOOD, P. R., MEATHERINGHAM, S. J., HARRINGTON, J. P., BOHLIN, R. C., FORD, H. C., STECHER, T. P., MARAN, S. P., 1997, ApJ, 474, 188

DREIZLER, S., WERNER, K., 1993, A&A, 278, 199

DUDZIAK, G., PÉQUIGNOT, D., ZIJLSTRA, A. A., WALSH, J. R., 2000, A&A, 363, 717

DWARKADAS, V. V., BALICK, B., 1998, ApJ, 497, 267

DWEK, E., 1998, ApJ, 501, 643

ELMEGREEN, B. G., 1998, in Abundance profiles: diagnostic tools for Galaxy History, Eds. D. Friedli et al., ASP Conference Series, Vol. 147, p. 279

ESCUDERO, A. V., COSTA, R. D. D., 2001, A&A, 380, 300

ESTEBAN, C., 2002, in Ionized Gaseous Nebulae, Eds. W. J. Henney, J. Franco, M. Martos, & M. Peña , RMxAASC, 12, 56

ESTEBAN, C., PEIMBERT, M., 1995, RMxAC, 3, 133

ESTEBAN, C., PEIMBERT, M., TORRES-PEIMBERT, S., 1999a, A&A, 342, L37

ESTEBAN, C., PEIMBERT, M., TORRES-PEIMBERT, S., GARCÍA-ROJAS, J., 1999b, RMxAA, 35, 65

ESTEBAN, C., PEIMBERT, M., TORRES-PEIMBERT, S., ESCALANTE, V., 1998, MNRAS, 295, 401

ESTEBAN, C., SMITH, L. J., VÍLCHEZ, J. M., CLEGG, R. E. S., 1993, A&A, 272, 299

ESTEBAN, C., VÍLCHEZ, J. M., 1992, ApJ, 390, 536

ESTEBAN, C., VÍLCHEZ, J. M., MANCHADO, A., EDMUNDS, M. G., 1990, A&A, 227, 515

ESTEBAN, C., VÍLCHEZ, J. M., MANCHADO, A., SMITH, L. J., 1991, A&A, 244, 205

ESTEBAN, C., VÍLCHEZ, J. M., SMITH, L. J., CLEGG, R. E. S., 1992, A&A, 259, 629

FAÚNDEZ-ABANS, M., MACIEL, W. J., 1986, A&A, 158, 228

FAÚNDEZ-ABANS, M., MACIEL, W. J., 1988, RMxAA, 16, 105

FERLAND, G. J., 2001, PASP, 113, 41

FERLAND, G. J. ET AL., 1996, in Analysis of Emission Lines, R. E. Williams & M. Livio (Cambridge: Cambridge Univ. Press), p. 83

FERLAND, G. J., KORISTA, K. T., VERNER, D. A., FERGUSON, J. W., KINGDON, J. B., VERNER, E. M., 1998, PASP, 110, 761

FERLAND, G., SAVIN, D. W., 2001, Spectroscopic Challenges of Photoionized Plasmas, ASP Conference Series, vol. 247

FICH, M., SILKEY, M., 1991, ApJ, 366, 107

FORD, H. C., 1983, in Planetary Nebulae, IAU Symposium no. 103, ed. D. R. Flower, p. 443

FORESTINI, M., CHARBONNEL, C., 1997, A&AS, 123, 241

FRANK, A., MELLEMA, G., 1994a, A&A, 289, 937

FRANK, A., MELLEMA, G., 1994b, A&A, 430, 800

FROESE FISHER, C., SAHA, H. P., 1985, Phys. Scri. 32, 181

GALAVÍS, M. E., MENDOZA, C., ZEIPPEN, C. J., 1997, A&AS, 123, 159

GALLI, D., STANGHELLINI, L., TOSI, M., PALLA, F., 1997, ApJ, 477, 218

GARCÍA-SEGURA, G., LANGER, N., MAC LOW, M.-M., 1996a, A&A, 316, 133

GARCÍA-SEGURA, G., MAC LOW, M.-M., LANGER, N., 1996b, A&A, 305, 229

GARCÍA-VARGAS, M. L., GONZÁLEZ-DELGADO, R. M., PÉREZ, E., ALLOIN, D., DÍAZ, A., TERLEVICH, E., 1997, ApJ, 478, 112

GARNETT, D. R., 1992, AJ 103, 1330

GARNETT, D. R., DINERSTEIN, H. L., 2001, ApJ, 558, 145

GIVEON, U., STERNBERG, A., Lutz, D., Feuchtgruber, H., Pauldrach, A. W. A., 2002, ApJ, 566, 880

GONÇALVES, D., CORRADI, R. L. M., MAMPASO, A., 2001, ApJ 547, 302

GONZÁLEZ DELGADO, R. M., LEITHERER, C., HECKMAN, T. M., 1999, ApJS, 125, 489

GÓRNY, S. K. G., STASIŃSKA, G., 1995, A&A, 303, 893

GÓRNY, S. K. G., TYLENDA, R., 2000, A&A, 362, 1008

GREGGIO, L., RENZINI, A., 1990, ApJ 364, 35

GREVESSE, N. & SAUVAL, A. J., 1998, in Solar Composition and Its Evolution - From Core to Corona, Eds. C. Fröhlich, M. C. E. Huber, S. K. Solanki and R. von Steiger, Kluwer Academic Publishers, Dordrecht, p. 161

GRUENWALD, R. B., VIEGAS, S. M., 1992, ApJS, 78, 153

GRUENWALD, R., VIEGAS, S. M., 1998, ApJ, 501, 221

GRUENWALD, R., VIEGAS, S. M., BROGUIÈRE, D., 1997, ApJ, 480, 283

GUERRERO, M. A., MANCHADO, A., 1996, ApJ, 472, 711

GUERRERO, M. A., MANCHADO, A., CHU, Y.-H., 1997, ApJ, 487

GUTIÉRREZ-MORENO, A., MORENO, H., 1988, PASP, 100.1497

HARRINGTON, J. P., 1968, ApJ, 152, 943

HARRINGTON, J. P., 1996, in Hydrogen-deficient Stars, Eds. C. S. Jeffery & U. Heber, ASP conf. ser. vol. 96, p. 193

HARRINGTON, J. P., BORKOWSKI, K. J., TSVETANOV, Z. I., 1997, in Planetary Nebulae, IAU Symposium no. 180, eds. H. J. Habing and H. J. G. L. M. Lamers, p. 235

HARRINGTON, J. P., FEIBELMAN, W. A., 1984, ApJ, 277, 716

HARRINGTON, J. P., LUTZ, J. H., SEATON, M. J., STICKLAND, D. J., 1980, MNRAS, 191, 13

HARRINGTON, J. P., MARIONNI, P. A., 1981, in The Universe at Ultraviolet Wavelengths: The First Two Yrs. of Intern. Ultraviolet Explorer p. 623-631 (SEE N81-25893 16-90)

HARRINGTON, J.P., SEATON, M.J., LUTZ, J.H., ADAMS, S., 1982, MNRAS 199, 517

HENNEY, W. J., 1998, ApJ, 503, 760

HENRY, R. B. C., 1989, MNRAS, 241, 453

HENRY, R. B. C., 1990, ApJ, 356, 229

HENRY, R. B. C., 1993, MNRAS, 261, 306

HENRY, R. B. C., KWITTER, K. B., BATES, J. A., 2000, ApJ, 531, 928

HERWIG, F., 2000, A&A, 360, 952

HERWIG, F., 2001, in Post-AGB Objects as a Phase of Stellar Evolution, Eds. R. Szczerba & S. K. Górny, Kluwer, p. 249

HILLIER, D. J., MILLER, D. L., 1998, ApJ, 496, 407

HOARE, M. G., ROCHE, P. F., GLENCROSS, WILLIAM M., 1991, MNRAS, 251, 584

HOLWEGER, H., 2001, in Joint SOHO/ACE workshop "Solar and Galactic Composition". Ed. R. F. Wimmer-Schweingruber, American Institute of Physics Conference proceedings vol. 598, p.23

HOU, J. L., PRANTZOS, N., BOISSIER, S., 2000, A&A, 362, 921

HOWARD, J. W., HENRY, R. B. C., MCCARTNEY, S., 1997, MNRAS, 284, 465

HUBENY, I., LANZ, T., 1995, ApJ, 439, 875

HUMMER, D. G., BERRINGTON, K. A., EISSNER, W., PRADHAN, A. K., SARAPH, H. E., TULLY, J. A., 1993, A&A, 279, 298

HUMMER, D. G., STOREY, P. J., 1992, MNRAS, 254, 277

HYUNG, S., ALLER, L. H., FEIBELMAN, W. A., 1994, PASP, 106, 745

HYUNG, S., ALLER, L. H., FEIBELMAN, W. A., LEE, W. B., DE KOTER, A., 2000, MNRAS, 318, 77

HYUNG, S., MELLEMA, G., LEE, S.-J., KIM, H., 2001, A&A, 378, 587

IBEN, I. JR., KALER, J. B., TRURAN, J. W., RENZINI, A., 1983, ApJ, 264, 605

IVANS, I. I., SNEDEN, C., KRAFT, R. P., SUNTZEFF, N. B., SMITH, V. V., LANGER, G. E., FULBRIGHT, J. P., 1999, AJ, 118, 1273

IZOTOV, Y. I., & THUAN, T. X., 1998, ApJ, 497, 227

IZOTOV,Y. I., THUAN T. X., & LIPOVETSKY, V. A., 1994, ApJ, 435, 647

JACOBY, G. H., 1979, PASP, 91, 754

JACOBY, G. H., CIARDULLO, R., 1999, ApJ, 515, 169

JACOBY, G. H., FORD, H. C., 1983, ApJ, 266, 298

JACOBY, G. H., MORSE, J. A., FULLTON, L. K., KWITTER, K. B., HENRY, R. B. C., 1997, AJ, 114, 2611

KALER, J. B., IBEN, I. JR., BECKER, S. A., 1978,ApJ, 224L, 63

KEYES, C. D., ALLER, L. H., FEIBELMAN, W. A., 1990, PASP, 102, 59

KINGDON, J. B., FERLAND, G. J., 1995, ApJ, 450, 691

KINGDON, J. B., FERLAND, G. J., 1996, ApJS, 106, 205

KINGDON, J. B., FERLAND, G. J., 1997, ApJ, 477, 732

KINGDON, J., FERLAND, G. J., FEIBELMAN, W. A., 1995, ApJ, 439, 793

KINGSBURGH, R. L., BARLOW, M. J., 1994, MNRAS, 271, 257

KÖPPEN, J., ACKER, A., STENHOLM, B., 1991, A&A, 248, 197

KÖPPEN, J., CUISINIER, F., 1997, A&A, 319, 98

KOESTERKE, L., 2001, Ap&SS, 275, 41

KOESTERKE, L., GRÄFENER, G., HAMANN, W.-R., 2000, in Thermal and Ionization Aspects of Flows from Hot Stars, ASP Conference Series, Vol. 204. Eds. H. Lamers & A. Sapar, p.239

KRISHNASWAMY-GILROY, K., SNEDEN, C., PILACHOWSKI, C. A., COWAN, J. J., 1988, ApJ, 327, 298

KWITTER, K. B., HENRY, R. B. C., 1996, ApJ, 473, 304

KWITTER, K. B., HENRY, R. B. C., 1998, ApJ, 493, 247

LACY, J. H., TOWNES, C. H., GEBALLE, T. R., HOLLENBACH, D. J., 1980, ApJ, 241, 132

LAME, N. J., HARRINGTON, J. P., BORKOWSKI, K., 1997, in Planetary Nebulae, IAU Symposium no. 180, Eds. H. J. Habing & H. J. G. L. M. Lamers, Dordrecht: Kluwer Academic Publishers, p. 252.

LAMERS, H. J. G. L. M., NOTA, A., PANAGIA, N., SMITH, L. J., LANGER, N., 2001, ApJ, 551, 764

LATTANZIO, J., FORESTINI, M., 1999, in Asymptotic Giant Branch Stars, IAU Symposium no. 191, Eds. T. Le Bertre, A. Lèbre & C. Waelkens, p. 31

LEISY, P., DENNEFELD, M., 1996, A&AS, 116, 95

LENNON, D. J., BURKE, V. M., 1994, A&AS, 103, 273

LESTER, D. F., DINERSTEIN, H. L., WERNER, M. W., WATSON, D. M., GENZEL, R., STOREY, J. W. V., 1987, ApJ, 320, 573

LEUENHAGEN, U., HAMANN, W.-R., 1998, A&A, 330, 265

LIU, X.-W., 1998, MNRAS, 295, 699

LIU, X.-W., 2002, in Ionized Gaseous Nebulae, Eds. W. J. Henney, J. Franco, M. Martos, & M. Peña, RMxAASC, 12, 70

LIU, X.-W., BARLOW, M. J., COHEN, M., DANZIGER, I. J., LUO, S.-G., BALUTEAU, J. P., COX, P., EMERY, R. J., LIM, T., PÉQUIGNOT, D., 2001a, MNRAS, 323, 343

LIU, X.-W., BARLOW, M. J., DANZIGER, I. J., STOREY, P. J., 1995a, ApJ, 450, L59

LIU, X.-W., LUO, S.-G., BARLOW, M. J., DANZIGER, I. J., STOREY, P. J., 2001b, MNRAS, 327, 141

LIU, X.-W., STOREY, P. J., BARLOW, M. J., CLEGG, R. E. S., 1995b, MNRAS, 272, 369

LIU, X.-W., STOREY, P. J., BARLOW, M. J., DANZIGER, I. J., COHEN, M., BRYCE, M., 2000, MNRAS, 312, 585

LUO, S.-G., LIU, X.-W., BARLOW, M. J., 2001, MNRAS, 326, 1049

LURIDIANA, V., CERVIÑO, M., BINETTE, L., 2001, A&A, 379, 1017

LURIDIANA, V., PEIMBERT, M., 2001, ApJ, 553, 633

LURIDIANA, V., PEIMBERT, M., LEITHERER, C., 1999, ApJ, 527, 110

LUTZ, D., FEUCHTGRUBER, H., GENZEL, R., KUNZE, D., ET AL., 1996, A&A, 315, L269

MACIEJEWSKI, W., MATHIS, J. S., EDGAR, R. J., 1996, ApJ, 462, 347

MACIEL, W. J., KÖPPEN, J., 1994, A&A, 282, 436

MACIEL, W. J., QUIREZA, C., 1999, A&A, 345, 629

MAEDER, A., 1990, A&AS, 84, 139

MANCHADO, A., POTTASCH, S. R., MAMPASO, A., 1988, A&A, 191, 128

MARIGO, P., 2001, A&A, 370, 194

MARIGO, P., BRESSAN, A., CHIOSI, C., 1996, A&A, 313, 545

MARSTON, A. P., 1997, ApJ, 475, 188

MARSTON, A. P., 1999, in Wolf-Rayet Phenomena in Massive Stars and Starburst Galaxies, IAU Symposium no. 193, Eds. K. A. van der Hucht, G. Koenigsberger & P. R. J. Eenens, p.306

MARTEN, H., SCHÖNBERNER, D., 1991, A&A, 248, 590

MARTEN, H., SZCZERBA, R., 1997, A&A, 325, 1132

MARTÍN-HERNÁNDEZ, N. L., PEETERS, E., MORISSET, C., TIELENS, A. G. G. M., COX, P., ROELFSEMA, P. R., BALUTEAU, J.-P., SCHAERER, D., MATHIS, J. S., DAMOUR, F., CHURCHWELL, E., KESSLER, M. F., 2002, A&A, 381, 606

MARTINS, L. P., VIEGAS, S. M., 2000, A&A, 361, 1121

MATHIS, J. S., 1983, ApJ, 267, 119

MATHIS, J. S., LIU, X.-W., 1999, ApJ, 521, 212

MATHIS, J. S., ROSA, M. R., 1991, A&A, 245, 625

MATHIS, J. S., TORRES-PEIMBERT, S., PEIMBERT, M., 1998, ApJ, 495, 328

MCGAUGH, S. S., 1991, ApJ, 380, 140

MCGAUGH, S. S., 1994, ApJ, 426, 135

MEATHERINGHAM, S. J., DOPITA, M. A., 1991a, ApJS, 75, 407

MEATHERINGHAM, S. J., DOPITA, M. A., 1991b, ApJS, 76, 1085

MEDINA, S., PEÑA, M., 2000, RMxAA, 36, 121

MELLEMA, G., 1995, MNRAS, 277, 173

MELLEMA, G., FRANK, A., 1995, MNRAS, 273, 401

MENDOZA, C., 1983, in Planetary nebulae, IAU Symposium no. 103, Ed. D. R. Flower, Dordrecht, D. Reidel, p. 143

MEYER, D. M., CARDELLI, J. A., SOFIA, U. J., 1997, ApJ, 490, L103

MEYER, D. M., JURA, M., CARDELLI, J.A., 1998, ApJ, 493, 222

MIDDLEMASS, D.,1988, MNRAS, 231, 1025

MIDDLEMASS, D., 1990, MNRAS, 244, 294

MILLER, G. J., CHU, Y.-H., 1993, ApJS, 85, 137

MONK, D. J., BARLOW, M. J., CLEGG, R. E. S., 1988, MNRAS, 234, 583

MONTEIRO, H., MORISSET, C., GRUENWALD, R., VIEGAS, S. M., 2000, ApJ, 537, 853

MOORE, B. D., HESTER, J. J., SCOWEN, P. A., 2000, AJ, 119

MORISSET, C., SCHAERER, D., MARTÍN-HERNÁNDEZ, N. L., PEETERS, E., DAMOUR, F., BALUTEAU, J.-P., COX, P., ROELFSEMA, P., 2002, A&A, 386, 558

NAHAR, S. N., 2002, in Planetary nebulae and their Role in the Universe, IAU Symposium no. 209, in press

NAHAR, S. N., PRADHAN, A. K., 1997, ApJS, 111, 339

NAHAR, S. N., PRADHAN, A. K., ZHANG, H. L., 2000, ApJS, 131, 375

NAJARRO, F., KRABBE, A., GENZEL, R., LUTZ, D., KUDRITZKI, R. P., HILLIER, D. J., 1997, A&A, 325, 700

NAPIWOTZKI, R., HEBER, U., KÖPPEN, J., 1994, A&A, 292, 239

NATTA, A. PANAGIA, N., 1981, ApJ, 248, 189

NOTA, A., LIVIO, M., CLAMPIN, M., SCHULTE-LADBECK, R., 1995, ApJ, 448, 788

NUSSBAUMER, H., & STOREY P., J., 1981, A&A, 99, 177

OCH, S. R., LUCY, L. B., ROSA, M. R., 1998, A&A, 336, 301

O'DELL, C. R., 2001, PASP, 113, 29O

O'DELL, C. R., HANDRON, K. D., 1996, AJ, 111, 1630

O'DELL, C. R., HENNEY, W. J., BURKERT, A., 2000, AJ, 119, 2910

OEY, M. S., DOPITA, M. A., SHIELDS, J. C., SMITH, R. C., 2000, ApJS, 128, 511

OEY, M. S., KENNICUTT, R. C., JR., 1993, ApJ, 411, 137

OEY, M. S., SHIELDS, J. C., 2000, ApJ, 539, 687

OLIVA, E., PASQUALI, A., RECONDITI, M., 1996, A&A, 305, L21

OLIVE, K. A., SKILLMAN, E. D., 2001, New Astronomy, 6, 119

OLIVE, K. A., STEIGMAN, G., SKILLMAN, E. D., 1997, ApJ, 483, 788

OSTERBROCK, D. E., 1989, Astrophysics of Gaseous Nebulae and Active Galactic Nuclei (Mill Valley: University Science Books)

OSTERBROCK, D., FLATHER, E., 1959, ApJ, 129, 26

OSTERBROCK, D. E., TRAN, H. D., VEILLEUX, S., 1992, ApJ, 389, 305

PAGEL, B. E. J., EDMUNDS, M. G., BLACKWELL, D. E., CHUN, M. S., SMITH, G., 1979, MNRAS, 189, 95

PALLA, F., BACHILLER, R., STANGHELLINI, L., TOSI, M., GALLI, D., 2000, A&A, 355, 69

PATRIARCHI, P., MORBIDELLI, L., PERINOTTO, M., BARBARO, G., 2001, A&A, 372, 644

PASQUALI, A., PERINOTTO, M., 1993, A&A, 280, 581

PAULDRACH, A. W. A., HOFFMANN, T. L., LENNON, M., 2001, A&A, 375, 161

PEETERS, E., MARTÍN-HERNÁNDEZ, N. L., DAMOUR, F., COX, P., ROELFSEMA, P. R., BALUTEAU, J.-P., TIELENS, A. G. G. M., CHURCHWELL, E., KESSLER, M. F., MATHIS, J. S., MORISSET, C., SCHAERER, D., 2002, A&A, 381, 571

PEIMBERT, M., 1967, ApJ, 150, 825

PEIMBERT, M., 1978, in Planetary nebulae, IAU Symposium no. 76, Ed. Y. Terzian, Dordrecht, Reidel, p. 215

PEIMBERT, M., 1990, Rep. Prog. Phys., 53, 1559

PEIMBERT, M., 1996, in Analysis of Emission Lines, Eds. R. E. Williams & M. Livio (Cambridge: Cambridge Univ. Press), p. 165

PEIMBERT, M., 2002, in Ionized Gaseous Nebulae, Eds. W. J. Henney, J. Franco, M. Martos, & M. Peña, RMxAASC, 12, 275

PEIMBERT, M., CARIGI, L., PEIMBERT, A., 2001 Astrophys. Space Sci. Suppl., 277, 147-156

PEIMBERT, A., PEIMBERT, M., LURIDIANA, V., 2002, ApJ 565, 668

PEIMBERT, M., RAYO, J. F., TORRES-PEIMBERT, S., 1978, ApJ, 220, 516

PEÑA, M., STASIŃSKA, G., ESTEBAN, C., KOESTERKE, L., MEDINA, S., KINGSBURGH, R., 1998, A&A, 337, 866

PEÑA, M., STASIŃSKA, G., MEDINA, S., 2001, A&A, 367, 983

PÉQUIGNOT, D., 1986, in Model Nebulae (Publication de l'Observatoire de Paris)

PÉQUIGNOT, D., ET AL., 2002, in Ionized Gaseous Nebulae, Eds. W. J. Henney, J. Franco, M. Martos, & M. Peña, RMxAASC, 12, 142

PÉQUIGNOT, D., BALUTEAU, J.-P., 1994, A&A, 283, 593

PÉQUIGNOT, D., PETITJEAN, P., BOISSON, C., 1991, A&A, 251, 680

PÉQUIGNOT, D., STASIŃSKA, G., ALDROVANDI, S. M. V., 1978, A&A, 63, 313

PÉQUIGNOT, D., WALSH, J. R., ZIJLSTRA, A. A., DUDZIAK, G., 2000, A&A, 361, L1

PÉREZ, E., 1997, MNRAS, 290, 465

PERINOTTO, M., BENCINI, C. G., PASQUALI, A., MANCHADO, A., RODRIGUEZ ESPINOSA, J. M., STANGA, R., 1999, A&A, 347, 967

PERINOTTO, M., KIFONIDIS, K., SCHOENBERNER, D., MARTEN, H., 1998, A&A, 332, 1044

PERINOTTO, M., PANAGIA, N., BENVENUTI, P., 1980, A&A, 85, 332

PILYUGIN, L. S., 2000, A&A, 362, 325

PILYUGIN, L. S., 2001, A&A, 369, 594

POTTASCH, S. R., 1984, Planetary Nebulae, Dordrecht, Reidel

RATAG, M. A., POTTASCH, S. R., DENNEFELD, M., MENZIES, J., 1997, A&AS, 126, 297

RAUCH, T., DEETJEN, J. L., DREIZLER, S., WERNER, K., 2000, in Asymmetrical Planetary Nebulae II: ¿From Origins to Microstructures, ASP Conference Series, Vol. 199. Eds. J. H. Kastner, N. Soker & S. Rappaport, p. 337

RELAÑO, M., PEIMBERT, M., BECKMAN, J., 2002, ApJ, 564, 704

RENZINI, A., VOLI, M., 1981, A&A, 94, 175

RICHER, M. G., 1993, ApJ, 415, 240

RICHER, M. G., MCCALL, M. L., 1995, ApJ, 445, 642

RICHER, M. G., MCCALL, M. L., STASIŃSKA, G., 1998, A&A, 340, 67

RICHER, M. G., STASIŃSKA, G., MCCALL, M. L., 1999, A&AS, 135, 203

RODRÍGUEZ-GASPAR, J. A., TENORIO-TAGLE, G., 1998, A&A, 331, 347

RUDOLPH, A. L., SIMPSON, J. P., HAAS, M. R., ERICKSON, E. F., FICH, M., 1997, ApJ, 489, 94

RUBIN, R. H., 1968, ApJ, 153, 761

RUBIN, R. H., 1989, ApJS, 69, 897

RUBIN, R. H., DUFOUR, R. J., WALTER, D. K., 1993, ApJ, 413, 242

RUBIN, R. H., SIMPSON, J. P., ERICKSON, E. F., HAAS, M. R., 1988, ApJ, 327, 377

RUBIN, R. H., SIMPSON, J. P., HAAS, M. R., ERICKSON, E. F., 1991, ApJ, 374, 564

RUBIN, R. H., SIMPSON, J. P., LORD, S. D., COLGAN, S. W. J., ERICKSON, E. F., HAAS, M. R., 1994, ApJ, 420, 772

RUSSELL, S. C., DOPITA, M. A., 1992, ApJ, 384, 508

SAMLAND, M., KÖPPEN, J., ACKER, A., STENHOLM, B., 1992, A&A, 264, 184

SANKRIT, R., HESTER, J. J., 2000, ApJ, 535, 847

SASSELOV, D., GOLDWIRTH, D., 1995, ApJ, 444, L5

SAUER, D., JEDAMZIK, K., 2002, A&A, 381, 361

SAVIN, D. W., 1999, ApJ, 523, 855

SCHAERER, D., 2000, in Stars, Gas and Dust in Galaxies: Exploring the Links, ASP Conference Proceedings, Vol. 221. Eds. D. Alloin, K. Olsen, & G. Galaz, p.99

SCHAERER, D., DE KOTER, A., 1997, A&A, 322, 598

SCHMIDT-VOIGT, M., KÖPPEN, J., 1987a, A&A, 174, 211

SCHMIDT-VOIGT, M., KÖPPEN, J., 1987b, A&A, 174, 223

SCHÖNING, T., BUTLER, K., 1998, A&AS, 128, 581

SEARLE, L., 1971, ApJ, 168, 327

SEATON, M., 1987, J. Phys. B, 20, 6363

SEMBACH, K. R., SAVAGE, B. D., 1996, ApJ, 457, 211

SHAVER, P. A., MCGEE, R. X., NEWTON, L. M., DANKS, A. C., POTTASCH, S. R., 1983, MNRAS, 204, 53

SHIELDS, G. A., 1974, ApJ, 193, 335

SHIELDS, G. A., 1978, ApJ, 219, 559

SHIELDS, G. A., 1983, in Planetary nebulae, Proceedings of the IAU Symposium 103, Ed. D. R. Flower, Dordrecht, D. Reidel Publishing Co., p. 259

SHIELDS, G. A., ALLER, L. H., KEYES, C. D., CZYZAK, S. J., 1981, ApJ, 248, 569

SHIELDS, J. C., FERLAND, G. J., 1994, ApJ, 430, 236

SHIELDS, J. C., KENNICUTT, R. C. JR., 1995, ApJ, 454, 807

SIMPSON, J.P., COLGAN, S. W. J., RUBIN, R. H., ERICKSON, E. F., HAAS, M. R., 1995, ApJ, 444, 721

SIMPSON, J. P., RUBIN, R. H., 1990, ApJ, 354, 165

SIMPSON, J. P., WITTEBORN, F. C., PRICE, S. D., COHEN, M., 1998, ApJ, 508, 268

SKILLMAN, E. D., 1989, ApJ, 347, 883

SMITH, L. J., 1996, in Wolf-Rayet stars in the framework of stellar evolution, Eds. J.M. Vreux, A. Detal, D. Fraipont-Caro, E. Gosset & G. Rauw, p.381

SMITS, D. P., 1996, MNRAS, 278, 683

SOFIA, U. J., CARDELLI, J. A., Guerin, K. P., Meyer, D. M. 1997, ApJ, 482, L105

SOFIA, U. J., MEYER, D. M., 2001, ApJ, 554, L221, SOFIA, U. J., MEYER, D. M. (erratum), 2001, ApJ, 558, 147

SPITZER, L., JR., 1948, ApJ, 107, 6

SPITZER, L. JR., 1978, Physical Processes in the Interstellar Medium, John Wiley & Sons, New York

STASIŃSKA, G., 1980a, A&A, 85, 359

STASIŃSKA, G., 1980b, A&A, 84, 320

STASIŃSKA, G., 1990, A&AS, 83, 501

STASIŃSKA, G., 1998, in Abundance Profiles: Diagnostic Tools for Galaxy History, ASP Conf. Ser. Vol. 147, Eds. D. Friedli, M. Edmunds, C. Robert & L. Drissen, p. 142

STASIŃSKA, G., 2002, in Ionized Gaseous Nebulae, Eds. W. J. Henney, J. Franco, M. Martos, & M. Peña, RMxAASC, 12, 62

STASIŃSKA, G., GÓRNY, S. K., TYLENDA, R., 1997, A&A, 327, 736

STASIŃSKA, G., IZOTOV, Y., 2001, A&A, 378, 817

STASIŃSKA, G., RICHER, M. G., McCALL, M. L., 1998, A&A, 336, 667

STASIŃSKA, G., SCHAERER, D., 1997, A&A, 322, 615

STASIŃSKA, G., SCHAERER, D., 1999, A&A, 351, 72

STASIŃSKA, G., SCHAERER, D., LEITHERER, C., 2001, A&A, 370, 1

STASIŃSKA, G., SZCZERBA, R., 1999, A&A, 352, 297

STASIŃSKA, G., SZCZERBA, R., 2001, A&A, 379, 1024

STASIŃSKA, G., TYLENDA, R., 1986, A&A, 155, 137

STASIŃSKA, G., TYLENDA, R., ACKER, A., STENHOLM, B., 1992, A&A, 266, 486

STEVENSON, C. C., McCALL, M. L., WELCH, D. L., 1993, ApJ, 408, 460

STORCHI-BERGMANN, T., CALZETTI, D., KINNEY, A. L., 1994, ApJ, 429, 572

STOREY, P. J., 1997, in Planetary Nebulae, IAU Symposium no. 180, eds. H. J. Habing and H. J. G. L. M. Lamers, p. 161

STOREY, P. J., HUMMER, D. G., 1995, MNRAS, 272, 41

TAMBLYN, P., RIEKE, G. H., HANSON, M. M., CLOSE, L. M., McCARTHY, D. W., JR., RIEKE, M. J., 1996, ApJ, 456, 206

TORRES-PEIMBERT, S., PEIMBERT, M., 1977, RMxAA, 2, 181

TORRES-PEIMBERT, S., PEIMBERT, M., PEÑA, M., 1990, A&A, 233, 540

TOVMASSIAN, G. H., STASIŃSKA, G., CHAVUSHYAN, V. H., ZHARIKOV, S. V., GUTIERREZ, C., PRADA, F., 2001, A&A, 370, 456

TYLENDA, R., 1979, Acta Astronomica, 29, 355

VAN DEN HOEK, L. B., GROENEWEGEN, M. A. T., 1997, A&AS, 123, 305

VAN HOOF, P. A. M., BEINTEMA, D. A., VERNER, D. A., FERLAND, G. J., 2000, A&A, 354, L41

VAN ZEE, L., SALZER, J. J., HAYNES, M. P., O'DONOGHUE, A. A., BALONEK, T. J., 1998, AJ, 116, 2805

VASSILIADIS, E., DOPITA, M. A., BOHLIN, R. C., HARRINGTON, J. P., FORD, H. C., MEATHER-INGHAM, S. J., WOOD, P. R., STECHER, T. P., MARAN, S. P., 1996, ApJS, 105, 375

VASSILIADIS, E., DOPITA, M. A., BOHLIN, R. C., HARRINGTON, J. P., FORD, H. C., MEATHER-INGHAM, S. J., WOOD, P. R., STECHER, T. P., MARAN, S. P., 1998, ApJS, 114, 237

VASSILIADIS, E., DOPITA, M. A., Morgan, D. H., Bell, J. F., 1992, ApJS, 83, 87

VIEGAS, S. M., CLEGG, R. E. S., 1994, MNRAS, 271, 993

VIGROUX, L., STASIŃSKA, G., COMTE, G., 1987, ApJ, 319, 358

VÍLCHEZ, J. M., ESTEBAN, C., 1996, MNRAS, 280, 720

VOLK, K., DINERSTEIN, H., SNEDEN, C., 1997, in Planetary Nebulae, IAU Symposium no. 180, Eds. H. J. Habing & H. J. G. L. M. Lamers, Dordrecht: Kluwer Academic Publishers, p. 284.

WALSH, J. R., DUDZIAK, G., MINNITI, D., ZIJLSTRA, A. A., 1997, ApJ, 487, 651

WALSH, J. R., WALTON, N. A., JACOBY, G. H., PELETIER, R. F., 1999, A&A, 346, 753

WEBSTER, B. L., 1988, MNRAS, 230, 377

WEIDEMANN, V., 1987, A&A, 188, 74

WEINGARTNER, J. C., DRAINE, B. T., 2001, ApJ, 548, 296

WEN, Z., O'DELL, C. R., 1995, ApJ, 438, 784

WERNER, K., 2001, Ap&SS, 275, 27

ZUCKERMAN, B., 1973, ApJ, 183, 863

Element Abundances In Nearby Galaxies

By DONALD R. GARNETT

Steward Observatory, University of Arizona, Tucson AZ 85721, USA

In these lectures I present a highly opinionated review of the observed patterns of metallicity and element abundance ratios in nearby spiral, irregular, and dwarf elliptical galaxies, with connection to a number of astrophysical issues associated with chemical evolution. I also discuss some of the observational and theoretical issues associated with measuring abundances in H II regions and gas and stellar surface densities in disk galaxies. Finally, I will outline a few open questions that deserve attention in future investigations.

1. Introduction

The measurement of element abundances in galaxies other than our own has a roughly forty-year history, beginning with early attempts to measure helium abundances in giant H II regions in the Magellanic Clouds and M33 (Aller & Faulkner 1962, Mathis 1962) and pioneering studies of heavy element abundances from forbidden lines in extragalactic H II regions (e.g. Peimbert & Spinrad 1970, Searle 1971, Searle & Sargent 1972). Since then this field has grown tremendously, with high quality oxygen abundance data in some 40 nearby spiral galaxies and more than 100 irregular and compact dwarf galaxies. The amount of data for other elements (C, N, Ne, S, and Ar) has also improved tremendously, thanks largely to improvements in visible-wavelength detectors and the launching of spacecraft observatories, such as IUE, HST, and ISO, which have opened up the UV and IR spectral regions for spectroscopy.

The direct importance of determining the distribution of metallicity and element abundance ratios in galaxies is the contribution these measurements make to chemical evolution, and by consequence the evolution of galaxies. The elements heavier than H and He in stars and the interstellar medium (ISM) are the accumulated product of previous generations of star formation. The overall metallicity (usually represented by O/H in H II regions/ISM, and by Fe/H in stars) is determined by the total amount of previous star formation. Element abundance ratios, particularly C/O, N/O, or s-process/Fe, track the relative contributions of low-mass stars and high-mass stars, incorporating information on the stellar initial mass function (IMF). The abundances can be affected by gas flows (infall, outflow, or internal flows). It is possible, with modeling, to infer important clues to the evolution of galaxies from abundance measurements.

Beyond galaxy evolution, abundance measurements provide important ancillary information relevant to other very important astrophysical problems, including:

• The dependence of the $I(CO)/N(H_2)$ conversion on environment and metallicity, which is critical for determining the amount of molecular gas in galaxies.

• The metallicity dependence of the Cepheid period-luminosity relation, currently under debate with respect to the determination of the Hubble constant and the extragalactic distance scale.

• Understanding the color evolution of galaxies. Colors of composite stellar populations depend on both age and metallicity; metallicity measurements thus reduce degeneracies in the interpretation of colors.

• Stellar mass loss rates, particularly the radiatively-accelerated winds of O and Wolf-Rayet stars, likely depend on the metallicity of the individual stars.

• The cooling function of interstellar gas. Cooling of interstellar gas is generally dominated by metals (ions of metals in X-ray and ionized gas, singly-ionized metals in neutral gas, and molecules other than H_2 in molecular clouds), so the thermal balance in the ISM is a function of metallicity, with obvious implications for the formation of stars and galaxies.

In any field of investigation, a few key questions arise which form a framework for specific studies. I formulate a few of them below.

(*a*) How do metallicity and element abundance ratios evolve within galaxies, and how do variations relate to the evolution of the gas content and stellar light?

(*b*) What galaxy properties determine the observed compositions of galaxies? How is metallicity affected by galaxy dynamics (interactions, gas flows, angular momentum evolution)?

(*c*) How did heavy elements get into the intergalactic medium (IGM)? Were they ejected from galaxies by supernova-driven winds, ejected in tidal streams during galaxy interactions and mergers, or did they come from the first, possibly pre-galactic, stars?

(*d*) How well do simulations of galaxy formation and evolution reproduce the observed metallicities and distribution of abundances in galaxies?

The purpose of these lectures is to review the results of a variety of element abundance studies in galaxies other than our own in the nearby universe. I will not try to be all-inclusive, as the field is vast. For example, I will not attempt to discuss abundances in elliptical galaxies in detail, as better experts have already written extensive reviews on the subject (e.g. Worthey 1998, Henry & Worthey 1999), nor will I say much about luminous IR starbursts or low surface brightness galaxies. Much of the methodology behind abundance measurements in stars and ionized gas will be covered in great detail in the lectures by Lambert, Langer and Stasińska; I will not spend much time on these subjects, but will highlight points of contention or uncertainty where appropriate. Likewise, Matteucci will discuss chemical evolution modeling in detail, so I will use the observational results to highlight areas where the data shed light on physical evolution of galaxies.

2. Observational Methods for Measuring Abundances

2.1. *Spectroscopy of H II Regions and Planetary Nebulae*

Most of the information we have on abundances in spiral and irregular galaxies have come from spectroscopy of H II regions. This is logical since H II regions are luminous and have high surface brightness (in the emission lines) compared to individual stars in galaxies. One can think of an H II region as an efficient machine for converting the extreme ultraviolet radiation of a hot, massive star into a few narrow emission lines, leading to a very luminous object in the optical/IR bands. As a result, observations of H II regions have typically provided our first look at abundances within galaxies. Indeed, emission lines are now being used to probe the ISM of galaxies at redshifts greater than 2, as will be discussed by Pettini in these proceedings.

Elements that are readily observed in the visible spectrum of H II regions include O, N, Ne, S, and Ar. With the exception of O, all of these elements may have important ionization states that emit only in the ultraviolet or infrared (for example, Ne^+, N^{+2}, S^{+3}). If we add UV spectroscopy we can study C and Si. Figure 1 shows an HST spectrum of one H II region in the SMC, showing the rich variety of forbidden emission lines and H, He recombination lines in the UV and optical spectrum. Other abundant heavy elements (such as Fe or Mg) may be observed in photoionized nebulae. One must always keep in mind that many elements in the ISM are strongly depleted onto grains,

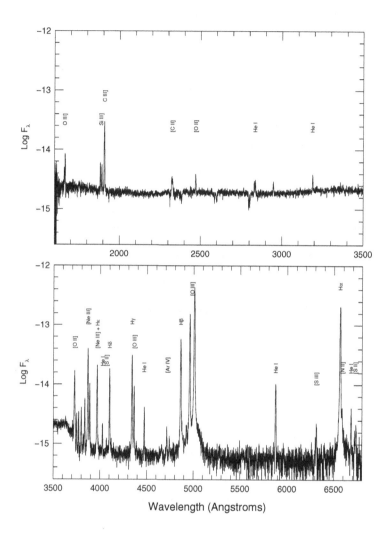

FIGURE 1. *Hubble Space Telescope* UV/optical spectrum of the H II region N88A in the Small Magellanic Cloud.

which affects the total abundance. This is an important but poorly known factor in many cases (such as O and C). Another caution is that H II regions show the composition of the present-day ISM, and are insensitive to the evolution of abundances with time.

Planetary nebulae (PNs) are essentially H II regions created by the ionization of a red giant envelope by the exposed hot stellar core, so spectroscopy of a PN provides information on a similar variety of elements as H II regions, with the same caveats. There are a number of significant differences, however. PNs are much less luminous than H II regions, so observations suitable for abundance measurements are restricted to the nearest galaxies; at the present time, measuring abundances in PNs is challenging in galaxies as close as M31 (Jacoby & Ciardullo 1999). Another difference is that the PN abundances are altered from the original stellar composition by nucleosynthesis – He, C, and N are often enriched, and even O may be affected. Thus, the use of PNs to measure abundances across galaxies must be pursued with caution. Nevertheless, measurements of PNs offer a means of measuring abundances across galaxies and their evolution over the range of ages of PN progenitors (a few tens to a few thousands of Myr). Stasińska will discuss PN abundance measurements in her lectures.

The observational and analytic techniques for determining abundances in H II regions have been discussed at great length by Skillman (1998) at the VIIIth Canary Island Winter School and by Stasińska in her lectures here. I will add a few remarks here on observing and deriving abundances.

2.1.1. *Observational Considerations*

It is easy to obtain a high-quality spectrum of an H II region in a nearby galaxy. It is not so easy to obtain a high-quality analysis afterward. Photon statistics is not the entire story in CCD spectroscopy. Additional random errors creep in during the flat-fielding and photometric calibration stages. It is difficult to flat-field a CCD frame to better than 1% even in imaging observations, where the most precise flat-fields involve matching the color of the target to that of the flat-field source (the night-sky for deep imaging – see Tyson 1986). Spectroscopists rarely observe such practices. In typical H II region spectroscopy, the flat-field is often obtained by combining an internal lamp to map the pixel-to-pixel sensitivity variations with a twilight sky observation to fit the slit vignetting. Both fill the slit in a different way than the object, which is an important consideration for the correction of interference fringing in the red. Flat-fields *repeatable* to 1% precision can be obtained over limited areas of a CCD spectrum, but the precision can be worse over regions where the lamp source is weak (in the blue part of the spectrum for example) or vignetting is strong.

The photometric calibration also contributes to the uncertainty of the measured spectrum. Flux standard stars are typically measured at widely spaced wavelengths (50 Å is common), and the sensitivity function of the instrument is determined by fitting a low-order polynomial or spline to the flux points. Such fits inevitably introduce low-order "wiggles" in the sensitivity function, which will vary from star to star. Based on experience, the best spectrophotometric calibration yield uncertainties in the *relative* fluxes of order 2-3% for widely-spaced emission lines; the errors may be better for ratios of lines closer than 20 Å apart. Absolute fluxes have much higher uncertainties, of course, especially for narrow-aperture observations of extended objects.

Another source of concern is the extended nature of H II regions and patchiness in interstellar reddening, which affects the measured line ratios. H II region spectra are often presented as integrated over the source. Reddening by dust is patchy everywhere we look, so the effects on the H II region spectrum must vary from point-to-point if we look at the spatial distribution. Although the spectrum of an H II region may be dominated by the areas with the highest surface brightness, it may be possible for a bright but obscured area to be given low weight, or for a region with a lower-quality spectrum to have an inflated surface brightness because of poorly-measured extinction. Thus the patchy nature of dust reddening must introduce additional uncertainty into measured line ratios. Spatially-resolved measurements are encouraged whenever possible.

The highly opinionated point here is that anyone who presents measured emission line strengths with uncertainties of 1% or less is probably not adding in all the error sources. Five percent uncertainties are probably more realistic for the brightest emission lines observed. Note that this level of precision is more than adequate for abundance measurements for most astrophysical problems.

2.1.2. *The Direct Method*

Direct abundance measurements can be made when one is able to measure the faint emission lines which are important diagnostics of electron temperature, T_e. The abundance of any ion relative to H^+ derived from the ratio of the intensity of a transition λ to the intensity of $H\beta$ is given by

$$\frac{N(X^{+i})}{N(H^+)} = \frac{I(\lambda)}{I(H\beta)} \frac{\epsilon(H\beta)}{\epsilon(\lambda)}, \tag{2.1}$$

where $\epsilon(\lambda)$ represents the volume emission coefficient for a given emission line λ. For collisionally-excited lines in the low-density limit, the analysis in section 5.9 of Osterbrock (1989) applies.

When T_e has been measured, the volume emission coefficient for a collisionally-excited line is given by

$$\epsilon(\lambda) = h\nu q_{coll}(\lambda) = \frac{hc}{\lambda} 8.63 \times 10^{-6}(\Omega/\omega_1)T_e^{-0.5}e^{-\chi/kT_e} \tag{2.2}$$

where Ω is the collision strength for the transition observed, ω_1 is the statistical weight of the lower level, and χ is the excitation energy of the upper level. Ω contains the physics in the calculation; it represents the electron-ion collision cross-section averaged over a Maxwellian distribution of electron velocities relative to the target ion at the relevant temperature. Thus Ω has a mild temperature dependence, which can introduce a trend in abundance ratios if not accounted for.

Note on collision strengths: the vast majority of these values are computed, not experimental. This does not mean that they have zero uncertainty! A recent example is given by the case of [S III] (Tayal & Gupta 1999). This new 27-state R-matrix calculation resulted in changes of approximately 30% in the collision strengths for optical and IR forbidden transitions from earlier calculations. This shows that even for commonly-observed ions the atomic data is still in a state of flux. Observers should take into account the probable uncertainty in atomic data when estimating errors in abundances.

Another thing to account for is the fact that ionized nebulae are not strictly isothermal. Because [O III] is usually the most efficient coolant, the thermal balance at any point in an H II region depends on the local abundance of O^{+2}, as well as the local radiation field. The *ion-weighted* electron temperature for a given ion can vary with respect to T(O III) in a predictable way (Garnett 1992), depending largely on the metallicity. Figure 2 shows a plot of measured electron temperatures for [O III], [S III], [O II], and [N II] compared with the relationships derived from model photoionized nebulae (solid lines). The measured temperatures show correlations which agree quite well with the model relations, although there is quite a bit of scatter in the [O II] temperatures, and there may be a slight offset between T[S III] and the predicted relation, which may be real or an observational artifact. These results indicate that the photoionization models provide a reliable predictor of the thermal properties of H II regions.

For recombination lines, the emission coefficient is given by

$$\epsilon(\lambda) = h\nu q_{rec}(\lambda) = \frac{hc}{\lambda} \alpha_{eff}(\lambda), \tag{2.3}$$

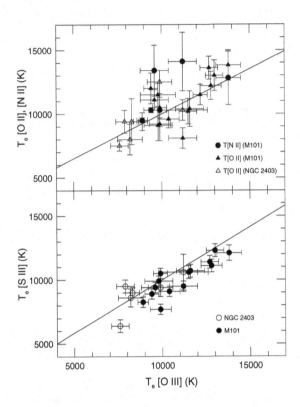

FIGURE 2. Comparison of electron temperatures derived from [O III], [O II], [N II], and [S III] measurements for H II regions in NGC 2403 and M101. The straight lines show the correlations predicted by photoionization models (Garnett 1992).

where $\alpha_{eff}(\lambda)$ is the "effective" recombination coefficient for the recombination line λ. α_{eff} incorporates the physics, including the cross-section for electron-ion recombination and the probability that a given recombination will produce the given emission line. α values for H vary as roughly T_e^{-1}; individual lines have mildly different T dependences, but recombination line ratios are only weakly dependent on T, and quite insensitive to n_e for densities less than 10^6 cm^{-3}.

Most astronomers are familiar with the bright H I Balmer and He I recombination lines in the optical spectrum of ionized nebulae. Heavier elements also emit a recombination spectrum, and O I, O II, C II, N I, N II and other permitted lines have been observed in PNs and the Orion Nebula. In principle, such recombination lines could yield more accurate abundances than the forbidden lines, because their emissivities all have roughly the same T dependence. In practice, the recombination lines scale roughly with element abundance, so even for O and C the RLs are typically fainter than 1% of Hβ, making them too faint to observe routinely in extragalactic H II regions. It is observed that recombination lines in some PNs give much higher abundances than the corresponding forbidden lines from the same ions (Liu et al. 1995, 2000; Garnett & Dinerstein 2001, 2002), and there is currently a raging debate over whether the recombination lines or the the forbidden lines provide more reliable abundances.

Measurements of infrared collisionally-excited fine-structure lines are gaining ground with the launch of the *ISO* spacecraft, and with the upcoming *SIRTF*, *SOFIA*, and *FIRST* missions. Recognizing that $\chi \approx$ 5-10 eV for UV forbidden lines, $\chi \approx$ 2-3 eV for optical forbidden lines, and $\chi <$ 0.2 eV for IR fine structure lines with $\lambda > 7\mu m$, we see that the exponential term in Equation 2.2 goes to nearly unity, and the IR lines have a weak temperature dependence. Thus it should be possible to determine accurate abundances free of concerns over temperature fluctuations. One caveat is that the very important [O III] and [N III] fine-structure lines are sensitive to density, suffering from collisional de-excitation at $n_e \approx$ 1000 cm^{-3}, so density fluctuations could introduce large uncertainties. Fine structure lines from Ne, S, and Ar in the 7-20μm range, however, are not so sensitive to density.

For extragalactic H II regions, the main limitations on IR observations so far have been small telescopes, high background, and short spacecraft lifetimes. Nevertheless, *ISO* is providing some information on H II regions in the Galaxy and other Local Group galaxies (and luminous starbursts), and the future missions promise even better data.

2.1.3. "Empirical" (Strong-Line) Calibrations

In many cases T_e can not be measured, either because the nebula is too faint or it is so cool that the temperature-sensitive diagnostic lines (for example [O III] λ4363) are too weak. Thus, there is interest in having an abundance indicator that uses the strong forbidden lines.

Pagel et al. (1979) identified the line intensity ratio

$$R_{23} = \frac{I([O\ II]\lambda 3727) + I([O\ III]\lambda\lambda 4959, 5007)}{H\beta} \tag{2.4}$$

as an indicator of O/H in H II regions. They noted, based on a sample of extragalactic H II regions, that the measured T_e, O/H, and R_{23} were all correlated. This works because of the relationship between O/H and nebular cooling: the cooling in the ionized gas is dominated by emission in IR fine-structure lines (primarily the [O III] 52μm and 88μm lines), so as O/H increases, the nebula becomes cooler. In response, the optical forbidden lines, especially the [O III] lines, become weaker as O/H increases (excitation goes down as T decreases).

The R_{23} vs. O/H relation is fairly well calibrated empirically (based on abundances using the direct method) for log O/H between -3.5 and -4.0 (Edmunds & Pagel 1984). For higher O/H, the strong-line method breaks down because few measurements of T_e exist; only two measurements have been made for H II regions with roughly solar O/H (Kinkel & Rosa 1994; Castellanos et al. 2002). In this regime, the relation has been calibrated using photoionization models (which I'll discuss later) that may have systematic errors. One other complication is that for log O/H < -3.8, the relation between R_{23} and O/H reverses, such that R_{23} decreases with decreasing abundance. The relation thus becomes double-valued, and at the turn-around region the uncertainties in O/H are much larger. This occurs because at very low metallicities the IR fine-structure lines no longer dominate the cooling because there are too few heavy elements. As a result the forbidden lines more directly reflect the abundances in the gas.

This double-valued nature of R_{23} has led some to seek other strong-line diagnostics. The ratio [O III]/[N II] (Alloin et al. 1979; Edmunds & Pagel 1984) has been promoted to break the degeneracy in R_{23}. This ratio does appear to vary monotonically with O/H, although the observational scatter generally is larger than for R_{23}. More recently, the emission line ratio

$$S_{23} = \frac{I([S\ II]\lambda\lambda6717, 6731) + I([S\ III]\lambda\lambda9069, 9532)}{H\beta} \qquad (2.5)$$

has been calibrated as an indicator of O/H by Díaz & Pérez-Montero (2000). S_{23} has the advantage of varying monotically over the range $-4.3 < \log$ O/H < -3.7 in which R_{23} becomes ambiguous. S_{23} does become double-valued for O/H > -3.4. Where this relation breaks down is uncertain at present because there are too few measurements. In addition, the ratio [N II]$\lambda6583$/Hα has been promoted as another possible measure of O/H (van Zee et al. 1998; Denicoló, Terlevich & Terlevich 2002). [N II]/Hα varies monotonically with O/H over the entire range over which it is calibrated, but the scatter is quite large, especially at low values of O/H in dwarf irregular galaxies. Note that S_{23} and [N II]/Hα are employed here as measures of the *oxygen* abundance, not sulfur or nitrogen and are calibrated by direct measurements of O/H. Thus, non-solar abundance ratios are not a concern.

At the same time, there are several limitations.

(*a*) None of these strong-line diagnostics is well calibrated for $\log$ O/H > -3.5. At higher metallicities, the calibration is largely derived from photoionization models.

(*b*) The accuracy of each of these calibrations is quite limited. For R_{23} the usual quoted uncertainty is ±0.2 dex, which is roughly the scatter; in the turnaround region, the uncertainty is significantly larger. The accuracy of S_{23} is probably about the same; although there are few data points to pin down the scatter at the present time. The scatter in [N II]/Hα is significantly larger, about ±0.3 dex; most of this scatter is real, not observational.

(*c*) The strong-line abundance relations are subject to systematic errors, because the forbidden-line strengths depend on the stellar effective temperature and ionization parameter as well as abundances. If a galaxy has a low star formation rate and only low-luminosity H II regions with cooler O stars, the empirical calibration could give a systematically different O/H than a galaxy with many of the most massive O stars and luminous giant H II regions.

2.1.4. *Photoionization Models*

Some have said that the use of photoionization models to estimate nebular abundances is the "last resort of scoundrels", so to speak. I admit that I have used photoionization models to commit offenses in the past. Since Grazyna Stasińska will cover the mechanics of photoionization modeling in detail in her lectures, I will confine my remarks to what I feel are major uncertainties and observational considerations that need to be addressed.

Gas-star geometry: This is an observational consideration, since the geometry influences the ionization parameter. The classical model of an H II region is a uniform sphere with a point source of ionizing photons. High-resolution images of real H II regions show that they are anything but this. The Orion Nebula is better modeled as a blister, with the brightest areas being the photoevaporating surface of molecular cloud. *Hubble Space Telescope* images of giant H II regions show them to resemble bubbles more than filled spheres, often with a surrounding halo of superbubbles. Young star clusters can often be found outside the main H II region associated with the superbubbles (Hunter et al. 1996). In many cases the ionizing cluster is not centrally condensed, but rather quite loose and extended (for example, NGC 604 and I Zw 18). Even the 30 Doradus nebula, which is dominated by the compact cluster R136a, also includes an extended distribution of O giants, supergiants and Wolf-Rayet stars, plus luminous embedded, possibly pre-main-sequence stars (see Walborn 1991, Bosch et al. 1999). To my knowledge, there has been no investigation of the effect of an extended distribution of ionizing sources on H II region spectra.

The effects of density and density variations are related to this problem. T_e, and thus the optical forbidden line strengths, are very sensitive to density because of collisional de-excitation of the far-IR fine structure cooling lines (Oey & Kennicutt 1993). This is most true for metal-rich H II regions, so, for example the relation between R_{23} and O/H from ionization models depends on the average density assumed. It is not clear yet that observed integrated densities from, say, the [S II] line ratio accurately reflect mean densities. It is highly likely that a range of densities is more representative of nebular structure, although a functional form is not yet known.

A great deal of work needs to be done in this area.

Wolf-Rayet stars: There are a number of myths about Wolf-Rayet stars and their influence on the H II regions. The first myth is that the presence of Wolf-Rayet stars indicates an age of at least 3 Myr for the ionizing OB association, which is the result obtained from stellar evolution and spectrum synthesis modeling. However, this idea is demonstrated to be not true specifically in the case of 30 Doradus, which contains numerous W-R stars, yet has a color-magnitude age of 1.9 Myr (Hunter et al. 1995). This contradiction is the result of a new population of hydrogen-rich W-R stars noted by de Koter, Heap & Hubeny (1997). Thus we need to reconsider our ideas about using W-R stars to constrain the ages of stellar populations.

The second myth is that W-R stars add a hard component of photons with energies greater than 54 eV to the ionizing radiation field. Again, this is a result from the combination of stellar atmosphere models for Wolf-Rayet stars and spectrum synthesis models. The reality is that only about 1 in 100 W-R stars emits significant amounts of radiation beyond 54 eV. Those W-R stars that are associated with nebular He II emission in nearby galaxies tend to be rare high excitation WN and WO types (Garnett et al. 1991). X-ray binaries are also implicated in nebular He II emission (Pakull & Angebault 1986). We do not yet understand the evolutionary status of these stars, so it is premature to predict them from the stellar evolution models. Indeed, a comparison of photoionization models with the spectral sequence of H II regions indicates that OB cluster models that include such hot W-R stars produce results that are not consistent with observed emission-line trends (Bresolin, Kennicutt, & Garnett 1999).

It is not clear that we know very well at all the ionizing spectral energy distribution of Wolf-Rayet stars, or of O stars for that matter. Stellar atmosphere models which incorporate stellar winds, departures from LTE and plane-parallel geometry, and realistic opacities are still in the development stage, and the effective temperature scale of O stars is still in flux (Martins, Schaerer, & Hillier 2002).

A great deal of work needs to be done in this area.

Dust: Dust has three major effects on the H II region spectrum. First, dust grains mixed with the ionized gas absorb Lyman-continuum radiation. Second, obscuration by dust is typically patchy; differential extinction between stars and gas can affect the emission line equivalent widths. Third, dust can affect the heating and cooling by emitting and recombining with photoelectrons.

The absorption cross-section for standard interstellar dust grains extends well into the EUV spectral region with a peak near 17 eV. Dust grains are thus quite capable of absorbing ionizing photons in the H II regions, and in fact can compete with H and He. When this occurs, the flux of Balmer line emission is reduced over the dust-free case. Figure 3 displays a set of ionization models showing the reduction in Hβ line emission over that expected from the number of ionizing photons for dusty H II regions. I have assumed

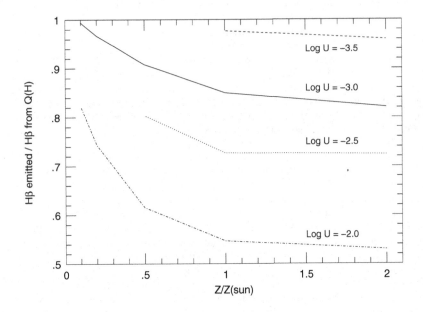

FIGURE 3. The ratio of emitted Hβ emission to that predicted from the stellar ionizing photon luminosity as a function of the nebular abundances, showing the effects of absorption of ionizing photons by dust grains. The models are for a stellar temperature of 40,000 K and assume a linear increase of the dust-to-gas ratio with metallicity. From Garnett (1999)

standard interstellar grains (Martin & Rouleau 1990), with a dust-to-gas ratio that varies linearly with metallicity over the range 0.1-2.0 solar O/H. The models show that grains can reduce the emitted Hβ flux by as much as 50%. The amount lost depends strongly on the ionization parameter, increasing for higher ionization parameters. A region with high U is likely to be a young one where the gas is close to the star cluster; thus more Hβ photons are missed, and EW(Hβ) reduced the most, for the youngest clusters.

Incidentally, the same phenomenon leads one to underestimate the number of ionizing photons. Therefore, claims of leakage of ionizing photons from H II regions, based on comparing N(Ly-c) from Balmer lines fluxes with that estimated from the OB star population, must be viewed with some skepticism.

Differential extinction between the stars and the gas can also affect EW(Hβ). Calzetti et al. (1994) found that the obscuration toward starburst clusters tended to be lower than that toward the ionized gas. They determined that, on average, A_V toward the stars was about one-half of that toward the gas. This is understandable if the stars have evacuated a cavity in the ionized gas through the combined effects of radiation pressure and stellar winds. The average derived obscuration for H II regions in spirals is $A_V \approx 1$ mag. If A_V(stars) is only 0.5 mag, then the observed EW(Hβ) will be about 40% lower than the intrinsic value.

These results suggest that dust effects can easily cause one to underestimate the intrinsic EW(Hβ), even for metallicities as low as 0.1 solar O/H. This would lead to a

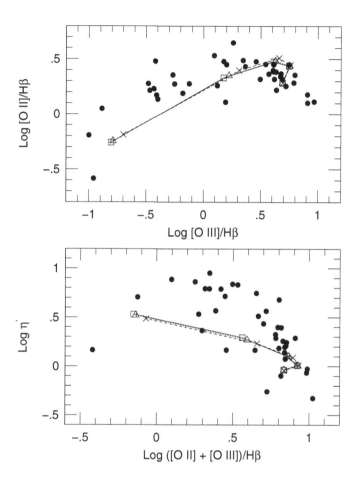

FIGURE 4. The effects of dust grains on forbidden-line strengths in H II regions. Three sequences of models are shown, with $T_{eff} = 40{,}000$ K and log $U = -3$. *Solid line plus squares:* dust-free models; *dashed line + crosses:* models with standard ISM grains; *dotted line plus triangles:* models with Orion-type grains. The effects of grains on the emission-line ratios are seen to be modest. From Garnett (1999)

systematic bias toward larger ages for the stellar population. One should therefore exercise caution in weighting EW(Hβ) as a constraint on the synthesis models.

By contrast, the effects of dust on the relative forbidden-line strengths are modest (Figure 4), except at high metallicities (Shields & Kennicutt 1995). One exception is in the case of very hot ionizing stars (for example the central stars of planetary nebulae), where ionization of grains can lead to additional photoelectric heating of the nebula (e.g., Ferland 1998, Stasińska & Szczerba 2001).

A great deal of work is needed in this area.

Note on atomic data for ionization calculations: the vast majority of the values used for photoionization and recombination cross-sections are computed, not experimental. This does not mean that their uncertainties are zero! In fact, the best values are probably not accurate to better than 15-20%. Thus, it is unreasonable to expect photoionization models to match real H II region spectra to an accuracy much better than this.

2.2. *Spectroscopy of Individual Stars*

Stellar spectroscopy has been a very valuable tool for studying the composition and evolution of stars in our Galaxy. Recent improvements in instrumentation and the construction of 8-10m telescopes has allowed this kind of work to be extended to other galaxies. It is not possible yet to do routine spectroscopy of F and G main sequence stars outside the Milky Way, so these studies have concentrated on A and B type supergiants or red giants. Nevertheless, detailed abundance studies of individual stars is not likely to extend far beyond the Local Group for some time because of telescope size limitations.

The supergiants are an important complement to spectroscopy of H II regions, since they sample similar spatial and temporal distributions. Furthermore, they overlap in many of the elements that can be studied: C, N, O, Ne, and so on. On the one hand, the supergiants provide information on elements such as Si, Fe and s-process elements that are depleted into grains in the ISM. On the other hand, the H II regions provide a valuable comparison for He, C, and N which may be affected by internal mixing and nucleosynthesis in the massive stars, which is covered by Norbert Langer's contribution. This is one of the important uncertainties in abundance studies for these stars; others include the degree to which conditions depart from LTE, and the effects of spherical geometry and stellar winds.

Spectroscopy of red giants is well established from Milky Way studies, as discussed by David Lambert. Red giants are valuable because they sample abundances over long time spans, from a 100 Myr to greater than 10 Gyr. A wide variety of elements can be studied, including α capture elements, Fe-peak elements, and neutron-capture elements. C, N, and O (and s-process elements in AGB stars) can be affected by internal mixing. Since red giants are fainter than H II regions or supergiants, detailed spectroscopy is limited to the Milky Way's satellite galaxies at present.

For metallicity distributions, one can examine lower spectral resolution diagnostics. The most useful of these has been the Ca II triplet indicator (Armandroff & Da Costa 1991), which uses the combined equivalent width of the Ca II triplet near 8500 Å, calibrated with metallicities of globular clusters, to infer the metallicity [Fe/H] (where the brackets denote the logarithmic abundance relative to that in the Sun). The main uncertainty of this method is that the Ca/Fe abundance ratio can vary depending on the star formation history, so the globular clusters may not provide the correct metallicity calibration for galaxies with a variety of star formation histories. Work needs to be done to calibrate the Ca II triplet with [Ca/H] rather than [Fe/H] to remove this ambiguity.

2.3. *Stellar Photometry and Color-Magnitude Diagrams*

Color-magnitude diagrams of galaxy populations can provide some information on the metallicity (or metallicity spread) of a stellar population, since features in the CMD, such as the color of the red giant branch, can vary with metallicity. Unfortunately, these features also vary with age of the populations, so there is a degeneracy between age and metallicity in the CMD (and in composite colors). Systematic uncertainties may

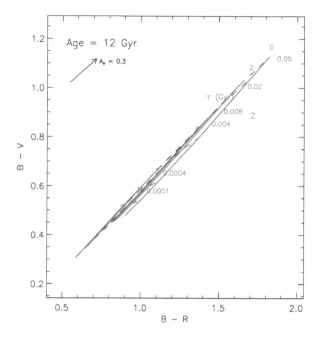

FIGURE 5. Synthetic B–R vs. B–V colors for disk galaxies based on population synthesis models. The colors are from Bruzual & Charlot models with an exponentially decaying star formation rate with timescale τ ranging from 0 Gyr to infinity and metallicity Z ranging from 0.0001 to 0.05; see Bell & de Jong (2000) for details of the models. Solid lines connect models of constant Z, while dashed lines connect models of constant τ. All galaxies start forming stars 12 Gyr ago. The effects of interstellar reddening by foreground screen with $A_V = 0.3$ magnitude is shown by the vector in the upper left corner. Diagram courtesy of Eric Bell.

also be introduced by variations in element abundance ratios and by reddening. Thus, color-magnitude diagrams are at best indicative of metallicities.

2.4. *Spectrum Synthesis of Stellar Populations*

Spectrum synthesis for deriving metallicities has been applied mostly to elliptical galaxies. Since I am not an expert on this, I refer the reader to Guy Worthey's review in Henry & Worthey (1999) and references therein for details. Needless to say, spectrum synthesis is an intricate and uncertain art; the results depend on the choices of spectral templates, element abundance ratios, and star formation histories. Worthey points out that 25 spectral indices are available to derive metallicities and ages for old stellar populations. (In 1986 there were 11 indices in the Lick spectral index system [Burstein, Faber, & González 1986].) Unfortunately, most of these vary in a degenerate way with age and metallicity of the stellar population. The best indices for breaking this degeneracy are (1) Hβ or a higher Balmer line, which are more sensitive to age; and (2) Fe4668 (which is actually a C_2 feature), more sensitive to metallicity. In addition, the Mg index (covering Mg b and Mg$_2$ features between 5150 and 5200 Å) provides information on Mg/Fe.

2.5. *Surface Photometry and Galaxy Colors*

Since the colors of stars are sensitive to both age and metallicity (as pointed out in Section 2.3 above) it is readily concluded that the colors of galaxies similarly depend on age and metallicity. The situation is a bit more complicated because in galaxies the colors represent composite stellar populations. A further complication is the ubiquitous presence of dust, which has a clumpy distribution mixed with the stars rather than a uniform screen. Nevertheless, the analysis of galaxy colors could provide a useful means of studying averaged ages and metallicities for stellar populations in very large samples of galaxies covering a wide range of redshifts, particularly galaxies that are too faint for spectroscopy.

As with stars, of course, a difficulty with using colors is that variations in age and metallicity cause similar variations in galaxy colors. This is especially true for optical colors, which have been known for some time to be almost completely degenerate with regard to variations in age and metallicity (see Figure 5). This degeneracy can be broken to a large extent by including IR photometry, particularly K-band surface photometry, as illustrated in Figure 6 from Bell & de Jong (2000). Bell & de Jong have exploited this property to derive *luminosity-weighted* mean ages and metallicities for a sample of low-inclination disk galaxies. Note that the luminosity weighting means that the derived properties do not represent the *present-day* metallicities, as the emission-line measurements do.

Note that because this method depends on synthesis models for the colors of the stellar population, it suffers the same limitations. The model colors depend on the star formation and chemical enrichment history of a given galaxy. At present very simple star formation histories are assumed: either an instantaneous burst or exponentially decaying continuous star formation (which approximates a constant star formation rate for very long decay timescales). These approximations may break down in cases of galaxies which have undergone multiple starbursts separated by long periods, or galaxies which have truncated star formation histories, possibly punctuated by starbursts as well. Dwarf galaxies, in particular, may not be well-reproduced by the synthesis models. Note also that the sensitivity of the color-color diagram decreases rapidly for very metal-poor or very old stellar populations.

Note also that the mean ages and metallicities of a given position are very sensitive to the dust correction, although the slopes of age and metallicity gradients are not. Finally, the uncertainties in photometry and sky subtraction grow as the surface brightness decreases, so inferred ages and metallicities become increasingly more uncertain in the outer parts of disks and in low surface brightness galaxies. In contrast, metallicity gradients derived from H II regions are more stable, as the luminosity of an H II region is largely independent of its position within a galaxy.

3. Abundances in Local Group Dwarf Elliptical Galaxies

Outside of the Magellanic Clouds, dwarf elliptical galaxies (dEs, sometimes also called dwarf spheroidals, or dSph) are the nearest companion galaxies to the Milky Way. The discovery of the tidally distorted Sagittarius dwarf elliptical (Ibata et al. 1994) brought the number of known dE companions for the Milky Way to nine. There may be more still hidden behind the Galaxy's obscuring dust layer. Recent studies have uncovered a host of dE companions of our sister galaxy M31 as well, and a few other more distant 'free-floating' dEs, such as the Cetus galaxy (Whiting et al. 1999) have been found

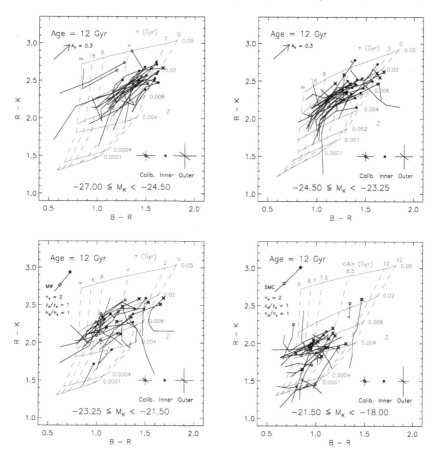

FIGURE 6. Synthetic B–R vs. R–K colors for disk galaxies from Bell & de Jong (2000); see this paper for complete details. Here the four panels show results for galaxies in different ranges of K-band absolute magnitude. The labels on the model grid are the same as in Figure 5; note that the bottom right panel shows the mean age < A > rather than the star formation timescale τ. The solid broken lines in each plot represent observed surface colors for disk galaxies in the Bell & de Jong sample; the attached open or filled circles are the central colors for each galaxy. The symbols in the bottom right of each panel show the calibration and sky subtraction error bars for the inner and outer annuli of a galaxy. Diagram courtesy of Eric Bell.

on Schmidt survey plates. Many of the properties of the Local Group dEs have been tabulated by Mateo (1998).

The dEs are deceptively simple stellar systems, with no young stars and apparently kinematically-relaxed stellar populations. Recent high-precision ground-based and HST CCD photometry have demonstrated that this is far from the truth. The dEs display a variety of complex multi-episode star formation histories. For example, Carina shows evidence for several distinct star formation events spread over several Gyr (Smecker-Hane et al. 1994); Sculptor and Fornax appear morphologically similar, but Sculptor appears to formed the bulk of its stars at an earlier time than Fornax (Tolstoy et al. 2001). Only

Ursa Minor appears to have something like a simple, monometallic stellar populations based on its color-magnitude diagram (Mighell & Burke 1999, but see below).

These star formation histories are of vital interest to understanding the evolution of the dEs, and they have raised some puzzles. Among these are the question of how, on the one hand, the dEs lost their gas, and how, on the other hand, they retained gas to experience multiple episodes of star formation! Combining the star formation history with the element enrichment history can, in principle, yield the information needed to understand the evolution of dwarf galaxies and their contribution to enrichment of the IGM.

3.1. *Metallicities*

In the absence of spectroscopy, metallicities for dEs have been derived from their CMDs by comparing the colors of the red giant branches (RGB) with those of globular clusters. This is not entirely satisfactory because of the age-metallicity degeneracy in RGB colors, and because element abundance *ratios*, which also affect RGB colors, may not be the same in the dEs as in the globulars. A few efforts have been able to estimate metallicities from low-resolution spectra obtained with 4-5 meter telescopes; these studies are summarized in Mateo (1998). The photometric and spectroscopic studies show the dEs to have quite low $<[Fe/H]>$, ranging from about -1.3 in Fornax to about -2.2 in Ursa Minor. Most of the dEs show evidence for a significant spread in [Fe/H], $\sigma([Fe/H]) = 0.2$-0.5 dex, based on the width in color of the RGB (disregarding possible age spread contributing to this).

With several 8-10 meter telescopes now available, medium-resolution ($R \approx 5,000$) spectroscopy of RGs in the Milky Way satellite dEs can be almost routinely done, while high-resolution spectroscopy ($R > 15,000$) is possible for the brightest giants. These facilities offer exciting new possibilities for understanding the evolution of the Local Group dEs. One example of what can be done is the VLT study of Tolstoy et al. (2001), who obtained Ca II triplet measurements for 37 red giants in Sculptor, 32 RGs in Fornax, and 23 RGs in the dI NGC 6822. Having measured metallicities for individual stars and existing CMDs for these galaxies, it was possible for Tolstoy et al. to assign ages to each star directly by comparison with isochrones of the proper metallicity, and subsequently to derive the time evolution of both the star formation rate and the metallicity. This is shown for Sculptor in Figure 7 taken from Tolstoy et al. The results are consistent with an initial burst of star formation between 11 and 15 Gyr ago with a metallicity [Fe/H] $\approx$ -1.8, a sharp subsequent decline in the SFR, followed by a slow decline in star formation until it stops approximately 5 Gyr ago. The metallicity evolution is very modest, with the youngest stars having a mean [Fe/H] of only -1.4. Their data for Fornax, on the other hand, show a low star formation rate over the time period 10-15 Gyr ago, then a sharply increased rate over the next few Gyr, and a higher mean [Fe/H] of -0.7 for the youngest stars.

The main source of uncertainty here is that the Ca/Fe ratio in these stars may not reflect that of the calibrator globular clusters stars - and the Ca/Fe ratio may actually vary with time within each galaxy! Nevertheless, with more data like this for the Local Group dwarfs it should be possible to derive very accurate evolution histories for these galaxies.

3.2. *Element Ratios*

Element abundance ratios are another important piece of information, since the various elements are synthesized in stars with different masses and lifetimes. The α-element/Fe ratio, in particular is a good diagnostic, because the α-capture elements are produced

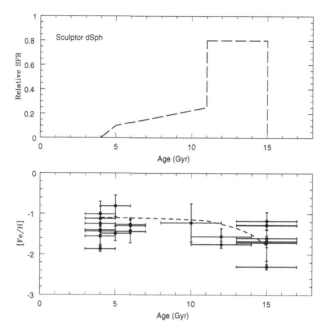

FIGURE 7. The star-formation and metallicity evolution history for Sculptor derived by Tolstoy et al. (2001) from Ca II triplet measurements and photometry of red giants. The upper panel shows a schematic plot of how the star formation rate may have varied over time. The lower panel shows the corresponding variation in metallicity over the same time period (*dashed line*). Overploted on the lower panel are the Ca II triplet measurements for individual Sculptor giants, with ages determined using isochrones.

mainly in very massive stars and expelled into the ISM by Type II or Ib,c supernovae, while Fe is mainly produced in Type I supernovae by longer-lived stars. Thus the α/Fe ratio is an indicator of how rapidly star formation and metal enrichment occurred within a system: high α/Fe indicates enrichment over short time scales, possibly in starburst events, while low α/Fe may indicate enrichment over much longer time scales, as in systems with roughly constant star formation rate over their lifetimes.

Of current interest is the question of whether the Galactic halo was formed in a monolithic collapse (Eggen, Lynden-Bell & Sandage 1962) or was aggregated from mergers of smaller sub-units (Searle & Zinn 1978). It is speculated that the nearby dE satellites may be representative of those sub-units.

It is clear from their CMDs that many of the Milky Way satellites are not like the halo population, since they contain stars that are much young than those in the halo. At the same time, we know that the Sagittarius dE is being tidally disintegrated by the Galaxy and will become part of the halo. Although systems like Fornax and Carina appear not to be representative of the current halo populations, much less complex systems like Ursa Minor and Draco could be similar to the structures out of which the halo formed.

The test of this possibility is that the ages and compositions of stars in the dEs are similar to those in the halo. It is known that halo stars show elevated [α/Fe] compared to disk stars, reflecting a dominant nucleosynthesis contribution from massive stars, while

[Ba/Eu] is low in halo stars, indicating that the s-process (which is the main source of Ba) has not had sufficient time to contribute to the abundances in halo stars, while the r-process (the main source of Eu) dominates in metal-poor stars. Indeed, very metal-poor halo giants show evidence for a purely r-process contribution to the abundances of heavy neutron capture elements (Sneden et al. 2000).

Little high-resolution spectroscopy of giants in even the nearest dEs have been carried out because of the faintness of the stars (16th-20th magnitude). However, the first such studies have become available due to the availability of the 10-m Keck telescopes (Shetrone, Coté & Sargent 2001). Shetrone et al. have obtained high-resolution spectra for 5-6 stars in each of the Draco, Ursa Minor, and Sextans galaxies, deriving abundances for a variety of elements. The comparison of element abundance ratios in these stars show a puzzling mixed bag of results. Although Shetrone et al. argue that [α/Fe] is low in the dEs compared to halo stars, closer examination shows that [Ca/Fe] and [Ti/Fe] do appear to be lower, but [Mg/Fe] and [Si/Fe] appear to agree with halo ratios. Meanwhile, [Ba/Eu] values in the dEs appear to be in good agreement with those in halo stars. The comparison between dEs and halo stars seems to be inconclusive at the present time, perhaps not surprising given the small samples of stars at present. The samples of stars for each galaxy certainly need to be enlarged to determine the evolution of element ratios in these galaxies. Nevertheless, the Shetrone et al. (2001) study illustrates the power of high-resolution spectroscopy with the new large telescopes. More work of this nature is highly encouraged.

4. Abundance Profiles in Spirals and Irregulars

Here I present an overview of the patterns of metallicity and element abundance ratios observed in spiral and irregular galaxies. I will discuss the results for both types of galaxies rather than separately; many aspects can be discussed for the combined groups, although there are a number of differences that could constitute the topic of an entire conference alone.

The observational data to be discussed represent a highly selected sample of abundances, gas masses, and stellar photometry from sources too numerous to mention here. (If you recognize your data in the following plots, feel free to take credit.) I will employ abundances derived almost exclusively from H II region spectra, since they contribute the largest set of abundance data for spirals and irregulars in the local universe.

4.1. *Gas and Stellar Masses*

The ultimate goal of any chemical evolution model is to account for the global and local metallicity within a galaxy, the gas and stellar mass distributions, and the stellar luminosity self-consistently. Thus, any discussion of abundances and chemical evolution should include a few words about observational determinations of gas and stellar masses and mass surface densities.

4.1.1. *Neutral and Molecular Gas*

Neutral gas is the largest component of gas in most spirals and irregulars, as determined from H I 21-cm hyperfine line measurements. The 21-cm line has been well-mapped in many nearby galaxies. With regard to determining gas fractions and surface densities, a few points should be kept in mind:

(*a*) *The size of the neutral gas disk is often much larger than the stellar disk in spirals and irregulars*, as much as 3-4 times the photometric radius R_{25} (e.g., Broeils & van Woerden 1994). One must obtain or use maps which cover the full extent of the H I disk, which can mean observing over degree-size scales for the nearest spirals. Determining the H I extent may be difficult in complicated interacting systems such as the M81 group (Yun et al. 1994).

(*b*) *Fully sampled maps are desirable.* Aperture synthesis maps, while providing the high spatial resolution needed for studies of kinematics and gas structure, can miss a large fraction of the H I emission on scales larger than the synthesized beam due to lack of short antenna baselines in the $u - v$ plane. (The closest spacings in an interferometer are, of course, one antenna diameter.) This is especially true for the highest resolution images. Although there are efforts to correct for the missing extended emission by including single-dish measurements, it is usually assumed that the extended emission is uniformly distributed. This assumption may not be correct.

(*c*) *The helium contribution to the mass is not negligible.* The helium accounts for 30-40% of the total gas mass, which must be included in the total gas mass and surface density.

Fortunately, many nearby galaxies have been well mapped in H I at kiloparsec scale resolution or better.

Molecular gas, mostly in the form H_2, is the important phase associated with star formation. Although H_2 may not dominate the total gas mass, it is often found to be the main component in the inner disk of Sbc or later type spirals, and so can be the main contribution to the gas surface density in such regions.

H_2 has no dipole moment and thus emits no strong dipole radiation of its own. The usual tracer of molecular gas is the abundant CO molecule, typically the ^{12}CO (J=1-0) millimeter-wave transition. The conversion from the measured I(CO) to the column density $N(H_2)$, X(CO), must be calibrated largely without the benefits of direct measurement of H_2 column densities. The result has been a long-standing controversy over the value of the CO-H_2 conversion factor and its dependence on metallicity. For example, Wilson (1995) has studied the CO-$N(H_2)$ relation in a variety of environments in Local Group galaxies, comparing $I(CO)$ with molecular cloud masses derived from the velocity dispersion assuming the clouds are in virial equilibrium. From her data Wilson found a roughly linear relation between X(CO) and 12 + log O/H corresponding to approximately a factor ten increase in X(CO) for factor ten decrease in O/H, from solar O/H to 0.1 times solar O/H. On the other hand, Israel (1997a,b) argues that virial equilibrium is a poor assumption for short-lived molecular clouds. He instead uses the FIR dust emission surface brightness and H I maps to determine the dust/N(H) ratio, then uses the FIR surface brightness to estimate $N(H_2)$ in regions where CO is detected. With this method Israel derives a variation in X(CO) with O/H which is much steeper than that obtained by Wilson: a factor of approximately 100 decrease in X for a factor 10 increase in O/H.

Both of these methods likely suffer from errors due to assumptions made. Virial equilibrium may very well be a poor assumption for molecular clouds. On the other hand, the FIR calibration depends on the assumption that the dust-to-gas ratio is the same in neutral gas and in molecular clouds; however, grains may be preferentially destroyed by shocks in the lower density neutral component. The FIR model is also very sensitive to the dust temperature. The relation between I(CO) and $N(H_2)$ likely depends on a variety of factors besides metallicity (Maloney & Black 1988). More work needs to be done to determine the best method to obtain H_2 masses.

4.1.2. *Stellar Mass Densities*

Masses and surface densities for the stellar component in galaxies are probably even more uncertain than molecular gas masses. Because of the flat rotation curves of disk

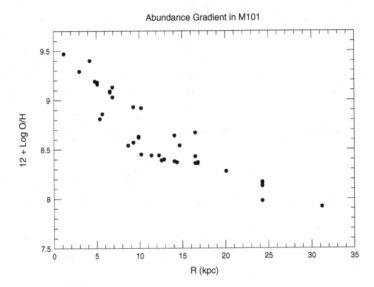

FIGURE 8. The gradient in O/H across the disk of the spiral galaxy M101 vs. galactocentric radius (Kennicutt & Garnett 1996).

galaxies, and the consequent inference that the galaxies are dominated by non-luminous, non-baryonic matter, the mass-to-light (M/L) ratio and mass surface density of the stellar component can not be derived from galaxy dynamics. Unless one can count the stars in a region directly (possible only for very nearby systems), it is necessary to infer M/L for the stellar component by indirect means. This is difficult because the luminosity and colors of composite stellar populations depend on both the star formation history and the metal enrichment history.

Nevertheless, recent work by Bell & de Jong (2001) indicates that the stellar M/L ratios of galaxies are rather robustly related to their colors. Bell & de Jong examined population synthesis models for galaxies assuming a variety of star formation histories. They found that M/L for the stellar component correlated very well with optical colors, although there is some scatter in the correlations. IR colors did not correlate as well with M/L because of the strong metallicity dependence of the IR luminosity of giants. The B-band M/L ratio shows a steep correlation with color, while the K-band M/L ratio shows a much less steep correlation (a factor three increase between B–R = 0.6 and B–R = 1.6, compared to a factor 10 increase in B-band M/L over the same color range). If the population synthesis models can reliably reproduce the colors and spectra of real galaxies, this method offers the possibility of greatly improved estimates of masses and mass surface densities for the stellar components of disk galaxies.

4.2. *Spatial Abundance Profiles*

The spatial distribution of abundances in galaxies depends coarsely on the Hubble type. Spectroscopic study of H II regions in unbarred or weakly barred spiral galaxies typically reveals a strong radial gradient in metallicity (Figure 8), as determined from O/H (Zaritsky, Kennicutt, & Huchra 1994; ZKH; Vila-Costas & Edmunds 1992; VCE). The derived O/H can drop by a factor of ten to thirty or fifty from the nucleus of a galaxy to the outer disk as demonstrated in galaxies with well-sampled data. In those spirals with spectroscopy of more than ten H II regions covering the full radial extent of the disk, there is little evidence that O/H gradients deviate from exponential profiles. Irregular galaxies, by contrast, show little spatial variation in abundances, to high levels of precision (Kobulnicky & Skillman 1996), indicating a well-mixed ISM. The data for strongly barred spiral galaxies shows evidence that their O/H gradients are more shallow than in unbarred spirals. I will discuss these galaxies in more detailed in a later section.

A quick glance at data on O/H in galaxies (e.g. Figure 8 of ZKH) does not immediately reveal any trends of metallicity among galaxies of different types. However, detailed examination of this data shows that there are significant correlations between abundances and abundance gradients in spirals and irregulars with galaxy structural properties. Here I shall review some of these correlations and some implications. Note also that for the most part this discussion applies only to high surface brightness "normal" spirals.

4.3. *Metallicity versus Galaxy Luminosity/Mass*

One well established correlation is the relation between metallicity and galaxy luminosity or (Garnett & Shields 1987, Lequeux et al. 1979). This is shown in the top panel of Figure 9, where I plot O/H determined at the half-light radius of the disc (R_{eff}) versus B-band magnitude M_B. This is the usual way of plotting the relation. [Note that the choice of what value of metallicity to use for spiral disks, where the metallicity is not constant, is somewhat arbitrary. I have used the value at one disk scale length in the past on the grounds that the disk scale length is a structural parameter determined by galaxy physics, whereas the photometric radius can be biased by observational considerations. The actual "mean" abundance in the disk ISM would be determined by convolving the abundance gradient with the gas distribution. As a simple compromise I have used the disk half-light radius, which is 1.685 times the disk scale length (de Vaucouleurs & Pence 1978), and so is still connected to galaxy structure. It should be noted also that a O/H - M_B correlation is derived whether one uses the central abundance, the abundance at one disk scale length, or some other fractional radius.] ZKH noted the remarkable uniformity of this correlation over 11 magnitudes in galaxy luminosity, for ellipticals and star-forming spiral and irregular galaxies.

To the extent that blue luminosity reflects the mass of a system, the metallicity-luminosity correlation suggests a common mechanism regulating the global metallicity of galaxies. What the mechanism might be is not very well understood at present. The most commonly invoked mechanism is selective loss of heavy elements in galactic winds (e.g., Dekel & Silk 1986). However, the metallicity-luminosity correlation for star-forming galaxies by itself does not imply that lower-luminosity galaxies are losing metals. Such a correlation could occur if there is a systematic variation in gas fraction across the luminosity sequence, either because the bigger galaxies have evolved more rapidly, or because the smaller galaxies are younger. In fact, there is evidence that both of these may be true. There is also evidence for fast outflows of hot X-ray gas from starburst galaxies such as M82 (Bregman, Schulman, & Tomisaka 1995). However, the question of whether this hot gas is escaping into the IGM or will be retained by any given galaxy depends not just on the gravitational potential, but also on details such as the vertical distribution of ambient gas and the radiative cooling which are not so well understood.

The question of loss of metals from galaxies is profound because of the existence of metals in low column density Lyman-α forest systems (Ellison et al. 1999), which are probably gas clouds residing outside of galaxies. Where the heavy elements came from in these systems is still a mystery; it is possible that they were seeded with elements

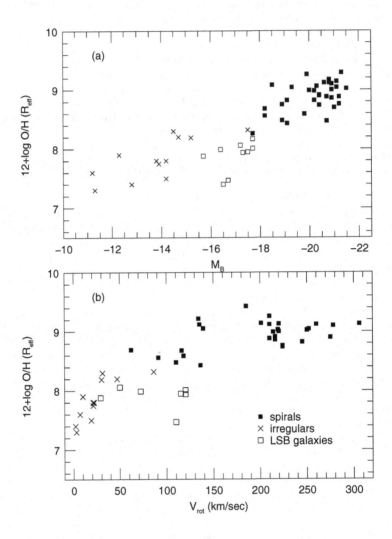

FIGURE 9. Top: The correlation of spiral galaxy abundance (O/H) at the half-light radius of disk from the galaxy nucleus vs. galaxy blue luminosity. Bottom: abundance versus maximum rotational speed V_{rot}.

from a generation of pre-galactic stars or with elements expelled in starbursts during galaxy formation. In the dense environments of galaxy clusters, ram pressure stripping by intracluster gas, tidal interactions, and galaxy mergers may also liberate material to

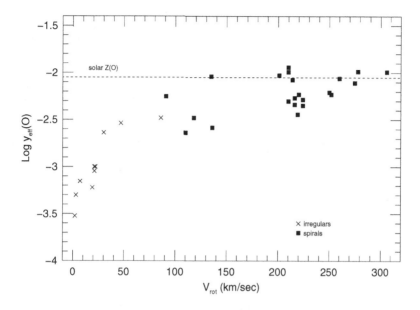

FIGURE 10. Effective yields y_{eff} for nearby spiral and irregular galaxies versus rotation speed V_{rot} (Garnett 2002). Filled squares represent the data for spirals while the crosses show the data for irregulars.

the intracluster medium (Mihos 2001). It is therefore a useful exercise to investigate what kinds of galaxies are potential candidates for ejecting heavy elements into the IGM.

I begin by looking at the metallicity-luminosity relation in a different way. It is often argued that the B-band luminosity is not a very good surrogate for mass, since the B-band light can be affected by recent star formation and dust. Therefore, in Figure 9(b) I plot the mean O/H for the galaxies in Fig. 9(a) versus rotation speed (obtained from resolved velocity maps [e.g., Casertano & van Gorkom 1991, Broeils 1992], not single-dish line widths), where for the spirals the rotation speed is taken to be the value on the flat part of the rotation curve. Interestingly, the Z-V_{rot} correlation turns over and flattens out for $V_{rot} \gtrsim 150$ km s-1, suggesting that spirals with rotation speeds higher than this have essentially the same average metallicity. Does this indicate a transition from galaxies that are likely to be losing metals to the IGM to galaxies that essentially retain the metals they produce?

This question can be examined further by studying metallicity as a function of gas fraction. In the context of the simple, closed box, chemical evolution, Edmunds (1990) derived a few simple theorems that show that outflows of gas and inflows of metal-poor gas cause galactic systems to deviate from the closed box model in similar ways. Specifically, defining the effective yield y_{eff}

$$y_{eff} = \frac{Z}{ln(\mu^{-1})},$$
(4.6)

outflows of any kind and inflows of metal-poor gas tend to make the effective yield smaller than the true yield of the closed box model. y_{eff} defined this way is an observable quantity, and provides a tool for studying gas flows in galaxies. Although the true yield is relatively uncertain, comparing effective yields for a sample of galaxies can provide information on the relative importance of gas flows from one galaxy to another.

Such a comparison is presented in Figure 10 (Garnett 2002), where I show data compiled on abundances, atomic and molecular gas, and photometry for 22 spiral and 10 nearby irregular galaxies. Figure 10 plots the effective yields derived for each galaxy using equation 3.1 and the global gas mass fraction versus galaxy rotation speed. The plot shows a very strong systematic variation in y_{eff} with V_{rot} in the dwarf irregular galaxies, with y_{eff} increasing asymptotically to a roughly constant value for the most massive galaxies. The uncertainties in the individual y_{eff} values are relatively large, because of relatively large uncertainties in M/L ratios for the stellar component and the CO - H_2 conversion for the molecular gas component; individual values of y_{eff} are probably not known to better than a factor of two. Nevertheless, the data show a factor of 30 systematic increase in y_{eff} from the least massive irregulars to the most massive spirals.

This result is striking verification that the yields derived for dwarf irregulars are significantly lower than in spiral galaxies, and shows that the variation is a systematic function of the galaxy potential. In strict terms, the trend in Figure 10 does not distinguish between infall of unenriched gas and outflows as the cause. However, the trend toward small y_{eff} in the least massive galaxies suggests that it is the loss of metals in galactic winds that drives the correlation. It would be of interest to use this correlation to estimate the total amount of gas lost in the small systems, and to determine the manner in which supernova energy feeds back into the ISM of the host galaxies.

Although not quite certain yet where one can say that galaxies are losing significant quantities of metals and which ones retain essentially all their metals, it appears likely that this boundary point is somewhere near $V_{rot} \approx 150$ km s^{-1}. Given this, one can surmise that the outflow of hot gas seen in the starburst galaxy M82 ($V_{rot} \approx 100$ km s^{-1}) may contribute significantly to enrichment of the IGM, while the hot gas flow seen in NGC 253 ($V_{rot} \approx 210$ km s^{-1}) is likely to remain confined to the galaxy.

4.4. Abundance Gradient Variations

Another commonly-noted correlation is illustrated in the top panel of Figure 11: the steepness of abundance gradients (expressed in dex/kpc) decreases with galaxy luminosity. However, more luminous galaxies have larger disk scale lengths, and so if one looks at the gradient per disk scale length (Figure 11, bottom panel), the correlation goes away. Interestingly, when one considers the errors in the computed gradients (25% is a typical uncertainty), then the scatter in measured gradient slopes may be consistent with purely observational scatter. Combes (1998) has suggested that a "universal" gradient slope per unit disk scale length may be explained by so-called "viscous disk" models (Lin & Pringle 1987); if the timescale for viscous transport of angular momentum is comparable to the star formation timescale (with the two timescales connected through the gravitational instability perhaps), such models naturally produce an exponential stellar disk. Chemical evolution models invoking viscous transfer have been able to produce abundance gradients (Clarke 1989; Tsujimoto et al. 1995), but it is not yet clear from the few models examined that they can yield very similar abundance gradients per unit disk scale length for a wide variety of spiral disks. This is something that deserves further study.

Another interesting possibility is presented by Prantzos & Boissier (2000). They con-

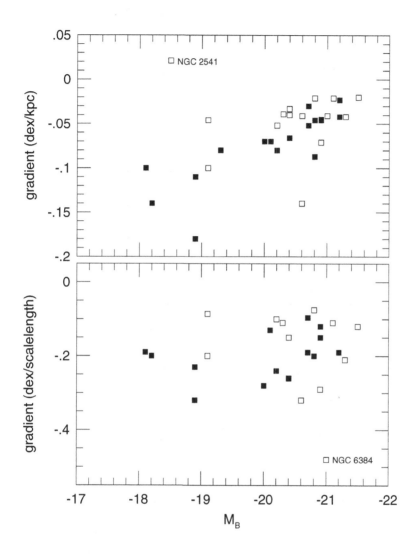

FIGURE 11. The correlation of abundance gradient vs. M_B, from Garnett et al. (1997a). The upper panel shows abundance gradients per kpc, while the lower panel shows gradients per unit disk scale length.

structed a sequence of chemical evolution models for disk galaxies by scaling the mass distribution (total mass, scale length) according to the scaling relations for cold dark matter halos of Mo, Mao & White (1998), in which the disk mass profile can be expressed using only two parameters: the maximum circular velocity (which corresponds to the

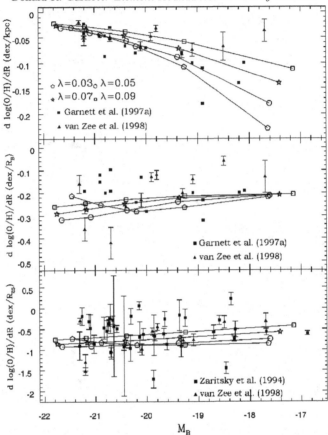

FIGURE 12. Comparison of generalized chemical evolution models from Prantzos & Boissier (2000) with observed composition gradients for spiral galaxies. Top: O/H gradient in dex/kpc vs. M_B. Middle: O/H gradient in dex per unit disk scale length vs. M_B. Bottom: O/H gradient over the photometric radius R_{25} vs. M_B. The curves show the model relations for constant λ for $\lambda = 0.03$, 0.05, 0.07, and 0.09.

halo mass) and the spin parameter λ (which corresponds to the angular momentum). A key assumption is that the scaled galaxies settle into exponential disks. Prantzos & Boissier computed models for galaxies under these assumptions, calibrated by reproducing measurements for the Milky Way. The basic results are illustrated in their Figure 4, reproduced here in Figure 12. The top panel plots the slope of the composition gradients in units of physical length (kiloparsecs) vs. M_B, the middle panel plots the gradients per unit disk scale length, while the bottom panel plots the gradient over the photometric radius R_{25}. In each case the model gradients are in good agreement on average with observed gradients, although the gradients per unit scale length and R_{25} are perhaps a bit steeper than observed. If this analysis holds up it would be quite remarkable, as it would have been difficult to imagine that the present-day ISM composition could be related to the initial properties of the disks in the distant past. This may reflect the assumption that the baryons start out with exponential mass distributions. Prantzos

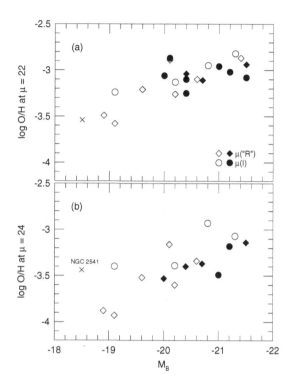

FIGURE 13. O/H at fixed value of galaxy surface brightness vs. M_B. Top: abundances at 22 mags arcsec^{-2}. Bottom: abundances at 24 mag arcsec^{-2}. More luminous spirals have higher abundances at a fixed surface brightness. Open symbols are from Garnett et al. (1997a); filled symbols represent additional data obtained from the literature.

& Boissier (2000) predict that there should be a small spread in observed gradients per kiloparsec in massive spirals and a large spread in small spirals, such that small spirals with large angular momentum should have shallower gradients (larger scale lengths) than those with lower angular momentum.

4.5. Metallicity vs. Surface Brightness

The uniformity of abundance gradients as a function of scale length suggests a close correlation between metallicity and disk surface brightness. Indeed, McCall (1982) and Edmunds & Pagel (1984) noted a remarkably tight correlation between O/H and disk surface brightness for late-type spirals. This has provided part of the motivation for models of self-regulated star formation, in which the radiation and mechanical energy produced by stars feeds back into the surrounding ISM and acts to inhibit further star formation. Models of this kind have been explored by Phillips & Edmunds (1991) and Ryder (1995), and appear to do a good job of reproducing the trends of both star formation rate and O/H with surface brightness. One caveat is that the interaction of the stellar energy output with the ISM is still poorly understood. Viscous disk models also provide a possible mechanism to tie the abundances to the underlying surface density distribution.

Edmunds & Pagel (1984) also noted that early-type spirals do not follow the same O/H-surface brightness correlation as the late types. Garnett et al. (1997a) put this on a more quantitative basis. Figure 13 displays the characteristic metallicity at two fixed values of disk surface brightness for a sample of spirals having either I- or R-band surface photometry; these bandpasses presumably sample the light from the old disk population better than B. The figure shows that metallicity-luminosity correlation appears to hold at all values of surface brightness across spiral disks. This result argues for two modes of enrichment in disk galaxies: a local mode, in which the metallicity is connected to the local mass density, and a global mode, in which an entire galaxy is enriched in a manner dependent on its total mass. One can imagine a global enrichment event which raises the metallicity of a galaxy to some level which depends on total mass, followed by sequential local enrichment which follows the mass density distribution. One caveat is that M/L, and thus the mass surface density at a given surface brightness, may vary systematically along the luminosity sequence in Figure 13. A more comprehensive study of mass surface density and gas fraction along this sequence should prove enlightening.

4.6. *Barred Spirals*

Bars are interesting because the gravitational potential of a bar is expected to induce a large-scale radial gas flow, possibly through radiative shocking and subsequent loss of angular momentum as the gas passes through the bar (Barnes 1991). The radial flow could significantly alter the metallicity distribution by mixing in gas from outer radii, thus weakening composition gradients, and is often argued to be the means to fuel the nuclei of active galaxies. The evidence so far accumulated indicates that barred spirals generally do have shallower composition gradients than weakly-/non-barred spirals (ZKH; Martin & Roy 1994). Martin & Roy (1994) have argued that the slope of the composition gradient correlates with both bar length and bar strength, defined as the ratio of bar length to width. This is illustrated in Figure 14, which shows the slope of the O/H gradient per kpc versus bar length a relative to the photometric radius (*top panel*) and versus bar strength $E_B = 10(1-b/a)$, where b is the bar width. Non-barred, barred, and irregular galaxies from the Martin & Roy sample are distinguished by different symbols, and the gradients have been adjusted for new distances to the galaxies. The plots show that (1) the non-barred spirals show a wide range of values for gradient slopes, although on average they are steeper than those for barred spirals; and (2) the O/H gradients for the barred spirals tend to get shallower with increasing bar length and bar strength. On the other hand, the trend depends how the gradients are scaled; if one plots the O/H gradient per unit scale length instead, the correlations disappear.

Curiously, some barred spirals (e.g., NGC 1365; Roy & Walsh 1997) show an O/H gradient within the bar, with flattening only outside the bar. One way to explain this is if strong star formation occurs in the bar, building up abundances faster than the radial flow can homogenize them (Friedli, Benz, & Kennicutt 1994).

4.7. *Spiral Bulges*

Abundances in bulges generally are obtained the same way as for ellipticals, by spectroscopy and modeling of absorption line indices from the integrated stellar population, and so suffer from the same uncertainties. Most of the line indices in the Lick system vary in the same way with age and metallicity, and so poorly distinguish between the two in model grids. A few, such as Hβ, Mg b, Fe5270, and Fe5335, do provide some ability

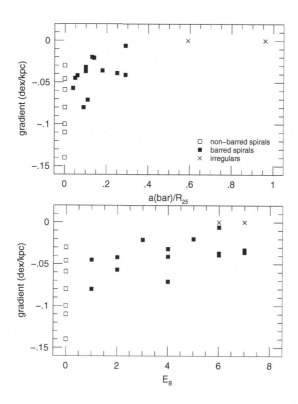

FIGURE 14. *Top:* O/H gradient per kpc vs. bar length a relative to the photometric radius R_{25}. Filled squares are barred spirals, unfilled squares non-barred spirals, and crosses are irregular galaxies. Bottom: O/H gradient vs. bar strength $E_B = 10(1-b/a)$, where b/a is the ratio of bar minor and major axes. Plot adapted from Martin & Roy 1994.

to separate age and metallicity, and have been used to obtain estimates of ages, [Fe/H], and [Mg/Fe] in spheroidal systems (Worthey 1998).

Most work on abundances in bulges have come from spectroscopy of the central regions. Jablonka, Martin, & Arimoto (1996) found that Mg_2 correlated with both bulge luminosity and stellar velocity dispersion in spirals with T = 0-5, but that the Fe5270 feature did not. Comparing with a grid of synthetic spectra with non-solar [α/Fe] they inferred that [Mg/Fe] increased with bulge luminosity as well. Similar results are obtained for ellipticals (Worthey 1998). Maps of spectral line indices obtained with integral field units (Peletier et al. 1999, de Zeeuw et al. 2002) show hints that Mg/Fe varies with radius, although only a small number of galaxies have been analyzed so far.

High-resolution spectroscopy of red giants in our own Galactic bulge recommends caution in interpreting line indices in integrated spectra of ellipticals and bulges. Giants in the Baade's Window region show high α/Fe ratios and a mean [Fe/H] ≈ -0.25, similar to [Fe/H] for solar neighborhood stars (McWilliam & Rich 1994). The mean metallicity is lower by about 0.3 dex compared to low-resolution spectroscopic and photometric determinations. This has several implications. Enhanced Ti/Fe ratios make the spectral types of the bulge giants later for the same IR colors compared to stars with solar abundance ratios. Enhanced α/Fe alter both stellar line indices and the location of isochrones in population synthesis models, which have been computed using solar abundance ratios so far. Thus, metallicities derived for spheroids may be overestimated by a factor of two or so. Most population synthesis Trager et al. (2000a, 2000b) attempt to correct the models for the effects of non-solar element abundances ratios. Such corrections are non-trivial, as both isochrones and line strengths are affected. No such corrections have been applied to bulges yet.

The formation of bulges is still mysterious. The candidate mechanisms are: monolithic collapse with rapid star formation (Eggen, Lynden-Bell, & Sandage 1962); mergers of roughly equal mass objects in hierarchical clustering models for galaxy formation (Baugh, Cole, & Frenk 1996: Kauffmann 1996); and secular growth from disk material, for example by mass transfer via bars (Combes et al. 1990; Hasan, Pfenniger, & Norman 1993). High α/Fe (that is, higher than the solar ratio) would tend to favor the models with rapid bulge formation from relatively metal-poor gas, because most of the Fe is expected to come from Type Ia supernovae. Solar or less α/Fe would tend to favor secular evolution from already enriched material, or star formation extended over times scales greater than 1 Gyr. The correlation of Mg/Fe with bulge luminosity and velocity dispersion suggests that a mix of formation mechanisms are at work; moreover, Andredakis et al. (1995) have found that bulge structure parameters correlate with Hubble type and bulge luminosity, such that large, luminous bulges tend toward $R^{1/4}$ profiles similar to ellipticals, while small bulges tend to have exponential profiles (although de Jong 1996 argues that bulges have exponential profiles in general). The trends in Mg/Fe and bulge shape together suggest that large bulges formed rapidly at early times, while small bulges may have formed more slowly via secular processes.

4.8. *Cluster Spirals and Environment*

Environment and interactions appear to play a significant role in the evolution of galaxies, particularly in dense environments. Interactions with satellites may be responsible for the significant fraction of lopsided spiral galaxies (Rudnick & Rix 1998). Disk asymmetry may affect the inferred spatial distribution of metals in the interstellar gas. For example, Kennicutt & Garnett (1996) noted an asymmetry in O/H between the NW and SE sides of M101, which may be related to the asymmetry in the structure of the disk. Zaritsky (1995) found a possible correlation between disk $B-V$ and the slope of the O/H gradient such that bluer galaxies tended to have steeper gradients. He suggested that accretion of metal-poor, gas-rich dwarf galaxies in the outer disk could steepen abundance gradients and make the colors bluer through increased star formation. On the other hand, the trend may simply reflect the fact that spirals with steep metallicity gradients tend to be lower-luminosity late Hubble types with bluer colors on average.

Rich clusters offer a variety of galaxy-galaxy and galaxy-ICM interactions. The cluster environment certainly affects the morphology of galaxies (Dressler 1980). It is also known that spirals near the center of rich clusters, for example, the Virgo cluster, show evidence for stripping of H I, especially from the outer disks (Warmels 1988; Cayatte et al. 1994). Such stripping is inferred to result from interaction of the galaxy ISM with the hot X-ray intracluster gas. The degree of H I stripping correlates with projected distance from the cluster core, although in Virgo the molecular content appears to be not affected (Kenney & Young 1989).

If field galaxies evolve through continuing infall of gas (Gunn & Gott 1972), then the truncation of H I disks in cluster spirals should have an effect on the chemical evolution. Specifically, infall of metal-poor gas reduces the metallicity of the gas at a given gas fraction. Truncation of such infall should then cause the chemical evolution of cluster spirals to behave more like the simple closed box model, and thus should have higher metallicities than comparable field spirals. This idea has led to several studies of oxygen abundances in Virgo spirals. The largest study so far is that of Skillman et al. (1996), who obtained data for nine Virgo spirals covering the full range of H I deficiencies. The results indicate that cluster spirals with the largest H I deficiencies have higher O/H abundances than field spirals with comparable M_B, rotation speeds, and Hubble types, while spirals with only modest or little H I stripping have abundances comparable to those of similar field spirals (Skillman et al. 1996, Henry et al. 1994, 1996). The samples studied so far have been small, and one must worry about possible systematic errors in abundance caused by the lack of measured electron temperatures. This is an area that could benefit from further study with larger galaxy samples and more secure abundance measurements.

5. Element Abundance Ratios in Spiral and Irregular Galaxies

The abundance ratios of heavy elements are sensitive to the initial mass function (IMF), the star formation history, and variations in stellar nucleosynthesis with, e.g., metallicity. In particular, comparison of abundances of elements produced in stars with relatively long lifetimes (such as C, N, Fe, and the s-process elements) with those produced in short-lived stars (such as O) probe the star formation history. Below, I review the accumulated data on C, N, S, and Ar abundances (relative to O) in spiral and irregular galaxies, covering two orders of magnitude in metallicity (as measured by O/H). The data are taken from a variety of sources on abundances for H II regions in the literature.

5.1. *Helium*

Helium, the second most abundant element, has significance for cosmology and stellar structure. Most ^{4}He was produced in the Big Bang, and the primordial mass fraction Y_p is a constraint on the photon/baryon ratio and thus on the cosmological model. The He mass fraction also affects stellar structure, but He is difficult to measure in stars and so must be inferred from other measurements. On the other hand, He I recombination lines are relatively easy to measure in H II regions, and so a large amount of data is available on He/H in ionized nebulae.

A great deal of effort has been spent in determining Y_p, and is covered in Gary Steigman's contribution, so I will be brief on this aspect. Peimbert & Torres-Peimbert (1974) initiated the current modern study of Y_p by making the simple assumption that the He mass fraction varies linearly with metallicity (or O/H), and thus used measurements of abundances in H II regions with a range of O/H to extrapolate to the pre-galactic He abundance at O/H = 0. Today there is very high signal/noise data on He abundances in approximately 40 metal-poor dwarf irregulars with O/H ranging from 2% to 10% solar, so the extrapolation to O/H = 0 can be estimated to high statistical precision.

The good news is that the best current estimates of Y_p agree to within 5%. This is amazing agreement for measurements derived from spectroscopy of distant galaxies, so we should all feel proud. Nevertheless, the differences in Y_p estimates are a source of consternation for theory of cosmological nucleosynthesis, as the two largest studies obtain values for Y_p which disagree at the 4-5σ significance level: Olive, Skillman & Steigman (1997) derived $Y_p = 0.234 \pm 0.003$, while Izotov & Thuan (1998) derived $Y_p = 0.244 \pm 0.002$ (statistical uncertainties only for both studies), from similar-sized H II region samples.

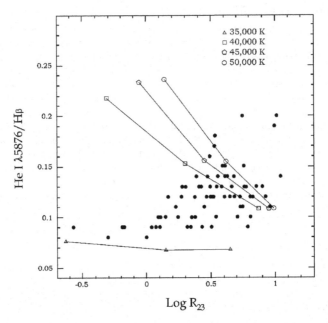

FIGURE 15. I(He I $\lambda5876$)/I(Hβ) vs. R_{23} from spectroscopy of H II regions in spiral galaxies (Bresolin et al. 1999), showing how the He I line strengths decrease for low R_{23} (high O/H). Overplotted are trends obtained from photoionization models with various values of T_{eff} for the ionizing stars, assuming that He varies with metallicity as $\Delta Y/\Delta = 2.5$.

Depending on which estimate is considered most reliable, Y_p either agrees with the best current estimate of D/H under standard Big Bang nucleosynthesis (for $Y_p = 0.244$, or it does not).

At present the battleground for Y_p is focused on sources of systematic error, and these are likely to yield the greatest improvements in He measurements, rather than measuring more data points. The areas that need work are:

• Corrections for He I absorption by the underlying OB association. These are fairly uncertain and affect the He I line ratios as well as the total He abundance. B main sequence and supergiants have the largest He I line strengths, and so need special attention; the B supergiant contribution is likely to be affected by stochasticity and uncertain stellar evolution.

• Density effects. He I lines in the optical spectrum, particularly the triplets, are subject to collisional excitation because the 2 ^{3}S level is metastable. The contribution of collisional excitation depends on both electron density and electron temperature. It is debated whether electron densities typically derived from [S II] line ratios are appropriate for He I. Detailed studies of density structure in a few good H II regions would provide useful information on this.

• Radiative transfer. Again, because the 2 ^{3}S level is metastable, transitions decaying into this level can build up large optical depths in H II regions. This leads to redistribution of line ratios among the triplets. This problem is coupled to the collisional excitation problem.

• Corrections for neutral He. He^0 can not be observed directly in H II regions; the

He^0 fraction must be inferred from ionization models. Since the ionization energy of He is 24.6 eV He^0 can exist in the H$^+$ zone. However, for ionizing radiation field with an effective temperature $T_{eff} > 40{,}000$ K the He$^+$ Strömgren radius is nearly coincident with the H$^+$ radius. This problem is largely solved by observing only high-ionization H II regions, with O$^+$/O < 0.15. Nevertheless, even in this case the He^0 corrections can be 1-2%, either positive or negative.

• Collisional excitation of H I. In high temperature metal-poor H II regions, collisional excitation of Hα could be significant (of order 5% or so). H I collisional excitation affects the He/H ratio and interstellar reddening estimates. The effect is very sensitive to the fraction of H^0 present in the highly-ionized zone, which is very uncertain.

Each of these error sources contribute perhaps 1-2% to the uncertainty in derived He abundances, but it is how they sum that determines the systematic error, which is not fully understood.

Helium abundances in spiral galaxies are less well-determined, because of more uncertain electron temperatures. The He abundance does have an effect on ionization structure, so it is of interest to know how He/H varies with metallicity in the inner disks of spirals. Does He/H continue to rise linearly with metallicity as in the metal-poor galaxies, or does it level off? This may be difficult to determine from H II region spectroscopy, as there appears to be a drop in the He I line strengths in the inner disks of spirals (Bresolin, Kennicutt, & Garnett 1999), contrary to what one would expect from ionization models with rising or even constant He/H (Figure 15). The low He I 5876 line strengths in the metal-rich regions observed so far are lower than expected even for primordial He/H. The trend of decreasing He I line strength is best explained if ionizing clusters in H II regions with metallicity above solar have radiation fields with characteristic temperatures of about 35,000 K. This is the strongest evidence available for a possible variation in the upper limit of the massive star IMF.

5.2. *Carbon*

Carbon abundances in H II regions have been difficult to determine with precision because the important ionization states (C$^+$, C^{+2}) have no strong forbidden lines in the optical spectrum. Only the UV spectrum shows collisionally excited lines from both C II and C III. A number of studies of carbon abundances in extragalactic H II regions were made with *IUE* (e.g., Dufour, Shields & Talbot 1982; Peimbert, Peña, & Torres-Peimbert 1986; Dufour, Garnett & Shields 1988), but for the most part the *IUE* observations suffered from low signal/noise and uncertainties due to aperture mismatches between UV and optical spectra.

The higher UV sensitivity of *HST* offered greatly improved measurements of UV emission lines from [C II] and C III], plus the opportunity to scale the C lines directly to [O II] and O III] lines in the UV, tremendously reducing the uncertainties due to reddening corrections and errors in T_e (Garnett et al. 1995a, 1999; Kobulnicky & Skillman 1998). The most recent data for C/O as a function of O/H in dwarf irregular and spiral galaxies from *HST* measurements are displayed in Figure 16. Some C (and O) is expected to be depleted onto interstellar dust grains. Sofia et al. (1997) showed that the gas-phase C abundance varies little with physical conditions in the local neutral ISM, suggesting a constant fraction of C in dust everywhere. They infer that C is depleted by about 0.2 dex. O should be depleted by no more than ≈ 0.1 dex everywhere (Mathis 1996). Thus, it is likely that our C/O values should all be increased by 0.1-0.2 dex, but we do not expect any systematic variation in the fractional depletions with metallicity.

Figure 16 shows a trend of steeply increasing C/O for log O/H > -4. This is in agreement with observations of C/O in disk stars in the Galaxy (Gustafsson et al. 1999).

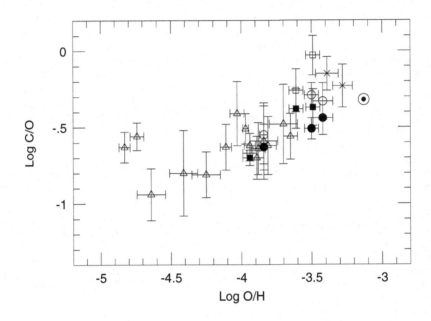

FIGURE 16. C/O abundance ratios (by number) from spectroscopy of H II regions in spiral and irregular galaxies (Garnett et al. 1995a, 1997b, 1999). *Open symbols*: irregular galaxies; *filled symbols*: spiral galaxy H II regions.

The C/O ratios in the most metal-poor galaxies are consistent with the predictions for massive star nucleosynthesis by Weaver & Woosley (1993; hereafter WW93) for their best estimate of the $^{12}C(\alpha,\gamma)^{16}O$ nuclear reaction rate factor. On the other hand, the amount of contamination by C from intermediate mass stars is poorly known in these galaxies.

The notable trend in Figure 16 is the apparent 'secondary' behavior of C with respect to O, despite the fact that C (i.e., ^{12}C) is primary. Tinsley (1979) demonstrated that such variations can be understood as the result of finite stellar lifetimes and delays in the ejection of elements from low- and intermediate-mass stars. If C is produced mainly in intermediate-mass stars, then the enrichment of C in the ISM is delayed with respect to O, which is produced in high-mass stars.

At the same time, C is also produced in high-mass stars, with a production yield that is fairly uncertain. In stars without mass loss, the relative yield of C with respect to O in massive stars is smaller than the solar ratio (WW93), which would demand that most C come from intermediate-mass stars. Maeder (1992), however, showed that stellar mass loss can affect the yields of C and O from massive stars. The effect of such mass loss is to remove He and C from the massive stars before they can be further processed into O. If the mass-loss rates depend on radiative opacity, and thus on metallicity, then the yields of C and O will depend on metallicity, with the C yield increasing with Z at the expense of O.

Figure 17 shows the data for the spiral galaxies M101 and NGC 2403 with the predic-

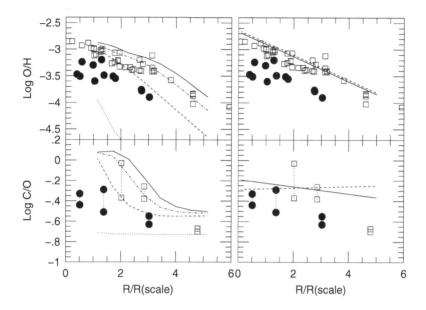

FIGURE 17. C/O and O/H gradients in M101 (open squares) and NGC 2403 (filled circles) plotted vs. radius normalized to the disk scale length (Garnett et al. 1999). *Left panels:* Galactic chemical evolution models from Carigi 1996, using massive star yields from Maeder 1992. The curves show the O/H and C/O gradients at different times, from 0.5 Gyr (*dotted curves*) to 13 Gyr (*solid curves*). *Right panels:* Galactic chemical evolution models from Götz & Köppen 1992 (*dashed line*) and Mollá et al. 1997 (*solid lines*) using massive star yields with no stellar winds. All models use the same intermediate-mass star yields from Renzini & Voli 1981.

tions of two sets of chemical evolution models overlaid. The left panels show a sequence of Galactic chemical evolution models using massive star nucleosynthesis models including stellar winds from Maeder (1992); the right panels shows two other Galactic chemical evolution models derived with massive star yields computed assuming no stellar mass loss. All of the models use the same intermediate-mass star yields. Although all of the models reproduce the O/H gradients reasonably well, only the models with Maeder yields seem able to reproduce the steep C/O gradients observed - with the caveat that these models were not tailored for the two galaxies in question. Comparison of solar neighborhood models with the observations of stars also tend to favor the nucleosynthesis models that take into account metallicity-dependent mass loss for massive stars (e.g., Prantzos et al. 1994; Carigi 2000).

The big uncertainties in all of this revolve around the theoretical yields. Problem number one is the $^{12}C(\alpha, \gamma)^{16}O$ reaction rate, which is still highly uncertain (Hale 1998). Problem two is uncertainty in convective mixing. ^{16}O is produced by α captures onto ^{12}C during helium burning. Mixing of fresh He into the convective zone can turn C into O rapidly (Arnett 1996, pp. 223-229). Finally, mass loss rates for stars in various evolutionary states and metallicities are also still uncertain. For intermediate-mass stars, differences in mixing and treatment of thermal pulses affect the C yields. The most recent models for nucleosynthesis in intermediate-mass stars still show large discrepancies in yields (Portinari et al. 1998; van den Hoek & Groenewegen 1997; Marigo et al. 1996, 1998). Until these problems are solved or we have empirically-derived C yields for stars of various masses, it will be difficult to reliably interpret the abundance trends.

5.3. *Nitrogen*

Nitrogen abundances in extragalactic H II regions are almost entirely derived from optical [N II] lines alone, because the other important species, N III, has strong emission lines only in the UV and FIR. Photoionization models generally predict that $N^+/O^+ = N/O$ under most conditions. Nevertheless, IR measurements of [N III]/[O III] in Galactic H II regions consistently find a steeper N/O gradient than that obtained from optical measurements of [N II]/[O II] (Lester et al. 1987; Martín-Hernández et al. 2002). This suggests that ionization corrections may be important (Garnett 1990). Direct comparison between [N II] and [N III] measurements in H II regions with varying properties is needed to understand the nitrogen ionization balance, so that the variation in N with metallicity can be studied accurately. Comparison with measurements in stars is also helpful, and it should be noted that measurements of abundances in B stars (e.g. Rolleston et al. 2000, Korn et al. 2002) yield O and N abundances in good agreement with the values for H II regions in the Milky Way and the LMC and the Local Group spiral M33 (McCarthy et al. 1995; Monteverde et al. 1997).

With this uncertainty in mind, Figure 18 shows how N/O varies with O/H from optical spectroscopy of H II regions in spiral and irregular galaxies. It has been known for some time that N seems to have two components: one component which follows O in a fixed ratio (log N/O $\approx$ −1.5) for log O/H < −3.7, as inferred from the constant N/O vs. O/H in metal-poor dwarf irregular galaxies (open circles), and a second component that increases faster than O at higher O/H as seen in spiral galaxies (filled circles and squares, crosses). The second component is produced via the classical CNO cycle during hydrogen burning in stars and requires the presence of C and O in the star from birth ("secondary N"), while the first component is postulated to come from the CN cycle on freshly-synthesized C (from He-burning) which has been convectively "dredged-up" into a hot H-burning zone at the base of the convective envelope, and does not require an initial seed of C or O (hence, "primary" N). The latter process is most commonly thought to occur in the asymptotic giant branch (AGB) stage of intermediate mass stars (Iben & Truran 1978), but has been found to occur in models of massive stars with increased convective overshooting or rotationally-induced mixing (e.g. Langer et al. 1997). It has been unclear which primary N source accounts most for the constant N/O in the dwarf galaxies. The massive star primary source does not appear to produce enough N to yield N/O $\approx$ 0.03. For the lower mass stars, the various AGB model calculations give rather discrepant results for the production of N (see Forestini & Charbonnel 1997; van den Hoek & Groenewegen 1997; Marigo 2001). The N production during the third dredge-up is very sensitive to the assumptions that determine the boundary of the convective zone and the overshoot. This is a highly hydrodynamic problem including explosive thermal pulse events and is difficult to model at present (Lattanzio 1998). It is likely that this will continue to be an important topic of study for the near future.

Some insight may be found by examining more distant objects. Recent studies of high-redshift Lyman-alpha absorption systems (plotted as stars in Figure 18) have found objects with N/S, N/Si, and N/O ratios much lower than in the dwarf galaxies (Pettini, Lipman & Hunstead 1995; Lu, Sargent & Barlow 1998). A wide range in inferred N/O is seen in the DLAs, but the lowest values are as much as a factor 10 smaller than the average for irregular galaxies. Although S and Si column densities are derived from

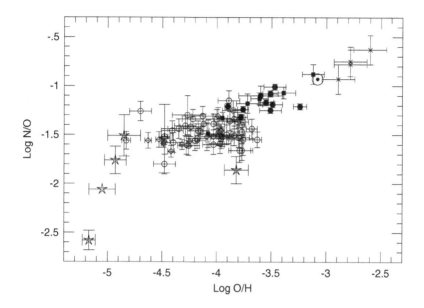

FIGURE 18. N/O abundance ratios in spiral and irregular galaxies. *Open circles*: Garnett 1990; *open diamonds*: Thuan et al. 1995; *filled circles*: Garnett et al. 1999; *filled squares*: Garnett & Kennicutt 1994, Torres-Peimbert et al. 1989; *plus signs*: Díaz et al. 1991; *stars*: high-redshift absorption line systems from Lu, Sargent & Barlow 1998. Note the very low N/O in some high-redshift systems.

S II and Si II, which can coexist with both ionized and neutral gas, ionization effects appear to insufficient to account for low N/S and N/Si (Vladilo et al. 2001). The results are consistent with the idea that the DLAs represent lines of sight through very young galaxies, with an age spread of a few hundred Myr, the timescale for enrichment of N from AGB stars. The higher N/O ratios seen in irregular galaxies would then be largely the product of AGB stars.

Some scatter is seen in N/O for the more metal-rich dwarf galaxies (Kobulnicky & Skillman 1998). This may be the result of localized enrichment by Wolf-Rayet stars. The most metal-poor dwarf galaxies seem to show very little scatter in N/O (Thuan et al. 1995). It is possible that this may simply reflect small number statistics (dwarf galaxies with log O/H < −4.5 and bright H II regions are rare). It is also possible to understand these galaxies if they are relatively old systems that experienced an episode of star formation in the past which enriched them to their present composition, and are experiencing a new starburst event after a long quiescent period.

5.4. *Neon, Sulfur and Argon*

Neon, sulfur, and argon are products of the late stages of massive star evolution. ^{20}Ne results from carbon burning, while S and Ar are products of O burning. As they are

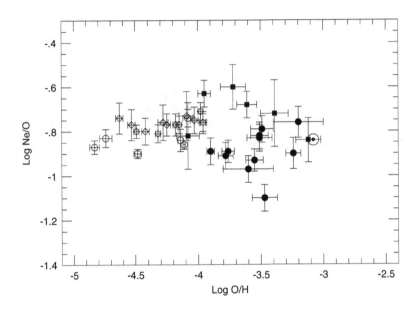

FIGURE 19. Ne/O abundance ratios in spiral and irregular galaxies. Symbols are the same as in Figure 18.

all considered part of the α-element group, their abundances are expected to track O/H closely.

Neon abundances in extragalactic H II regions are derived primarily from optical measurements of [Ne III], although spacecraft measurements of the IR [Ne II] and [Ne III] fine-structure lines are becoming available. A representative sample of Ne/O values for H II regions with measured T_e in spiral and irregular galaxies is shown in Figure 19. The scatter increases for the H II regions with higher O/H because of the more uncertain electron temperatures. It is apparent that the Ne abundance tracks O quite closely, in agreement with results from planetary nebulae (Henry 1989).

Figure 20 shows data for sulfur and argon. For log O/H < −3.5, S/O and Ar/O are essentially constant with O/H, and fall within the range predicted by WW93. For log O/H > −3.5, however, there is evidence for declining S/O and Ar/O as O/H increases. The cause of the decline is not clear. It is possible that the ionization corrections for unseen S^{+3} have been underestimated in the more metal-rich H II regions. More observational study, especially IR spectroscopy, of metal-rich H II regions is needed, to rule out ionization or excitation effects.

Because S and Ar are produced close to the stellar core, the yields of S and Ar may be sensitive to conditions immediately prior to and during the supernova explosion, such as explosive processing or fall-back onto the compact remnant (WW93). If real, however, the declining S/O and Ar/O cannot be accounted for by simple variations in the stellar mass function and hydrostatic nucleosynthesis (Garnett 1989); some variation in massive

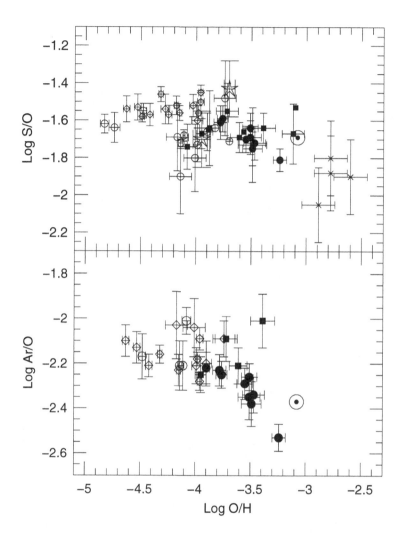

FIGURE 20. *Top:* S/O abundance ratios in spiral and irregular galaxies. *Bottom:* Ar/O abundance ratios in spiral and irregular galaxies. Symbols are the same as in Figure 18.

star and/or supernova nucleosynthesis at high metallicities (perhaps due to strong stellar mass loss) may be needed.

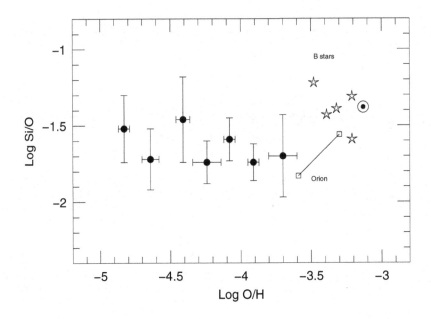

FIGURE 21. Si/O abundance ratios in irregular galaxies, from Garnett et al. 1995b (filled circles). The open squares represent two different measurements for the Orion nebula, while the stars show averages for four samples of Galactic B stars (see Garnett et al. 1995b for details).

5.5. *Other Elements*

Few other elements have been measured systematically in extragalactic H II regions, over a wide range of O/H.

Silicon can be measured in H II regions through the UV Si III] doublet at 1883, 1892 Å. Figure 21 shows Si/O measurements from Garnett et al. (1995b) for a small sample of metal-poor galaxies, along with data for several samples of B stars and the Orion nebula. Si/O appears to be roughly constant but smaller than the average for the Sun and the solar neighborhood B stars. Silicon is certainly depleted onto grains in the ISM, and the results in Figure 21 are consistent with a Si depletion of about -0.2 to -0.4 dex in the H II region sample. This is probably appropriate for not very dense, ionized gas (Sofia, Cardelli & Savage 1994).

Iron has a variety of emission lines from [Fe II] and [Fe III] in the optical spectrum. The [Fe III] 4658 Å is often observed in extragalactic H II regions. Izotov & Thuan (1999) measured [Fe III] in several metal-poor emission-line galaxies and derived Fe abundances. They obtained Fe/O ratios that were similar to the values found for metal-poor stars in the Galactic halo, and used this to argue that the their emission-line galaxies were very young. However, the ionization corrections for Fe are very uncertain, and Fe is highly depleted onto grains in the ISM. If Si is depleted by 0.2 to 0.4 dex, one can use the depletion analysis of Sofia et al. (1994) to estimate that Fe is depleted by about 0.7 to 1.1 dex. It is improbable that O/Fe ratios in H II regions can be used to interpret the enrichment history of the ISM without better understanding of the depletion factors.

6. Open Questions and Concluding Remarks

In conclusion, I'd like to enumerate a few questions regarding the chemical evolution of galaxies that seem to need further investigation.

• What are the primary mechanisms determining the shape and slope of abundance gradients in spiral galaxies? We have seen that chemical evolution models tend to predict that composition gradients should get shallower in the inner disks spirals, but this has not been observed in real galaxies. Does viscous evolution play a role in maintaining an exponential gradient, by transferring new gas into the inner disk? This is essentially a hydrodynamical problem. It is also important to determine how closely composition gradients follow an exponential profile, since the H II region abundance scale is still uncertain for high metallicity. IR spectroscopy will make a significant contribution to constraining the abundance calibrations in the metal-rich regime.

• How homogeneous are abundances in galaxies at a given place and time? Conflicting studies have argued for significant (± 0.2-0.3 dex) variations in abundances on small scales, or for a very homogeneous composition (< 0.1 dex variations). The question is relevant to the time scales for cooling and mixing of stellar ejecta with the ambient ISM. Most chemical evolution models assume instantaneous mixing, but is this a good approximation? Detailed studies of abundance variations across small (< 1 kpc) regions of galaxies will tell us how homogeneous the ISM composition is.

• Is infall of gas presently occurring in galaxies, and how does the rate evolve with time? This is a big unknown, since most chemical evolution models use relatively slow, ongoing infall of metal-poor gas to suppress the fraction of metal-poor G dwarfs. Observational evidence for classical infall is sketchy. The high-velocity clouds seen at high latitudes may represent infall, or may be part of a Galactic fountain flow, or may be associated with interaction between the Galaxy and its satellite galaxies.

• What galaxies may be losing metals to the IGM, and how much do they contribute? Is there a threshold mass above which galaxies retain metals?

• Does galaxy environment influence composition? The Virgo cluster studies need to be followed up by larger samples in a wider variety of cluster environments. The spiral-rich Ursa Major cluster and the Coma cluster are two obvious choices for continued study.

• Is there zero metallicity gas (or stars for that matter) in galaxies? Pre-enrichment by an initial stellar Population III also provides a solution to the G-dwarf problem and to the origin of metals seen in Lyα forest clouds. What is the composition of the huge gas reservoirs in the outer parts of spirals and irregulars?

For abundance work in ionized nebulae, we need to understand better the effects of dust on the thermal and ionization balance. The calculated ionizing spectra from massive stars are still in a state of flux, as more physics and opacity are included; this is an area that will continue to require attention as computing power grows. The effects of inhomogeneous structure on the observed emission-line spectrum of H II regions also needs to be addressed.

Stellar nucleosynthesis also needs continuing attention. The biggest remaining problem in theoretical stellar evolution and nucleosynthesis continues to be the treatment of convective mixing, which affects both structure and nucleosynthesis. This is also a hydrodynamical problem requiring improved computing power, and should provide a source of entertainment (and argument) for some time.

We are seeing great improvements in the study of abundances in nearby galaxies, particularly with the new space-based UV and IR observatories, which are greatly improving the data for elements besides oxygen. With these new data, we are in a better position to connect the present-day abundance patterns in galaxies with those observed in high-redshift gas clouds. Eventually, emission-line spectroscopy of distant galaxies should greatly expand our information on heavy element abundances at early times, and allow us to trace the evolution of metallicity in the universe in greater detail. This will be an important complement to the absorption line work on DLAs, as the connection between the emission-line gas and the stellar component is much more clear, and the disk component can be sampled more completely than with absorption studies.

For nearby galaxies one challenge for observers is to compile a homogeneous reference set of abundance data, to provide a statistically significant sample for outlining the relationships between abundances, galaxy mass, and Hubble type, and for understanding the effects of environment on abundance profiles. Some Hubble types are poorly represented in the database, particularly very early Hubble types (Sa-Sab), and very late types (Sd). Basic structural data for nearby galaxies also need improvement. The amount and distribution of molecular gas is one area for improvement. Stellar mass-to-light ratios and the stellar mass surface density distribution is another. Wide-field imaging in the infrared should help reduce the uncertainty in the mass of the stellar component.

One might have noticed that many of the theoretical questions mentioned above come down to hydrodynamics. Understanding star formation, the evolution of galaxies, and the structure of stars all involve hydrodynamics at fundamental levels, so I believe that improved hydrodynamical modeling of all of these phenomena will be the key to a better understanding of galaxy evolution. Understanding the mechanisms which connect abundances in galaxies to galaxy structure should provide a continuing challenge to galaxy evolution theorists.

I am grateful to the organizers of this Winter School for the opportunity to meet and interact with the young scientists (on both the galactic and stellar side) whose research papers I have been enjoying in the past couple of years. Special thanks go to Eric Bell for producing Figure 5 for me, and for informative discussions of the properties of galaxy colors and population synthesis models. The review presented here has also benefited by numerous conversations over the past years with Daniela Calzetti, Mike Edmunds, Gary Ferland, Claus Leitherer, John Mathis, Andy McWilliam, and Verne Smith. Finally, I must also acknowledge my various collaborators on galaxy abundances (Reginald Dufour, Rob Kennicutt, Greg Shields, Evan Skillman, Manuel Peimbert, and Silvia Torres-Peimbert) who have contributed greatly to many of the results I have presented in these lectures. My work on abundances in galaxies has been supported the past four years by NASA grant NAG5-7734.

REFERENCES

ALLER, L. H. & FAULKNER, D. J., 1962, PASP 74, 219

ALLOIN, D., COLLIN-SOUFRIN, S., JOLY, M., & VIGROUX, L., 1979, A&A 78, 200

ANDREDAKIS, Y. C., PELETIER, R. F., & BALCELLS, M. 1995, MNRAS 275, 874

ARMANDROFF, T. E. & DA COSTA, G. S. 1991, AJ 100 1329

ARNETT, D., 1996, Supernovae and Nucleosynthesis, (Princeton University Press)

BARNES, J., 1991, IAU Symposium 126, Dynamics of Galaxies and Their Molecular Cloud Distributions, eds. F. Combes and F. Casoli (Dordrecht: Kluwer), p. 363

BAUGH, C. M., COLE, S., & FRENK, C. S., 1996, MNRAS 283, 1361

BELL, E. F., & DE JONG, R. S., 2000, MNRAS 312, 497

BELL, E. F., & DE JONG, R. S., 2001, ApJ 550, 212

BOSCH, G., TERLEVICH, R., MELNICK, J., & SEHMAN, F., 1999, A&AS 137, 21

BREGMAN, J. N., SCHULMAN, E., & TOMISAKA, K., 1995, ApJ 439, 155

BRESOLIN, F., KENNICUTT, R. C., JR., & GARNETT, D. R., 1999, ApJ 510, 104

BROEILS, A. H., 1992, PhD thesis, University of Groningen

BROEILS, A. H., & VAN WOERDEN, H., 1994, A&AS 107, 129

BURSTEIN, D., FABER, S. M., & GONZÁLEZ, J. J., 1986, AJ 91, 1130

CALZETTI, D., KINNEY, A. L., & STORCHI-BERGMAN, T. 1994, ApJ 429, 582

CARIGI, L., 1996, RMxAA 32, 179

CARIGI, L., 2000, RMxAA 36, 171

CASERTANO, S., & VAN GORKOM, J. H., 1991, AJ 101, 1231

CASTELLANOS, M., DÍAZ, A. I., & TERLEVICH, E., 2002, MNRAS, 329, 315

CAYATTE, V., KOTANYI, C., BALKOWSKI, C., & VAN GORKOM, J. H., 1994, AJ 107, 1003

CLARKE, C. J. 1989, MNRAS, 238, 283

COMBES, F., 1998, in Abundance Profiles: Diagnostic Tools for Galaxy History, eds. D. Friedli, M. Edmunds, C. Robert, and L. Drissen (San Francisco: ASP), p. 300

COMBES, F., DEBBASCH, F., FRIEDLI, D., & PFENNIGER, D., 1990, A&A 233, 82

DE KOTER, A., HEAP, S. R., & HUBENY, I., 1997, ApJ, 477, 792

DE JONG, R. S. 1996, A&AS, 118, 557

DE VAUCOULEURS, G., & PENCE, W. D., 1978, AJ, 83, 1163

DE ZEEUW, P. T., ET AL., 2002, MNRAS, 329, 513

DEKEL, A., & SILK, J. 1986, ApJ, 303, 39

DENICOLÓ, G., TERLEVICH, R., & TERLEVICH, E. 2002, MNRAS 330, 69

DÍAZ, A. I., & PÉREZ-MONTERO, E. 2000, MNRAS 312, 130

DÍAZ, A. I., TERLEVICH, E., VÍLCHEZ, J. M., PAGEL, B. E. J., & EDMUNDS, M. G. 1991, MNRAS 253, 245

DRESSLER, A. 1980, ApJ, 236, 351

DUFOUR, R. J., GARNETT, D. R. & SHIELDS, G. A., 1988 ApJ 332, 752

DUFOUR, R. J., SHIELDS, G. A. & TALBOT, R. J., JR., 1982, ApJ, 252, 461

EDMUNDS, M. G., 1990, MNRAS, 246, 678

EDMUNDS, M. G. & PAGEL, B. E. J., 1984, MNRAS, 211, 507

EGGEN, O., LYNDEN-BELL, D. & SANDAGE, A., 1962, ApJ, 136, 748

ELLISON, S. L., SONGAILA, S., SCHAYES, J. & PETTINI, M., 1999, AJ, 120, 1175

FERLAND, G. J., 1998, Ringberg Workshop on the Orion Complex (astro-ph/9808107)

FORESTINI, M. & CHARBONNEL, C., 1997 A&AS 123, 241

FRIEDLI, D., BENZ, W. & KENNICUTT, R. C. JR., 1994 ApJ 430, L105

GARNETT, D. R., 1989 ApJ 345, 282

GARNETT, D. R., 1990 ApJ 363, 142

GARNETT, D. R., 1992 AJ 103, 1330

GARNETT, D. R., 1999 in Spectrophotometric Dating of Stars and Galaxies, eds. I. Hubeny, S. Heap, and R. H. Cornett (San Francisco: ASP), p. 61

GARNETT, D. R., 2002, in preparation

GARNETT, D. R., & DINERSTEIN, H. L., 2001, RMxAASC 10, 13

GARNETT, D. R., & DINERSTEIN, H. L., 2002, ApJ 558, 145

GARNETT, D. R., DUFOUR, R. J., PEIMBERT, M., TORRES-PEIMBERT, S., SHIELDS, G. A., SKILLMAN, E. D., TERLEVICH, E., & TERLEVICH, R. J.,, 1995b, ApJ 449, L77

GARNETT, D. R., & KENNICUTT, R. C. JR., 1994, ApJ 426, 123

GARNETT, D. R., KENNICUTT, R. C. JR., CHU, Y.-H., & SKILLMAN, E. D. 1991, ApJ 373, 458

GARNETT, D. R., & SHIELDS, G. A., 1987, ApJ, 317, 82

GARNETT, D. R., SHIELDS, G. A., SKILLMAN, E. D., SAGAN, S. P., AND DUFOUR, R. J., 1997a, ApJ 489, 63

GARNETT, D. R., SHIELDS, G. A., PEIMBERT, M., TORRES-PEIMBERT, S., SKILLMAN, E. D., DUFOUR, R. J., TERLEVICH, E., AND TERLEVICH, R. J., 1999, ApJ 513, 168

GARNETT, D. R., SKILLMAN, E. D., DUFOUR, R. J., PEIMBERT, M., TORRES-PEIMBERT, S., TERLEVICH, E., TERLEVICH, R. J., AND SHIELDS, G. A., 1995a, ApJ 443, 142

GARNETT, D. R., SKILLMAN, E. D., DUFOUR, R. J., AND SHIELDS, G. A., 1997b, ApJ 481, 174

GÖTZ, M., & KÖPPEN, J., 1992 A&A 260, 455

GUNN, J. E., & GOTT, J. R., 1972 ApJ 176, 1

GUSTAFSSON, B., KARLSSON, T., OLSSON, E., EDVARDSSON, B., AND RYDE, N., 1999, A&A 342, 426

HALE, G. M. 1998, Stellar Evolution, Stellar Explosions, and Galactic Chemical Evolution, ed. A. Mezzacappa (Bristol: Institute of Physics), p. 17

HASAN, H., PFENNIGER, D., & NORMAN, C. 1993, ApJ 409, 91

HENRY, R. B. C., 1989, MNRAS 241, 453

HENRY, R. B. C., BALKOWSKI, C., CAYATTE, V., EDMUNDS, M. G., & PAGEL, B. E. J., 1996, MNRAS 293, 635

HENRY, R. B. C., PAGEL, B. E. J., & CHINCARINI, G. L. 1994, MNRAS 266, 421

HENRY, R. B. C., & WORTHEY, G., 1999, PASP 111, 919

HUNTER, D. A., BAUM, W. A., O'NEIL, E. J., JR., & LYNDS, R. 1996, ApJ 456, 174

HUNTER, D. A., SHAYA, E. J., HOLTZMAN, J. A., LIGHT, R. M., O'NEIL, E. J. JR., & LYNDS, R. 1995, ApJ 448, 179

IBATA, R. A., GILMORE, G., & IRWIN, M. J. 1994 Nature 370, 194

IBEN, I. JR. & TRURAN, J. W. JR., 1978 ApJ 220, 980

ISRAEL, F. P., 1997a A&A 317, 65

ISRAEL, F. P., 1997b A&A 328, 471

IZOTOV, Y. I., & THUAN, T. X., 1998 ApJ 500, 188

IZOTOV, Y. I., & THUAN, T. X., 1999 ApJ 511, 639

JABLONKA, P., MARTIN, P, & ARIMOTO, N. 1996, AJ 112, 1415

JACOBY, G. H., & CIARDULLO, R. 1999, ApJ 515, 169

KAUFFMANN, G., 1996, MNRAS, 281, 475

KENNEY, J. D. P., & YOUNG, J. S., 1989, ApJ, 344, 171

KENNICUTT, R. C., JR. & GARNETT, D. R., 1996, ApJ, 456, 504

KINKEL, U., & ROSA, M. R., 1994, A&A, 282, 37

KOBULNICKY, H. A. & SKILLMAN, E. D., 1996, ApJ, 471, 211

KOBULNICKY, H. A. & SKILLMAN, E. D., 1998, ApJ, 497, 601

KORN, A. J., KELLER, S. C., KAUFER, A., LANGER, N., PRZYBILLA, N., STAHL, O. & WOLF, B., 2002, A&A, 385, 143

LANGER, N., FLIEGNER, J., HEGER, A. & WOOSLEY, S. E., 1997 Nucl. Phys. A621, 457

LATTANZIO, J. C. 1998, Stellar Evolution, Stellar Explosions, and Galactic Chemical Evolution, ed. A. Mezzacappa (Bristol: Institute of Physics), p. 299

LESTER, D. F., DINERSTEIN, H. L., WERNER, M. W., WATSON, D. M., GENZEL, R. L., & STOREY, J. W. V., 1987, ApJ 320, 573

LEQUEUX, J., RAYO, J. F., SERRANO, A., PEIMBERT, M., & TORRES-PEIMBERT, S., 1979 A&A 80, 155

LIN, D. N. C., & PRINGLE, J. E., 1987, ApJ, 320, L87

LIU, X.-W., STOREY, P. J., BARLOW, M. J., & CLEGG, R. E. S. 1995, MNRAS 272, 369

LIU, X.-W., STOREY, P. J., BARLOW, M. J., DANZIGER, I. J., COHEN, M., & BRYCE, M. 2000, MNRAS 312, 585

LU, L., SARGENT, W. L. W. & BARLOW, T. A., 1998 AJ 115, 55

MAEDER, A., 1992 A&A 264, 105

MALONEY, P., & BLACK, J. H., 1988, ApJ, 389, 401

MARIGO, P., 2001, A&A, 370, 194

MARIGO, P., BRESSAN, A., & CHIOSI, C., 1996, A&A, 313, 545

MARIGO, P., BRESSAN, A., & CHIOSI, C., 1998, A&A, 331, 580

MARTIN, P. G., & ROULEAU, F., 1990, in Extreme Ultraviolet Astronomy, eds. R. F. Malina and S. Bowyer (Oxford: Pergamon), p. 341

MARTIN, P., & ROY, J.-R., 1994, ApJ, 424, 599

MARTÍN-HERNÁNDEZ, N. L., PEETERS, E. et al., 2002, A&A, 381, 606

MARTINS, F., SCHAERER, D., & HILLIER, D. J. 2002, A&A 382, 999

MATEO, M., 1998, ARAA 36, 435

MATHIS, J. S., 1962, ApJ 136, 374

MATHIS, J. S., 1996, ApJ 472, 643

McCALL, M. L., 1982, PhD thesis, University of Texas at Austin

McCARTHY, J. K., LENNON, D. J., VENN, K. A., KUDRITZKI, R.-P., PULS, J., & NAJARRO, F., 1995, ApJ 455, L135

McWILLIAM, A. & RICH, R. M., 1994, ApJS 91, 749

MIGHELL, K. J. & BURKE, C. J., 1999, AJ 118, 366

MIHOS, J. C., 2001, in GALAXY DISKS AND DISK GALAXIES, ASP Conference Series Vol. 230, eds. J. G. Funes and E. M. Corsini, p. 491

MO, H., MAO, S., & WHITE, S. D. M., 1998, MNRAS, 295, 319

MOLLÁ, M., FERRINI, F., & DÍAZ, A. I., 1997, ApJ, 475, 519

MONTEVERDE, M. I., HERRERO, A., LENNON, D. J., & KUDRITZKI, R.-P., 1997, ApJ, 474, 107

OEY, M. S., & KENNICUTT, R. C. JR. 1993 ApJ 411, 137

OSTERBROCK, D. E. 1989, Astrophysics of Gaseous Nebulae and Active Galactic Nuclei (Mills Valley, CA: University Science Books)

OLIVE, K. A., SKILLMAN, E. D., & STEIGMAN, G., 1997 ApJ 483, 788

PAGEL, B. E. J., EDMUNDS, M. G., BLACKWELL, D. E., CHUN, M. S., & SMITH, G., 1979, MNRAS 189, 95

PAKULL, M. W., & ANGEBAULT, L. P., 1986, Nature 322, 511

PEIMBERT, M., PEÑA, M., & TORRES-PEIMBERT, S., 1986 A&A 158, 266

PEIMBERT, M., & SPINRAD, H., 1970 ApJ 159, 809

PEIMBERT, M., & TORRES-PEIMBERT, S., 1974 ApJ 193, 327

PELETIER, R. F. et al., 1999, MNRAS, 310, 863

PETTINI, M., LIPMAN, K. & HUNSTEAD, R. W., 1995, ApJ, 451, 100

PHILLIPS, S., & EDMUNDS, M. G., 1991, MNRAS, 251, 84

PORTINARI, L., CHIOSI, C., & BRESSAN, A., 1998, A&A 334, 505

PRANTZOS, N., & BOISSIER, S., 2000, MNRAS, 313, 338

PRANTZOS, N., VANGIONI-FLAM, R., & CHAUVEAU, S., 1994, A&A, 285, 132

RENZINI, A., & VOLI, M., 1981, A&A 94, 175

ROLLESTON, W. R. J., SMARTT, S. J., DUFTON, P. L., & RYANS, R. S. I., 2000, A&A 363, 537

ROY, J.-R. & WALSH, J. R., 1997, MNRAS, 288, 715

RUDNICK, G. & RIX, H.-W., 1998, AJ, 116, 1163

RYDER, S. D., 1995, ApJ, 444, 610

SEARLE, L., 1971, ApJ, 168, 327

SEARLE, L. & SARGENT, W. L. W., 1972, ApJ, 173, 25

SEARLE, L. & ZINN, R., 1978, ApJ, 225, 357

SHETRONE, M. D., COTÉ, P. & SARGENT, W. L. W., 2001, ApJ, 548, 592

SHIELDS, J. C. & KENNICUTT, R. C. JR., 1995, ApJ, 454, 807

SKILLMAN, E. D., 1998, Stellar Astrophysics for the Local Group, eds. A. Aparicio, A. Herrero, and F. Sánchez (Cambridge University Press), p. 457

SKILLMAN, E. D., KENNICUTT, R. C., JR., SHIELDS, G. A., & ZARITSKY, D., 1996, ApJ, 462, 147

SMECKER-HANE, T. A., STETSON, P. B., HESSER, J. E., & LEHNERT, M. D., 1994 AJ 108, 507

SNEDEN, C., COWAN, J. J, IVANS, I. I., FULLER, G. M., BURLES, S., BEERS, T. C. & LAWLER, J. E., 2000 ApJ 533, L139

SOFIA, U. J., CARDELLI, J. A, GUERIN, K. P. & MEYER, D. M., 1997 ApJ 482, L105

SOFIA, U. J., CARDELLI, J. A. & SAVAGE, B. D., 1994 ApJ 430, 650

STASIŃSKA, G. & SZCZERBA, G. P., 2001, A&A 379, 1024

TAYAL, S. S. & GUPTA, G. P., 1999 ApJ 526, 544

THUAN, T. X., IZOTOV, Y. I. & LIPOVETSKY, V. A., 1995 ApJ 445, 108

TINSLEY, B. M., 1979 ApJ 229, 1046

TOLSTOY, E., IRWIN, M. J., COLE, A. A., PASQUINI, L., GILMOZZI, R. & GALLAGHER, J. S., 2001 MNRAS 327, 918

TORRES-PEIMBERT, S., PEIMBERT, M. & FIERRO, J., 1989 ApJ 345, 186

TRAGER, S. C., FABER, S. M., WORTHEY, G. & GONZÁLEZ, J. J., 2000a, AJ 119, 1645

TRAGER, S. C., FABER, S. M., WORTHEY, G. & GONZÁLEZ, J. J., 2000b, AJ 120, 165

TSUJIMOTO, T., YOSHII, Y., NOMOTO, K. & SHIGEYAMA, T., 1995 A&A 302, 704

TYSON, J. A., 1986 J. Opt. Soc. Amer. A 3, 2131

VAN DEN HOEK, L. B. & GROENEWEGEN, M. A. T., 1997 A&AS 123, 305

VAN ZEE, L., SALZER, J., HAYNES, M., O'DONOGHUE, A. & BALONEK, T., 1998 AJ 116, 2805

VILA-COSTAS, M. B. & EDMUNDS, M. G., 1992, MNRAS, 259, 121 (VCE)

VLADILO, G., CENTURIÓN, M., BONIFACIO, P. & HOWK, J. C., 2001 ApJ 557, 1007

WALBORN, N. R. 1991, IAU Symposium 148: The Magellanic Clouds, eds. R. Haynes and D. Milne (Dordrecht: Kluwers), p. 145

WARMELS, R. H. 1988, A&AS 72, 427

WEAVER, T. A. & WOOSLEY, S. E., 1993 Phys. Rep. 227, 65 (WW93)

WHITING, A. B., HAU, G. K. T. & IRWIN, M. J., 1999 AJ 118, 2767

WILSON, C. D., 1995, Ap,J 448, L97

WORTHEY, G., 1998, PASP, 110, 888

YUN, M. S., HO, P. T. P. & LO, K. Y., 1994, Nature 372, 530

ZARITSKY, D., 1995, ApJ, 448, L17

ZARITSKY, D., KENNICUTT, R. C. JR. & HUCHRA, J. P., 1994, ApJ, 420, 87 (ZKH)

Chemical Evolution Of Galaxies And Intracluster Medium

By FRANCESCA MATTEUCCI

Department of Astronomy, University of Trieste
Via G.B. Tiepolo 11
34100 Trieste, Italy

In this series of lectures I discuss the basic principles and the modelling of the chemical evolution of galaxies. In particular, I present models for the chemical evolution of the Milky Way galaxy and compare them with the available observational data. ¿From this comparison one can infer important constraints on the mechanism of formation of the Milky Way as well as on stellar nucleosynthesis and supernova progenitors. Models for the chemical evolution of elliptical galaxies are also shown in the framework of the two competing scenarios for galaxy formation: monolithic and hierachical. The evolution of dwarf starbursting galaxies is also presented and the connection of these objects with Damped Lyman-α systems is briefly discussed. The roles of supernovae of different type (I, II) is discussed in the general framework of galactic evolution and in connection with the interpretation of high redshift objects. Finally, the chemical enrichment of the intracluster medium as due mainly to ellipticals and S0 galaxies is discussed.

1. Basic parameters of chemical evolution

Galactic chemical evolution is the study of the evolution in time and space of the abundances of the chemical elements in the interstellar gas in galaxies. This process is influenced by many parameters such as the initial conditions, the star formation and evolution, the nucleosynthesis and possible gas flows.

Here I describe each one separately:

• Initial conditions- One can assume that all the initial gas out of which the galaxy will form is already present when the star formation process starts or that the gas is slowly accumulated in time. Then one can assume that the initial chemical composition of this gas is primordial (no metals) or that some pre-enrichment has already taken place (e.g. Population III stars). As we will see in the following, different assumptions are required for different galaxies.

• The birthrate function- Stars form and die continuously in galaxies, therefore a recipe for star formation is necessary. We define the stellar birthrate function as the number of stars formed in the time interval dt and in the mass range dm as:

$$B(m,t) = \psi(t)\varphi(m)dtdm \tag{1.1}$$

where:

$$\psi(t) = SFR \tag{1.2}$$

is the star formation rate (SFR), and:

$$\varphi(m) = IMF \tag{1.3}$$

is the initial mass function (IMF). The SFR is assumed to be only a function of time and the IMF only a function of mass. This is clearly an oversimplification but is necessary in absence of a clear knowledge of the star formation process.

• Stellar evolution and nucleosynthesis- Nuclear burnings take place in the stellar interiors during the star lifetime and produce new chemical elements, in particular metals.

These metals, together with the pristine stellar material is restored into the interstellar medium (ISM) at the star death. This process clearly affects crucially the chemical evolution of the ISM. In order to take into account the elemental production by stars we define the "yields", in particular the *stellar yields* (the amount of elements produced by a single star) and the *yields per stellar generation* (the amount of elements produced by an entire stellar generation).

• Supplementary Parameters- Infall of extragalactic gas, radial flows and galactic winds are important ingredients in building galactic chemical evolution models.

2. The stellar birthrate

2.1. *Theoretical recipes for the SFR*

Several parametrizations, besides the simple one of a constant $\psi(t)$, are used in the literature for the SFR:

• Exponentially decreasing:

$$SFR = \nu e^{-t/\tau_*} \tag{2.4}$$

with $\tau_* = 5 - 15$ Gyr in order to produce realistic values which can be compared with the present time SFR in the Milky Way (Tosi, 1988).

• The Schmidt (1963) law:

$$SFR = \nu \sigma_{gas}^k \tag{2.5}$$

is the most widely adopted formulation for the SFR. It was originally formulated by Schmidt (1959;1963) as a function of the volume gas density with $k = 2.0$. He measured the space density of stars in different regions of the Galaxy in relation to the number density of neutral hydrogen, measured by means of the 21 cm emission The formulation as a function of the surface gas density σ_{gas} is normally preferred for studying the Milky Way disk and galactic disks in general. In principle, Schmidt's formulation as a function of the volume gas density and that with σ_{gas} are equivalent when $k = 1$. Kennicutt, in a series of papers (1983;1989;1998a,b) tried to assess the dependence of massive star formation on the surface gas density in disk galaxies, by comparing H_α emission with the data on the distribution of HI and CO and found that the SFR can be well represented by a Schmidt law with $k = 1.4 \pm 0.15$. The quantity ν is the efficiency of star formation and is expressed in units of $time^{-1}$.

• A more complex formulation, which depends also upon the total surface mass density, was suggested by Dopita & Ryder (1994) and can be written as:

$$SFR = \nu \sigma_{tot}^{k_1} \sigma_{gas}^{k_2} \tag{2.6}$$

with σ_{tot} being the total surface mass density. This kind of SFR is related to the feedback mechanism between the energy injected into the ISM by supernovae (SNe) and stellar winds and the local potential well.

• In alternative to the Schmidt law, Kennicutt proposed also the following star formation law, which fits equally well the observational data:

$$SFR = 0.017\Omega_{gas}\sigma_{gas} \propto R^{-1}\sigma_{gas} \tag{2.7}$$

with Ω_{gas} being the angular rotation speed of the gas.

2.2. *The tracers of star formation*

The main tracers of star formation in galaxies are:

• Counts of luminous supergiants in nearby galaxies under the assumption that their number is proportional to the SFR.

• The H_α and H_β flux from HII regions, which are ionized by young and hot stars, under the assumption that such flux is proportional to the SFR (Kennicutt 1998b):

$$SFR(M_\odot yr^{-1}) = 7.9 \cdot 10^{-42} L_{H_\alpha}(ergs^{-1}) \tag{2.8}$$

• From the integrated UBV colors and spectra of galaxies one can estimate the relative proportions of young and old stars and derive the ratio between the present time SFR and the average SFR in the past.

• The frequency of type II SNe as well as the distribution of SN remnants and pulsars can be used as tracers of the SFR. These tracers have been used for deriving the SFR in the Galactic disk.

• The radio emission from HII regions can also be a tracer of the SFR.

• The ultraviolet continuum and the infrared continuum (star forming regions are surrounded by dust) are also connected to the SFR as in the following expression:

$$SFR(M_\odot yr^{-1}) = 0.9 \cdot 10^{-6} \frac{L(UV)}{L_{bol_\odot}} \tag{2.9}$$

derived by means of a two-slope IMF by Donas et al. (1987).

• Finally, the SFR can be derived from the distribution of molecular clouds (Rana 1991).

All of these formulations for the SFR need the assumption of an IMF and viceversa, the derivation of the IMF needs the assumption of a star formation history, as described in the next section. The local SFR, derived under the assumption of a particular IMF, suitable for the solar neighbourhood, gives the following range of values:

$$SFR = 2 - 10 M_\odot pc^{-2} Gyr^{-1} \tag{2.10}$$

(see Timmes et al. 1995)

2.3. *The IMF: Various Parametrizations*

The IMF is a probability distribution function and is normally approximated by a power law, namely:

$$\varphi(m) = am^{-(1+x)} \tag{2.11}$$

which is the number of stars with masses in the interval m, $m + dm$. The IMF can be one-slope (Salpeter 1955; x=1.35) or multi-slope (Scalo 1986,1998; Kroupa et al. 1993). The IMF is usually normalized as:

$$\int_0^\infty m\varphi(m)dm = 1 \tag{2.12}$$

The IMF is derived locally from the present day mass function (PDMF) which in turn is obtained by counting the Main Sequence stars per interval of magnitude. Then the star counts are transformed into number of stars per pc^2 and then a mass-luminosity relation is adopted to pass from the luminosity to the mass.

2.4. *Derivation of the IMF*

As already mentioned, the IMF is derived by the observed PDMF, which is the current mass distribution of Main-Sequence (MS) stars per unit area, $n(m)$.

For stars $(0.1 < m/m_\odot \le 1)$ with lifetimes $\tau_m \ge t_G$ (with t_G being the Galactic age), $n(m)$ can be written as:

$$n(m) = \int_0^{t_G} \varphi(m)\psi(t)dt \tag{2.13}$$

where t_G is the Galactic lifetime. These stars are all still on the MS. If $\varphi(m)$ is constant in time, as it is usually assumed, then:

$$n(m) = \varphi(m) < \psi > t_G \tag{2.14}$$

where $< \psi >$ is the average SFR in the past.

For stars with $\tau_m << t_G$ ($m \geq 2M_\odot$), we see on the MS only those born after the time ($t = t_G - \tau_m$). The PDMF is therefore:

$$n(m) = \int_{t_G - \tau_m}^{t_G} \varphi(m)\psi(t)dt \tag{2.15}$$

Again, if $\varphi(m)$ is constant in time:

$$n(m) = \varphi(m)\psi(t_G)\tau_m \tag{2.16}$$

under the assumption that $\psi(t_G) = \psi(t_G - \tau_m)$, where $\psi(t_G)$ is the SFR at the present time, t_G, and τ_m is the lifetime of a star of mass m.

The IMF, $\varphi(m)$, in the mass interval $1 - 2 \ M_\odot$ depends on the ratio:

$$b(t_G) = \frac{\psi(t_G)}{< \psi >} \tag{2.17}$$

It has been shown by Scalo (1986) that a good fit between the two portions of the $\varphi(m)$, namely below $1M_\odot$ and above $2M_\odot$, requires:

$$0.5 \leq b(t_G) \leq 1.5 \tag{2.18}$$

which means that the SFR in the local disk should have varied in time less than a factor of 2 during the whole disk lifetime.

2.5. The Infall Rate: Various Parametrizations

The presence of infall of extragalactic gas in the chemical evolution of galaxies is demanded by the G-dwarf metallicity distribution in the solar vicinity and by the existence of high velocity clouds infalling towards the galactic disk. The origin of this infalling gas on the Galaxy is not yet entirely clear and measurements of the metallicity of such gas are necessary to decide whether it originates in the galactic disk (galactic fountain) or if it has an extragalactic origin.

Several parametrizations for the infall rate have been used so far:
- Constant in space and time
- Variable in space and time such as:

$$IR = A(R)e^{-t/\tau(R)} \tag{2.19}$$

with $\tau(R)$ is constant or varying along the disk and $A(R)$ is derived by fitting the present time distribution of $\sigma_{tot}(R, t_G)$.

3. Nucleosynthesis

During the Big Bang the light elements (D, ^3He, ^4He and ^7Li) were produced. On the other hand, all the elements heavier than ^7Li, with the exception of Be and B, are produced inside stars. The light elements ^6Li, Be and B are instead manufactured by spallation processes in the ISM due to the interaction between cosmic rays and interstellar atoms.

3.1. Nucleosynthesis in the Big Bang

I give here a brief summary of the main steps in the Big Bang nucleosynthesis.

When the temperature in the universe was $T = 10^{12}$K, only weak interactions causing conversions between protons and neutrons occurred, namely:

$$p + e \leftrightarrow n + \nu, \tag{3.20}$$

$$p + \tilde{\nu} \leftrightarrow n + e^+. \tag{3.21}$$

The nucleosynthesis started when $T = 10^9$K and lasted until $T = 10^8$K. The first element to be formed was D (a nucleus composed by a neutron plus a proton) and subsequently ^3He from the reaction:

$$D + p \rightarrow^3 He + \gamma \tag{3.22}$$

followed by:

$$^3He + n \rightarrow^4 He + \gamma \tag{3.23}$$

Then also very small fractions of ^7Li (10^{-9} by mass) and ^7Be (10^{-11} by mass) were produced.

One of the major achievements in cosmology is that it can account simultaneously for the primordial abundances of H, D, ^3He, ^4He and ^7Li but only for a low density universe. The comparison between the observed primordial abundances and the Big Bang nucleosynthesis calculations can allow to impose constraints upon the baryon to photon ratio (η) in the universe. In particular, for a baryon to photon ratio $\eta \sim 3 \cdot 10^{-10}$ the baryonic density parameter of the universe is (Peacock, 1999):

$$0.010 \leq \Omega_b h^2 \leq 0.015 \tag{3.24}$$

3.2. Stellar Nucleosynthesis

Before discussing stellar nucleosynthesis we need to define the crucial stellar mass ranges.

- Brown Dwarfs ($M < M_L$, $M_L = 0.08 - 0.09 M_\odot$). They never ignite H and their lifetimes are larger than the age of the universe.
- Low mass stars ($0.5 \leq M/M_\odot \leq M_{HeF}$) ($M_{HeF}$=1.85-2.2$M_\odot$ depending on stellar models) ignite He explosively and become C-O white dwarfs (WD). If $M < 0.5M_\odot$ they become He WD. The lifetimes range from several 10^9 years up to several Hubble times.
- Intermediate mass stars ($M_{HeF} \leq M/M_\odot \leq M_{up}$) ignite He quiescently. M_{up} is the limiting mass for the formation of a C-O degenerate core and is in the range 5-9$M_\odot$, depending on stellar evolution models. Their lifetimes range from several 10^7 to 10^9 years. These stars die as C-O WDs if they are not in binary systems. If in binary systems, stars in this mass range can give rise to Type Ia SNe (see later).
- Massive stars ($M > M_{up}$)
Stars in the mass range $M_{up} \leq M/M_\odot \leq 10 - 12$ become e-capture SNe (Type II SNe) and leave neutron stars as remnants. Stars in the range $10 - 12 \leq M/M_\odot \leq M_{WR}$ ($M_{WR} \sim 20 - 40M_\odot$) end their lives as core-collapse SNe (Type II) and leave a neutron star or a black hole as a remnant. Stars in the range $M_{WR} \leq M/M_\odot \leq 100$ probably become Type Ib SNe. The lifetimes of these stars are in the range from several 10^7 to $\sim 10^6$ years.
- Very Massive Stars ($M > 100M_\odot$), if they exist, explode by means of "pair creation" and are called pair-creation SNe. In fact, at $T \sim 2 \cdot 10^9$ K a large portion of the

gravitational energy goes into creation of pairs (e^+, e^-), thus the star becomes unstable and explodes. These SNe leave no remnant and have lifetimes $< 10^6$ years.

• Supermassive objects $(400 \leq M/M_\odot \leq 7.5 \cdot 10^5)$, if they exist, either explode due to explosive H-burning or collapse directly to black holes. The only available nucleosynthesis calculations for these stars are from Woosley et al. (1984).

All the elements with mass number A from 12 to 60 have been formed in stars during the quiescent burnings occurring during their lifetime. The main nuclear burnings are H, He, C, Ne, O and Si. Stars transform H into He and then He into heaviers until the Fe-peak elements, where the binding energy per nucleon reaches a maximum. At this point nuclear fusion reactions cannot occur anymore and the Fe nucleus starts contracting. When the central density reaches the atomic nuclear density, matter becomes uncompressible and a core-bounce occurs with the consequent formation of a shock wave and ejection of the star mantle. However, a problem exists for the explosion of stars with large Fe cores since most of the collapse gravitational energy is used to photodisintegrate Fe, thus weakening the shock wave and preventing the mantle ejection. To overcome this problem it has been suggested the existence of some mechanisms able to rejuvenate the shock wave, such as neutrino-heating from the collapsing neutron star, rotation and magnetic fields.

Here is a summary of the main nucleosynthesis stages: the first element to be burned is H, which is transformed into He through the proton-proton chain or the CNO-cycle, then ^4He is transformed into ^{12}C through the triple α reaction. Elements heavier than ^{12}C are then produced by synthesis of α-particles thus producing the so-called α-elements (O, Ne, Mg, Si, S, Ca and Ti). The last main burning in stars is the ^{28}Si-burning which produces ^{56}Ni which then β-decays into ^{56}Co and ^{56}Fe. Si-burning can be quiescent or explosive (depending on the temperature) but it always produces Fe. Explosive nucleosynthesis occurs in the inverse order (Si, O, Ne, C, He, H) relative to quiescent nucleosynthesis, depending on the fact that it starts in the center and propagates outwards following the passage of the shock wave. The main products of explosive nucleosynthesis are the Fe-peak elements. It is worth noting that low and intermediate mass stars never ignite C and thus end their lives as C-O white dwarfs. Stars with masses below 10– 12 $M_\odot$ (depending on stellar models) explode during O-burning and end up as e-capture SNe. Only massive stars can ignite all six nuclear fuels until they form an Fe-core. S- and r-process elements (elements with $A > 60$ up to Th and U) are formed by means of slow or rapid (relative to the β- decay) neutron capture by Fe seed nuclei. In particular, s-processing occurs during quiescent He-burning both in massive and low and intermediate mass stars, whereas r-processing occurs during SN explosions.

3.3. *Supernova Progenitors*

Supernovae, planetary nebulae (PNe) and, to a minor extent, stellar winds are the means to restore the nuclearly enriched material into the ISM, thus giving rise to the process of chemical evolution. There are two main Types of SNe (II, I) then divided in subclasses: SNe IIL, IIP and SNe Ia, Ib, Ic. As already mentioned before, SNe II, which are believed to be the end state of stars more massive than $10M_\odot$ exploding after a Fe core is formed (core-collapse SNe), produce mainly α-elements (O, Ne, Mg, Si, S, Ca) plus some Fe. The amount of Fe produced by type II SNe is one of the most uncertain quantities since it depends upon the so-called mass cut (how much Fe remains in the collapsing core and how much is ejected) and on explosive nucleosynthesis.

Type Ia SNe are believed to originate from the C-deflagration of a WD reaching the Chandrasekhar mass (1.44 $M_\odot$) after accretion of material from a young companion in a close binary system. C-deflagration occurs as a consequence of such accretion and an explosion ensues destroying the whole star. No remnant is then left behind. They produce a large amount ($\sim 0.6 - 0.7 M_\odot$) of ^{56}Ni (Fe) plus traces of C to Si elements. C-deflagration is the explosive burning which best reproduces the observed abundance pattern in Type Ia SN remnants. The best model for this kind of explosion is model W7 by Nomoto, Thielemann & Yokoi (1984). The amount of Fe produced by the other Type I SNe is smaller than that produced by the Type Ia ones at least a factor of two or more. Therefore, Type Ia SNe should be considered as the responsible for the Fe enrichment in the universe.

3.4. *Element production*

Here is a summary of element production:

• Big Bang → light elements H, D, ^3He, ^4He, ^7Li. Deuterium is only destroyed inside stars to form ^3He. ^3He is also mainly destroyed. The only stars producing some ^3He are those with masses $< 2.5 M_\odot$. Recent prescriptions for the yields of ^3He are from Forestini & Charbonnel (1997) and Sackmann & Boothroyd (1999). Lithium: ^7Li is produced during the Big Bang but also in stars: massive AGB stars, SNe II, carbon-stars and novae. Some ^7Li should also be produced in spallation processes by galactic cosmic rays (see Romano et al. 2001).

• Spallation Processes → ^6Li, Be and B.

• Type II SNe → α-elements (O, Ne, Mg, Si, S, Ca), some Fe, s-process elements ($A < 90$) and r-process elements. Yields are from Woosley & Weaver (1995) and Thielemann et al. (1996).

• Type Ia SNe → Fe-peak elements. Yields are from Nomoto et al. (1984) and Nomoto et al. (1997).

• Low and intermediate mass stars → ^4He, C, N, s-process ($A > 90$) elements. Yields are from Renzini & Voli (1981), Marigo et al. (1996), van den Hoek & Groenewegen (1997) and Gallino et al. (1998).

3.5. *Stellar yields*

In order to include the results from nucleosynthesis into chemical evolution models we need to define the stellar yields. The stellar yield of an element i is defined as the mass fraction of a star of mass m which has been newly created as species i and ejected:

$$p_{im} = (\frac{M_{ej}}{m})_i \qquad (3.25)$$

In order to compute p_{im} we need to know some fundamental quantities from stellar evolution and nucleosynthesis:

• M_α is the mass of the He-core (where H is turned into He)

• M_{CO} is the mass of the C-O core (where the He is turned into heaviers)

• M_{rem} is the mass of the remnant (WD, neutron star, black hole)

These masses are related to each other by the following relations:

$$M_\alpha - M_{He} = M_{CO}$$

where M_{He} is the newly formed and ejected ^4He and:

$$M_{CO} - M_r = M_C + M_O + M_{heaviers}$$

The values of these different quantities are given by stellar evolution and nucleosynthesis calculations.

4. Modelling chemical evolution

4.1. *Analytical models*

The simplest model of chemical evolution is the *Simple Model* for the chemical evolution of the solar neighbourhood. The basic assumptions of the Simple Model are:
- the system is one-zone and closed, namely there are no inflows or outflows,
- the initial gas is primordial (no metals),
- $\varphi(m)$ is constant in time,
- the gas is well mixed at any time.

In the following we will adopt the formalism of Tinsley (1980) and define:

$$\mu = \frac{M_{gas}}{M_{tot}} \tag{4.26}$$

as the fractional mass of gas, with:

$$M_{tot} = M_* + M_{gas} \tag{4.27}$$

where M_* is the mass in stars (dead and alive). Possible non-baryonic dark matter is not considered.

The mass of stars can be expressed as:

$$M_* = (1 - \mu)M_{tot}. \tag{4.28}$$

The abundance by mass of an element i is defined by:

$$X_i = \frac{M_i}{M_{gas}} \tag{4.29}$$

where M_i is the mass in the form of the specific element i. It is well known that the abundances must satisfy the condition, $\sum_i X_i(t) = 1$, where the summation is over all the chemical elements.

The initial conditions are:

$$M_{gas}(0) = M_{tot} \quad ; X_i(0) = 0, \tag{4.30}$$

where i refers to metals. The equation for the evolution of the gas in the system can be written as:

$$\frac{dM_{gas}}{dt} = -\psi(t) + E(t) \tag{4.31}$$

where $E(t)$ is the rate at which dying stars restore both the enriched and unenriched material into the ISM. $E(t)$ can be written as:

$$E(t) = \int_{m(t)}^{\infty} (m - M_{rem})\psi(t - \tau_m)\varphi(m)dm \tag{4.32}$$

where $m - M_{rem}$ is the total mass ejected from a star of mass m, and τ_m is the lifetime of a star of mass m. When $E(t)$ is substituted into the gas equation one obtains an integer-differential equation which can be solved analytically only by assuming Instantaneous Recycling Approximation (I.R.A.). In this approximation one assumes that all stars less massive than $1M_\odot$ live forever whereas all stars more massive than $1M_\odot$ die instantaneously. In other words, I.R.A. allows us to neglect the stellar lifetimes and solve analytically equation (4.31). Under I.R.A., we can define the returned fraction:

$$R = \int_1^{\infty} (m - M_{rem})\varphi(m)dm \tag{4.33}$$

which is called fraction because is divided by $\int_1^{\infty} m\varphi(m)dm = 1$, which depends on the normalization of the IMF. We also define the yield per stellar generation as:

$$y_i = \frac{1}{1-R} \int_1^{\infty} m p_{im} \varphi(m) dm \tag{4.34}$$

where p_{im} is the stellar yield previously defined. Then, by substituting R and y_i into equation (4.32) we obtain:

$$E(t) = \psi(t) R \tag{4.35}$$

and:

$$\frac{dM_{gas}}{dt} = -\psi(t)(1-R) \tag{4.36}$$

The equation for the evolution of the chemical abundances can be written as:

$$\frac{d(X_i M_{gas})}{dt} = -X_i \psi(t) + E_i(t) \tag{4.37}$$

where:

$$E_i(t) = \int_{m(t)}^{\infty} [(m - M_{rem}) X_i(t - \tau_m) + m p_{im}] \cdot \psi(t - \tau_m) \varphi(m) dm \tag{4.38}$$

contains both the unprocessed and the newly produced element i. It is worth noting that this equation is valid for metals and not for elements which are wholly or partly destroyed in stars. Under the assumption of I.R.A. the above eq. becomes:

$$E_i(t) = \psi(t) R X_i(t) + y_i(1-R)\psi(t) \tag{4.39}$$

When substituted in (4.37) the equation can be solved analytically with the previous initial conditions and the solution is:

$$X_i = y_i ln(\frac{1}{\mu}) \tag{4.40}$$

the famous solution for the Simple Model. The yield which appears in the above solution is known as *effective yield*, simply defined as the yield $y_{i_{eff}}$ that would be deduced if the system were assumed to be described by the Simple Model:

$$y_{i_{eff}} = \frac{X_i}{ln(1/\mu)} \tag{4.41}$$

The meaning of the effective yield can be understood with the following example: if $y_{i_{eff}} > y_i$(true yield) then the actual system has attained a higher abundance for the element i at a given gas fraction μ.

4.2. Failure of the Simple Model

The Simple Model predicts too many stars with metallicity lower than [Fe/H]= -1.0 dex relative to observations. This is known as "G-DWARF PROBLEM". However, the G-dwarf is no more a problem since several solutions have been suggested.

Possible solutions to the G-dwarf problem include:

-Slow formation of the solar vicinity by gas infall

-Variable IMF

-Pre-enriched gas

Generally, the slow infalling gas is preferred since it is the most realistic suggestion and produces results in very good agreement with the observations, as we will see in the following.

4.3. Analytical models with gas flows

The equation for the evolution of abundances in presence of gas flows transforms into:

$$\frac{d(X_i M_{gas})}{dt} = -X_i(t)\psi(t) + E_i(t) + X_{A_i}A(t) - X_i(t)W(t) \tag{4.42}$$

where $A(t)$ is the accretion rate of matter with abundance of the element i X_{A_i} and $W(t)$ is the rate of loss of material from the system. The case $A(t) = W(t) = 0$ obviously corresponds to the Simple Model. The case $A(t) = 0$, $W(t) \neq 0$ corresponds to the outflow model. The easiest way of defining $W(t)$ in order to solve the equation analytically is to assume:

$$W(t) = \lambda(1 - R)\psi(t) \tag{4.43}$$

where $\lambda \geq 0$ is the wind parameter. The analytical solution for the equation of metals, which can be integrated between 0 e $X_i(t)$ and between $M_{gas}(0) = M_{tot}$ and $M_{gas}(t)$, is:

$$X_i = \frac{y_i}{(1 + \lambda)}ln[(1 + \lambda)\mu^{-1} - \lambda] \tag{4.44}$$

For $\lambda = 0$ we recover the solution of the Simple Model.

The case of $A(t) \neq 0$ and $W(t) = 0$ corresponds to the accretion model. The easiest way to choose the accretion rate is:

$$A(t) = \Lambda(1 - R)\psi(t) \tag{4.45}$$

with Λ being a positive constant different from zero. The solution of the equation of metals for a primordial infalling material ($X_{A_i} = 0$) and $\Lambda \neq 1$ is :

$$X_i = \frac{y_i}{\Lambda}[1 - (\Lambda - (\Lambda - 1)\mu^{-1})^{-\Lambda/(1-\Lambda)}] \tag{4.46}$$

as shown by Matteucci & Chiosi (1983). If $\Lambda = 1$ (extreme infall model) the solution is:

$$X_i = y_i[1 - e^{-(\mu^{-1}-1)}] \tag{4.47}$$

where the quantity $\mu^{-1} - 1$ represents the ratio between the accreted mass and the initial mass.

5. Equations with Type Ia and II SNe

In general, if one wants to compute in detail the evolution of the abundances of elements produced and restored into the ISM on long timescales, the I.R.A. approximation is a bad approximation. Therefore, it is necessary to consider the stellar lifetimes in the chemical evolution equations and solve them with numerical methods.

If G_i is the mass fraction of gas in the form of an element i, we can write:

$$\dot{G}_i(t) = -\psi(t)X_i(t)$$

$$+ \int_{M_L}^{M_{Bm}} \psi(t - \tau_m)Q_{mi}(t - \tau_m)\phi(m)dm$$

$$+ A \int_{M_{Bm}}^{M_{BM}} \phi(m)$$

$$\cdot [\int_{\mu_{min}}^{0.5} f(\mu)\psi(t - \tau_{m2})Q_{mi}(t - \tau_{m2})d\mu]dm$$

$$+ B \int_{M_{Bm}}^{M_{BM}} \psi(t - \tau_m)Q_{mi}(t - \tau_m)\phi(m)dm$$

$$+ \int_{M_{BM}}^{M_U} \psi(t - \tau_m) Q_{mi}(t - \tau_m) \phi(m) dm$$

$$+ X_{A_i} A(t) - X_i W(t) \tag{5.48}$$

where B=1-A. The parameter A represents the unknown fraction of binary stars giving rise to type Ia SNe and is fixed by reproducing the observed present time SN Ia rate. Generally, values in the range $A = 0.05 - 0.09$ (according to the IMF) reproduce well the observed SN Ia rate in the Galaxy. The chemical abundances are defined as: $X_i = \frac{G_i}{G}$, where $G = \mu = \frac{M_{gas}}{M_{tot}} = \frac{\sigma_{gas}}{\sigma_{tot}}$. The total mass M_{tot} (or surface mass density σ_{tot}) refers to the mass of stars (dead and alive) plus gas at the present time.

$W_i(t) = X_i W(t)$ is the galactic wind rate for the element i, whereas $A_i(t) = A(t) X_{A_i}$, with $A(t)$ being the accretion rate for the element i.

M_{Bm} is the total minimum and M_{BM} the total maximum mass allowed for binary systems giving rise to Type Ia SNe (Matteucci & Greggio 1986). For the model of a C-O WD plus a red-giant companion for the progenitors of Type Ia SNe, $M_{BM} \leq 16 M_\odot$. M_{Bm} is more uncertain and often has been taken to be $3 M_\odot$ to ensure that the primary star (the initially more massive in the binary system) would be massive enough to guarantee that after accretion from the companion the C-O white dwarf eventually reaches the Chandrasekhar mass and ignites carbon. This formulation of the SN Ia rate was originally proposed by Greggio & Renzini (1983). Greggio (1996) presented revised criteria for the choice of M_{Bm}. In particular, the suggested condition for the explosion of the system is:

$$M_{WD} + \epsilon M_{2,e} \geq M_{Ch} \tag{5.49}$$

(where $M_{2,e}$ is the envelope mass of the evolving secondary, M_{WD} is the mass of the white dwarf and M_{Ch} is the Chandrasekhar mass).

For models involving Sub-Chandrasekhar white dwarf masses, which have been suggested to explain subluminous Type Ia SNe Greggio obtains: $M_{WD} \geq 0.6 M_\odot$ and $\epsilon M_{2,e} \geq 0.15$

The masses $M_L = 0.8$ and $M_U = 100 M_\odot$ define the lowest and the highest mass, respectively, contributing to the chemical enrichment. The function $\tau_m(m)$ describes the stellar lifetimes. The quantity $Q_{mi}(t - \tau_m)$ contains all the information about stellar nucleosynthesis for elements either produced or destroyed inside stars or both, and is defined as in Talbot & Arnett (1973).

5.1. *Type Ia SN rates*

The single degenerate (SD) scenario is based on the original suggestion of Whelan & Iben (1973), namely C-deflagration in a WD reaching the Chandrasekhar mass after accreting material from a star which becomes red giant and fills its Roche lobe. An alternative to the SD scenario is represented by the double degenerate (DD) scenario, where the merging of two C-O WDs, due to gravitational wave radiation, creates an object exceeding the Chandrasekhar mass and exploding by C-deflagration (Iben & Tutukov 1984). Negative results from observational searches for very close binary systems made of massive enough WDs (Bragaglia et al. 1990) has made the DD scenario less attractive and people to concentrate more on the SD scenario.

A recent model has been suggested by Hachisu et al. (1996, 1999) and is based on the classical scenario of Whelan & Iben (1973) but with a metallicity effect. It predicts that no Type Ia systems can form for [Fe/H]< -1.0. This model seems to have some difficulty in explaining the low [α/Fe] ratios observed in Damped Lyman-α systems (DLAs) which show that even at low metallicities is present the effect of Type Ia SNe. Recently, Matteucci & Recchi (2001) showed that there are some problems with this scenario also in explaining the observed [O/Fe] vs. [Fe/H] relation in the solar neighbourhood and concluded that the best model is still the SD one in the formulation proposed by Greggio & Renzini (1983). They also showed that the typical timescale for Type Ia SN enrichment, defined as the time t_{SNIa} when the SN rate reaches the maximum, varies strongly from galaxy to galaxy and that it is not correct to adopt a universal $t_{SNIa}=1$ Gyr, as often quoted in the literature. Their results indicate that for an elliptical galaxy with high SFR, $t_{SNIa} = 0.3 - 0.5$ Gyr, for a spiral Galaxy like the Milky Way, $t_{SNIa} = 4 - 5$ Gyr and for an irregular galaxy with a continuous but very low SFR, $t_{SNIa} > 5$ Gyr. This fact has very important consequences on the chemical evolution of galaxies of different morphological types, as we will see in the following.

6. The formation and evolution of the Milky Way

The first application of models of chemical evolution is the Milky Way (MW) Galaxy for which we have most of the available data. Before describing the chemical evolution of the Galaxy I recall some of the ideas proposed for the formation of the MW. Eggen, Lynden-Bell & Sandage (1962) suggested a rapid collapse for the formation of the Galaxy lasting $\sim 3 \cdot 10^8$ years implying that no spread in the age of Globular Clusters should be observed. They based their suggestion on the finding that halo stars have high radial velocities whitnessing the initial fast collapse. Later on, Searle & Zinn (1978) proposed a central collapse but also that the outer halo formed by merging of large fragments taking place over a considerable timescale > 1 Gyr. More recently, Berman & Suchov (1991) proposed the *hot Galaxy picture*, an initial strong burst of star formation which inhibited further star formation for a few gigayears, while a strong galactic wind was created. Subsequently, the remainder of the proto-Galaxy, contracted and cooled to form the major stellar components observed today. At the present time, the most popular idea on the formation of the Galaxy is that the inner halo formed rather quickly on a time scale of 0.5-1 Gyr, whereas the outer halo formed more slowly by mergers of fragments or accretion from satellites of the MW (see Matteucci 2001).

6.1. Models for the Milky Way

• SERIAL FORMATION APPROACH: halo, thick and thin disk formed in sequence, as a continuous process (e.g. Matteucci & François 1989).

• PARALLEL FORMATION APPROACH: the various Galactic components start forming at the same time and from the same gas but evolve at different rates (e.g. Pardi, Ferrini & Matteucci 1995). At variance with the previous scenario it predicts overlapping of stars belonging to the different components.

• TWO-INFALL APPROACH: the evolution of the halo and disk are totally independent and they form out of two separate infall episodes (overlapping in metallicity is also predicted) (e.g. Chiappini, Matteucci & Gratton, 1997; Chang et al. 1999, Alibés et al. 2001).

• STOCHASTIC APPROACH: in the early halo phases, mixing was not efficient, thus pollution from single SNe (Tsujimoto et al. 1999; Argast et al. 2000; Oey 2000) can be seen in very metal poor stars. This approach predicts a large spread in the abundance ratios at very low [Fe/H], even larger than observed.

6.2. The two-infall model

Here I describe in more detail the two-infall approach (Chiappini et al. 1997) since it gives the best agreement with observations for the halo and disk.

The basic equations for the evolution of the MW in this case contain only an accretion term (no outflow):

$$\frac{dG_i(r,t)}{dt} = -X_i\psi(r,t)+$$

$$(1-\alpha)\int_{M_L}^{M_{Bm}}\psi(r,t-\tau_m)Q_{m_i}(t-\tau_m)\varphi(m)dm+$$

$$\alpha n\int_{M_L}^{M_{Bm}}\psi(r,t-\tau_m-\Delta t)Q_{m_i}(t-\tau_m)\varphi(m)dm+$$

$$A\int_{M_{BM}}^{M_Bm}SNIa+$$

$$(1-A)\int_{M_{BM}}^{M_Bm}SNeII+\int_{M_{BM}}^{M_U}SNeII+$$

$$(\frac{dG_i}{dt})_{inf}$$

These equations are the same as (5.48) plus a term which contains the contribution from novae, which are perhaps important producers of ^7Li, ^{15}N and ^{13}C (D'Antona & Matteucci, 1991; Romano et al. 2001). The nova contribution contains the parameter $\alpha=0.0155$ which represents the fraction of WDs which are in binary systems giving rise to nova events. The value of α is chosen to reproduce the present time nova rate $R_{novae}\sim 26yr^{-1}$, after assuming that each nova has $\sim 10^4$ outbursts all over its lifetime. The quantity $\Delta t\sim 1$ Gyr is the assumed time-delay for the starting of nova activity since the formation of the binary system.

The infall rate term is defined as:

$$A_i(t)=\frac{dG_i(r,t)}{dt}=\frac{A(r)(X_A)_ie^{-t/\tau_H}}{\sigma_{tot}(r,t_G)}+\frac{B(r)(X_A)_ie^{-(t-t_{max})/\tau_D(r)}}{\sigma_{tot}(r,t_G)} \tag{6.50}$$

where τ_H is the timescale for the inner halo formation (0.5-1 Gyr) and $\tau_D(r)$ is the thin disk timescale varying with galactocentric distance (inside-out formation):

$$\tau_D(r)=0.875r-0.75 \tag{6.51}$$

The SFR is given by:

$$\psi(t)=\nu\sigma_{tot}(r,t)^{k_1}\sigma_{gas}(t)^{k_2} \tag{6.52}$$

with $k_1=0.5$ and $k_2=1.5$. The behaviour of this SFR for the halo-thick disk and the thin-disk phase, respectively, is show in Figure 1. A threshold density ($\sigma_{th}=7M_\odot pc^{-2}$) in the SFR is assumed.

6.3. Applications to the Local Disk

• *The G-dwarf metallicity distribution* is shown in Figure 2 where the data are compared with the predictions of the two infall model assuming a time scale for the formation of the local disk of 8 Gyr. (Chiappini et al. 1997; Boissier & Prantzos 1999; Chang et al. 1999; Chiappini et al. 2001).

• *The relative abundance ratios as functions of the relative metallicity (relative to the Sun) [X/Fe] vs. [Fe/H]*, they are interpreted as due to the time-delay between Type Ia and II SNe. In fact, for the α-elements the slowly declining [α/Fe] ratio at low [Fe/H] (halo phase) is normally interpreted as due to the pollution from massive stars, whereas

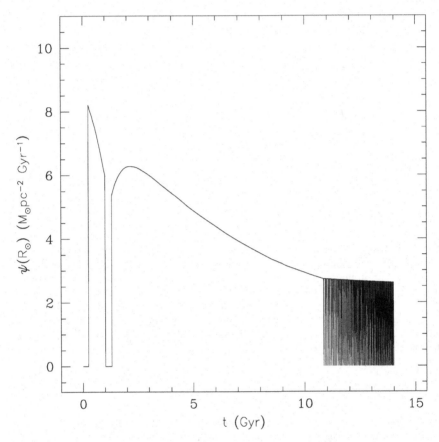

FIGURE 1. Predicted star formation rate in the halo-thick disk phase and in the thin disk phase during the evolution of the Milky Way. Notice that because of the existence of a threshold in the gas density the SFR halts in between the two major episodes of infall and oscillates in the last phases of the evolution of the disk. The models are from Chiappini et al. (2001).

the abrupt change in slope occurring at $\sim [Fe/H] = -1.0$ dex is due to the Type Ia SNe restoring the bulk of iron. From the $[\alpha/Fe]$ vs. $[Fe/H]$ diagram one can infer the timescale for the formation of the halo ($\tau_h \sim 1.5$-2.0 Gyr, Matteucci & François, 1989; Chiappini et al. 1997), just by means of the age-$[Fe/H]$ relationship, which indicates the time at which the metallicity of the turning point is reached. In Figure 3 we do not show the usual plot but $[Fe/O]$ vs. $[O/H]$ since in this plot are evident some features which do not appear in the classical diagram. In particular, the data show evidence for a gap around $[O/H] = -0.3$ dex, corresponding to $[Fe/H] \sim -1.0$ dex (Gratton et al. 2000). This gap is well reproduced by the model which predicts a hiatus (no more than 1 Gyr) in the SFR between the end of the thick-disk phase and the beginning of the formation of the thin disk. This hiatus produces, in fact, a situation where O is no longer produced whereas Fe is produced, and this is revealed by the increase of $[Fe/\alpha]$ at constant $[\alpha/H]$. This effect has been observed also for $[Fe/Mg]$ vs. $[Mg/H]$ by Fuhrmann (1998). The model shown

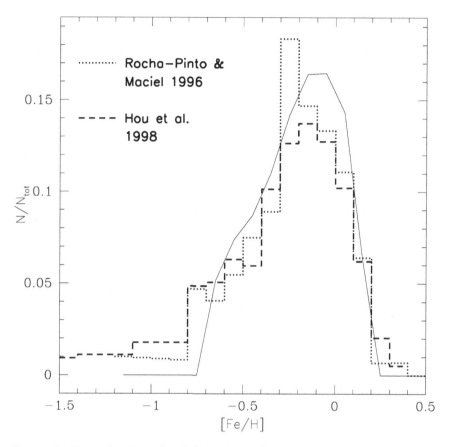

FIGURE 2. Observed and predicted G-dwarf metallicity distribution. The data are from Rocha-Pinto & Maciel (1996) and Hou et al. (1998). The model predictions (continuous line) are from Chiappini et al. (2001). The model assumes that the timescale for the disk formation in the solar neighbourhood is 8 Gyr.

in Figure 3 is a two-infall model with a threshold density for the star formation and is just the existence of such a threshold which produces the hiatus in the SFR, evident in Figure 1.

6.4. *Applications to the whole disk*

• *Abundance Gradients* are known to exist along the Galactic disk from data from various sources (HII regions, PNe, B stars). These data suggest that the gradient for oxygen is ∼ −0.07 dex/kpc in the galoctocentric distance range 4-14 kpc. It is not yet clear if the slope is unique or if there is a change in slope as a function of the galactocentric distance. Similar gradients are found for N and Fe (see Matteucci 2001 and references therein).

• *Gas Distribution.* HI is roughly constant over a range of 4-10 kpc along the Galactic

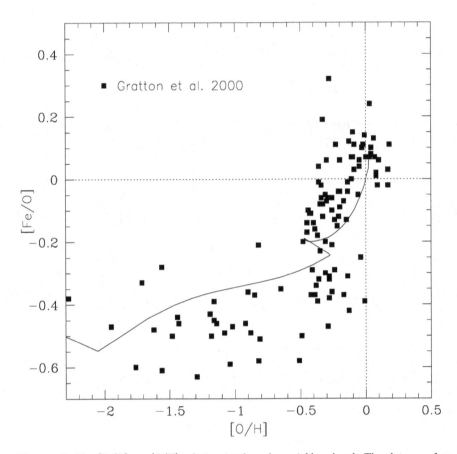

FIGURE 3. The [Fe/O] vs. [O/H] relation in the solar neighbourhood. The data are from Gratton et al.(2000) and the model (continuous line) from Chiappini et al.(2001). In this figure is evident the existence of a gap in the data at around [O/H]=-0.3 dex, corresponding to [Fe/H] ~ -1.0 dex, which is also predicted by the model.

disk while H_2 follows the light distribution. No models can explain the two distributions. The total gas increases towards the center with a peak at 4-6 kpc.

• *The SFR Distribution* is obtained from various tracers (Lyman-α continuum, pulsars, SN remnants, molecular clouds) and shows that the SFR increases with decreasing galactocentric distance reaching a peak at 4-6 kpc in correspondence of the gas peak.

In order to fit gradients, SFR and gas one has to assume that the disk formed inside-out, in agreement with a previous suggestion by Larson (1976) and that the SFR should be a strongly varying function of the galactocentric distance. In Figure 4 is shown a comparison between an inside-out model and the abundance gradient, the gas, star and SFR distributions.

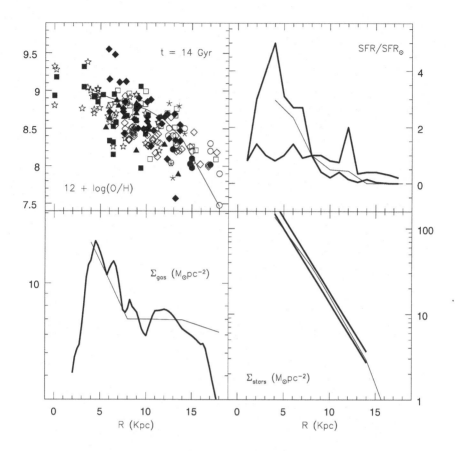

FIGURE 4. Comparison between data and model predictions from Chiappini et al. (2001) (continuous lines). In the first panel from the top we show the oxygen abundance gradient. In the second panel the variation of the $SFR/SFR_\odot$, in the third panel the gas distribution and in the fourth panel the distribution of stars along the Galactic disk.

6.5. *The Role of Radial Flows in the evolution of the Galactic Disk*

The gas infalling onto the disk can induce radial inflows by transferring angular momentum to the gas in the disk. Angular momentum transfer can be due to the gas viscosity in the disk and induce inflows in the inner parts of the disk and outflows in the outer parts. All viscous models suggest that metallicity gradients can be steepened by radial (in)flows, especially if an outer star formation cut-off is assumed (Clarke 1989; Yoshii & Sommer-Larsen 1989). All models agree that the velocity of radial (in)flows should be low ($v < 2$ km/sec). Observationally is not clear if these radial flows exist. Edmunds & Greenhow (1995), by means of analytical models of galactic chemical evolution concluded that there is no simple *one-way* effect of radial flows on abundance gradients. Portinari & Chiosi (2000) suggested that radial inflows may represent a possible explanation of the peak of the gas at 4-6 kpc. In our opinion radial flows, if they exist, are never the main cause for the formation of abundance gradients.

6.6. *The Role of the IMF in the evolution of the Galactic Disk*

Abundance gradients, in principle can be obtained by a variation of the IMF along the disk. One can assume either that more massive stars form in external regions or the contrary, that more low mass stars form in external regions. It has been shown convincingly that neither of the two options work (Carigi 1996; Chiappini et al. 2000), and that there is no evidence in favor of such variations along the Galactic disk. Therefore, we can exclude the variation of the IMF as the main cause of abundance gradients and conclude that the best agreement with the observed properties of the Galactic disk is obtained by assuming a constant IMF.

6.7. *Scenarios for Bulge Formation*

Various scenarios have been proposed insofar for the formation of the Galactic bulge but only one seems to reproduce the observed abundance pattern. Here I recall the main scenarios:

• Accretion of extant stellar systems which eventually settle in the center of the Galaxy.

• Accumulation of gas at the center of the Galaxy and subsequent evolution with either fast or slow star formation.

• Accumulation either rapid or slow of metal enriched gas from the halo or thick disk in the Galaxy center.

• Formation occurs out of inflow of metal enriched gas from the thin-disk.

The metallicity distribution of stars in the bulge as well as the $[\alpha/Fe]$ vs. $[Fe/H]$ relations help in selecting the most probable scenario and suggest that a fast accumulation of gas in the Galactic center accompanied by fast star formation is the best scenario. In this framework, the bulge must have formed contemporarily to the inner halo on a similar timescale. In Figures 5 and 6 we show some model predictions compared with the metallicity distribution of bulge stars and the predicted $[\alpha/Fe]$ vs. $[Fe/H]$ ratios for the bulge together with the predictions for the solar neighbourhood. The model for the bulge presented in the two figures (Matteucci, Romano & Molaro, 1999) assumes a much stronger star formation rate than in the solar neighbourhood (by a factor of 10) with the same nucleosynthesis prescriptions and a timescale for bulge formation of 0.5 Gyr as opposed to 8 Gyr in the solar vicinity. In Figure 5 it is evident that the best model to reproduce the stellar metallicity distribution requires a Salpeter (1955) IMF which is sligthly flatter than that used for the solar neighbourhood (Scalo, 1986). The predicted $[\alpha/Fe]$ ratios in Figure 6 indicate that the bulge stars should show overabundances of α-elements for most of the $[Fe/H]$ range, as it seems also suggested by observations (Mc William & Rich, 1994; Barbuy, 1999). This is a consequence of the time-delay between Type Ia and II SNe, coupled with a very fast evolution in the bulge as compared to the solar vicinity. In fact, in this case high values of $[Fe/H]$ are reached in the gas before a substantial number of SNe Ia has the time to restore the bulk of iron. The contrary occurs in a system with lower star formation than in the solar neighbourhood, such as the external regions of the disk and irregular galaxies. In these systems we expect that the overabundance of α-elements relative to Fe is maintained only for a short interval of $[Fe/H]$ (see Pagel, 1997: Matteucci 2001).

7. Disks of Other Spirals

Abundance gradients

Abundance gradients in dex/kpc are known to exist also in the disk of other spirals showing that they are steeper in smaller disks, but this correlation disappears if the

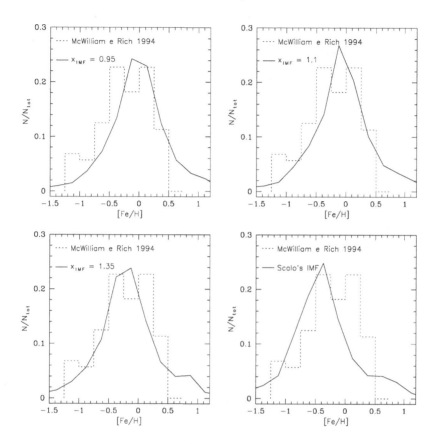

FIGURE 5. Comparison between data (dotted lines) and model predictions for the metallicity distribution of bulge stars. The data are from McWilliam & Rich (1994) and the models (continuous lines) from Matteucci et al. (1999). Each panel corresponds to a model with a different IMF slope, as indicated.

gradients are expressed in units of dex/scalelength, thus indicating the existence of a universal slope per unit scalelength (e.g. Garnett, 1998). Another remarkable characteristic about abundance gradients in other spirals is that they appear to be flatter in galaxies with central bars, suggesting that the dynamical effect of the bar can influence the evolution of the disk.

The SFR

The SFR is measured mainly from H_α emission (Kennicutt, 1998b) and implies a correlation with the total surface gas density (HI+H_2) as discussed in section 2.1.

Gas distributions

Gas distributions, especially the HI distribution is known for a fair sample of spirals. There are indications of differences between field and cluster spirals (e.g. Skillman et al. 1996).

Integrated colors

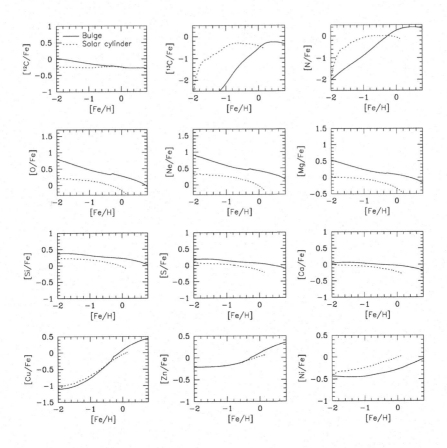

FIGURE 6. Predicted [α/Fe] vs. [Fe/H] relations for the bulge (continuous lines) and for the solar neighbourhood (dotted lines). The models are from Matteucci et al. (1999).

Studies of integrated colors of spiral disks reflect the observed abundance gradients and are well reproduced by an inside-out formation of disks similar to what assumed for the Milky Way (Josey & Arimoto 1992; Jiménez et al. 1998; Prantzos & Boissier 2000).

8. Conclusions on the Milky Way and other spirals

The comparison between observations and models for the Milky Way suggests that:

• The disk of the Galaxy formed mostly by infall of primordial or very metal poor gas accumulating faster in the inner than in the outer regions (inside-out scenario, i.e. $\tau_D(R)$ decreases with decreasing R).

• In the framework of the inside-out scenario, the SFR should be a strongly varying function of the galactocentric distance. In particular, it should either depend on the total surface mass density (feed-back mechanism) or on the angular circular velocity of gas. Both formulations for the SFR are supported by observations. Under the assumption of a strongly varying SFR, the observed abundance gradients can be nicely reproduced.

• Radial flows probably are not the main cause of gradients but can help in reproducing the gas profile.

• A constant IMF (in time and space) should be preferred, since variable IMFs have been tested and they cannot reproduce all of the observational constraints of the solar neighbourhood and the whole disk at the same time.

• The properties of the disks of other spirals also indicate an inside-out disk formation.

• In the framework of semi-analytical models of galaxy formation (Mo, Mao & White, 1998), the evolution of galaxy disks can be described by means of scaling laws calibrated on the Galaxy with V_c and λ as parameters (Jiménez et al. 1998; Prantzos & Boissier, 2000), where V_c is a measure of the mass of the dark halo and λ is a measure of the specific angular momentum of the halo.

• Dynamical processes such as the formation of a central bar can influence the evolution of disks and deserve more attention in the future.

• Abundance ratios (e.g. $[\alpha/\text{Fe}]$) in stars al large galactocentric distances can give us a clue to interpret the formation of the disk and the halo (inside-out or outside-in). In fact, in a typical inside-out scenario we predict that the $[\alpha/\text{Fe}]$ ratios would decrease with galactocentric distance, due to the weaker and sometimes intermittent SF regime, whereas in an outside-in scenario we would expect the contrary due to the faster evolution of the more external region which quickly consume the gas before Type Ia SNe have time to restore the bulk of Fe.

• The predictions of chemical evolution models can be tested in a cosmological context to study the galaxy surface brightness and size evolution as a function of redshift. Roche et al. (1998) have already done that and suggested that a size and luminosity evolution, as suggested by the inside-out scenario fits better the observations.

• Inside-out formation of the Galaxy is suggested also by the fact that the globular clusters of the inner halo are coeval ($\Delta t \sim 0.5$ Gyr, Rosenberg et al. 1999), whereas the age difference seems to increase in the outer halo.

• The metallicity distribution of stars in the Bulge as well as the observed $[\alpha/\text{Fe}]$ ratios suggest a very fast formation of the Bulge (~ 0.5-0.8 Gyr), due a the very fast SFR, as predicted by succesfull chemical evolution models and by Elmegreen (1999).

9. Elliptical Galaxies

Elliptical galaxies are mainly found in galaxy clusters, they show a large range of luminosities and masses and are made by stars as old as those in globular clusters. No cold gas is observed in these galaxies but hot X-ray gas halos are present. Because of their large masses and metal content they are probably the most important factories of metals in the universe.

9.1. *Observational properties*

Here I recall the main observational features of elliptical galaxies:

• The existence of Color-Magnitude and color-velocity dispersion (σ_o) relations (colors become redder with increasing luminosity and mass; e.g. Bower et al. 1992) is interpreted as a metallicity effect, namely as the fact that the metallicity decreases outwards similarly to what happens in spiral disks.

• The existence of a Mg_2–σ_o (where Mg_2 is a metallicity index) relation reinforces the previous point (Bender et al. 1993; Bernardi et al. 1998; Colless et al. 1999; Kuntschner et al. 2001).

• Inside ellipticals, Mg_2 correlates also with the escape velocity v_{esc}, as first shown by Franx & Illingworth (1990), indicating that the magnesium index is larger where the escape velocity is larger.

• M/L_B increases by a factor of ~ 3 from faint to bright ellipticals implying a tilt of the fundamental plane of ellipticals (Bender et al. 1992). The fundamental plane is the particular plane occupied by these galaxies in the space defined by the stellar velocity dispersion, the effective radius and the surface brigthness.

• Abundance gradients inside ellipticals have been measured by means of metallicity indices such as Mg_2 and $< Fe >$ (Carollo et al. 1993; Davies et al. 1993; Kobayashi & Arimoto 1999). These gradients correspond roughly to a gradient in [Fe/H] of the stellar component of $\Delta[Fe/H]/\Delta logr \sim -0.3$.

The average metallicity of the stellar component in ellipticals is $< [Fe/H] >_* \sim -0.3$dex (from -0.8 to +0.3).

• By comparing synthetic metallicity indices with the observed ones some authors (Worthey et al. 1992; Weiss et al. 1995; Kuntschner et al. 2001) have suggested that the average stellar $< [Mg/Fe] >_*$ is larger than zero (from 0.05 to +0.3 dex) in nuclei of giant ellipticals. Moreover, there is indication that $< [Mg/Fe] >_*$ increases with increasing σ_o (M_{gal}) and luminosity (Worthey et al. 1992; Jorgensen 1999; Kuntschner et al. 2001).

In particular the relation found by Kuntschner et al. (2001) is: $[Mg/Fe]=0.30(\pm0.06)log\sigma_o - 0.52(\pm0.15)$.

9.2. *Formation of Ellipticals*

Several mechanisms have been suggested for the formation and evolution of elliptical galaxies, in particular:

• Early monolithic collapse of a gas cloud or early merging of lumps of gas where dissipation plays a fundamental role (Larson 1974; Arimoto & Yoshii 1987; Matteucci & Tornambè 1987). In this scenario the star formation stops soon after a galactic wind develops and the galaxy evolves passively since then.

• Bursts of star formation in merging subsystems made of gas (Tinsley & Larson 1979). In this picture star formation stops after the last burst and gas is lost via stripping or wind.

• Early merging of lumps containing gas and stars in which some dissipation is present (Bender et al. 1993).

• Merging of early formed stellar systems in a wide redshift range and preferentially at late epochs (Kauffmann et al. 1993). A burst of star formation can occur during the major merging where $\sim 30\%$ of the stars can be formed (Kauffmann 1996).

The main difference between the monolithic collapse scenario and the hierarchical merging relies in the time of galaxy formation, occurring quite early in the former scenario and continuously in the latter scenario. As we will see, there are arguments either in favour of the monolithic or the hierarchical scenario.

9.3. *Formation of Ellipticals at low z*

Here I recall some of the main arguments in favor of the formation of ellipticals at low redshifts:

• Relative large values of the H_β index measured in a sample of nearby ellipticals which could indicate prolonged star formation activity up to 2 Gyr ago (González 1993; Trager et al. 1998).

• The tight relations in the fundamental plane are due to a conspiracy of age and metallicity in the sense that it should exist an age-metallicity anticorrelation implying that the more metal rich galaxies are also younger (Ferreras et al. 1999; Trager et al. 2000).

• The apparent paucity of high luminosity ellipticals at $z \sim 1$ compared to now claimed by a series of authors (Kauffmann et al. 1993; Zepf, 1997; Menanteau et al. 1999).

9.4. *Formation of Ellipticals at high z*

Here I recall the arguments in favor of a formation of ellipticals at high redshift:

• The tightness of the color-central velocity dispersion relation found for Virgo and Coma galaxies (Bower et al. 1992). If the formation of ellipticals were a continuous process we should expect a much larger spread in the galaxy colors for a given central velocity dispersion. In particular, the argument goes like that: from the observed color scatter one can derive $t_H - t_F \sim 2$Gyr (where t_H is the Hubble time and t_F the time of galaxy formation). If $t_H = 15$ Gyr then the youngest ellipticals must have formed ~ 13 Gyr ago at $z \geq 2$ (Renzini, 1994).

• The thinness of the fundamental plane for ellipticals in the same two clusters, in particular the M/L vs. M relation (Renzini & Ciotti 1993).

• The tightness of the color-magnitude relation for ellipticals in clusters up to $z \sim 1$ (Kodama et al. 1998; Stanford et al. 1998)

• The modest passive evolution measured for cluster ellipticals at intermediate redshift (van Dokkum & Franx 1996; Bender et al. 1996).

• Lyman-break galaxies at $z \geq 3$ where the SFR$\sim 50 - 100 M_\odot yr^{-1}$ could be the young ellipticals (Steidel et al. 1996; 1998).

• The strongly evolving population of Luminous Infrared Galaxies suggesting that they are progenitors of massive spheroidals (Blain et al. 1999; Elbaz et al. 1999).

9.5. *Models for ellipticals based on galactic winds*

Monolithic models assume that ellipticals suffer a strong star formation and quickly produce galactic winds when the energy from SNe injected into the ISM equates the potential energy of the gas. Then, star formation is assumed to halt after the development of a galactic wind. In this framework, the evolution of ellipticals crucially depends on the time at which a galactic wind occurs, t_{GW}. For this reason, it is extremely important to understand the SN feedback and star formation process. The condition for the onset of a wind can be written as:

$$(E_{th})_{ISM} \geq E_{Bgas} \tag{9.53}$$

The thermal energy of gas due to SNe and stellar wind heating is:

$$(E_{th})_{ISM} = E_{th_{SN}} + E_{th_w} \tag{9.54}$$

with

$$E_{th_{SN}} = \int_0^t \epsilon_{SN} R_{SN}(t^{'}) dt^{'} \tag{9.55}$$

and

$$E_{th_w} = \int_0^t \int_{12}^{100} \varphi(m) \psi(t^{'}) \epsilon_w dm dt^{'} \tag{9.56}$$

for the contribution from SNe and stellar winds, respectively. The quantity R_{SN} represents the SN rate (II and Ia). The quantities $\epsilon_{SN} = \eta_{SN} \epsilon_o$ with $\epsilon_o = 10^{51}$erg where ϵ_o is the typical SN blast wave energy, and $\epsilon_w = \eta_w E_w$ with $E_w = 10^{49}$erg (typical energy injected by a $20 M_\odot$ star taken as representative), are the efficiencies for the energy transfer from SNe II and Ia into the ISM. These efficiencies, η_w and η_{SN}, can be assumed as free parameters or be calculated from the results of the evolution of a SN remnant in the ISM and the evolution of stellar winds. The SN feedback is, in fact, a crucial parameter. The formulation of Cox (1972) for the efficiency of energy injection from SNe derives from following the evolution of the shock wave produced by the explosion into an ISM with constant density:

$$\epsilon_{SN} = 0.72\epsilon_o \ erg \qquad (9.57)$$

for $t_{SN} \leq t_c = 5.7 10^4 \epsilon_o^{4/17} n_o^{-9/17}$ years, where t_c is the cooling time, and t_{SN} is the time elapsed from the SN explosion, and:

$$\epsilon_{SN} = 2.2\epsilon_o (t_{SN}/t_c)^{-0.62} \ erg \qquad (9.58)$$

for $t_{SN} > t_c$.

With these prescriptions only few % of ϵ_o are deposited into the ISM (see Bradamante et al. 1998). However, multiple SN explosion should change the situation. Unfortunately, very few calculations of this type are available. An important point to consider is also that SNe Ia, which explode after Type II SNe, should provide more energy into the ISM than Type II SNe since they explode in an already formed cavity (see Recchi, Matteucci & D'Ercole, 2001).

The total mass of the galaxy is expressed as $M_{tot}(t) = M_*(t) + M_{gas}(t) + M_{dark}(t)$ with $M_L(t) = M_*(t) + M_{gas}(t)$ and the binding energy of gas is:

$$E_{Bgas}(t) = W_L(t) + W_{LD}(t) \qquad (9.59)$$

and:

$$W_L(t) = -0.5G\frac{M_{gas}(t)M_L(t)}{r_L} \qquad (9.60)$$

represents the potential well due to the luminous matter, whereas:

$$W_{LD}(t) = -Gw_{LD}\frac{M_{gas}(t)M_{dark}}{r_L} \qquad (9.61)$$

is the potential well due to the interaction between dark and luminous matter, where $w_{LD} \sim \frac{1}{2\pi}S(1 + 1.37S)$ with $S = r_L/r_D$ (Bertin et al. 1992).

The SFR is usually assumed to be:

$$SFR = \nu M_{gas} \qquad (9.62)$$

where $\nu \propto M_L^{-\gamma}$ ($\gamma = $-0.11, Arimoto & Yoshii 1987), owing to the fact that the star formation efficiency is just the inverse of the timescale for star formation:

$$\nu = \tau_{SF}^{-1} \qquad (9.63)$$

and that:

$$\tau_{SF} \propto \tau_{coll} \propto \tau_{ff} \qquad (9.64)$$

with τ_{coll} and τ_{ff} being the collapse and the free-fall timescale, respectively. Since dynamical timescales are longer for more massive galaxies, the efficiency of star formation should decrease with galactic mass. The efficiency ν coupled with the increase of the potential well as M_L increases leads to the fact that more massive galaxies form stars for a longer period before suffering a galactic wind. This fact has been invoked for explaining the observed mass-metallicity relation (Larson, 1974). However, this is at variance with the observed [Mg/Fe] vs. σ_o relation, since in this scenario the more massive ellipticals should show the lowest [Mg/Fe].

9.6. *Failure of Larson's Model*

There are different ways of obtaining that $[Mg/Fe]$ in the stellar component increases when the galactic mass M_L increases. These ways are:

- Different timescales for star formation (Worthey et al. 1992) in the sense that star formation should be more efficient in more massive galaxies ($\nu \propto M_L^\gamma$). In this case, the situation could be such that a galactic wind occurs earlier in more massive systems, *the inverse wind scenario* (Matteucci 1994).

- A variable IMF from galaxy to galaxy favoring more massive stars (Mg producers) in more massive galaxies (Worthey et al. 1992; Matteucci, 1994).

- Different amounts and/or concentrations of dark matter as functions of M_L In particular, less dark matter should be present in the most massive systems (Matteucci, Ponzone & Gibson, 1998). As a consequence of this, again galactic winds occur earlier in more massive objects.

- A selective loss of metals: more massive systems loose more Fe relative to α-elements than less massive ones (Worthey et al. 1992).

As mentioned before, Matteucci (1994) proposed a model that she called *inverse wind model*, where the efficiency of star formation is an increasing function of galactic mass, thus implying a shorter period of star formation in massive ellipticals. In fact, the efficiency of star formation is chosen in such a way that in massive ellipticals the galactic wind occurs before than in less massive ones. This produces the increase of the [Mg/Fe] ratio as a function of galactic mass. This approach bears a resemblance with the merging scenario of Tinsley & Larson (1979) where the efficiency of star formation was assumed to increase with the total mass of the system. In the inverse wind model a very massive elliptical of $M_L = 10^{12} M_\odot$ starts developing a wind before 1 Gyr from the beginning of star formation, whereas a small ellipticals form stars for a longer period. As a result, the average $< [Mg/Fe] >_*$ is larger in massive than in small ellipticals. The same effect can be obtained by varying the IMF in such a way that more massive ellipticals tend to form more massive stars. It is worth noting that this second hypothesis could also explain the tilt of the fundamental plane in M/L_B. On the other hand, the hypothesis of the variable dark matter does not produce relevant effects, as shown by Matteucci et al. (1998). The only assumption which has not been tested quantitatively is the selective loss of metals. The main problem with all of these alternatives is, in any case, the lack of a good physical justification depending on the poorly known processes of star formation and SN feeback, and in the future some effort should be devoted to these fields.

It is worth noting that the hierarchical clustering scenario for galaxy formation cannot produce a solution to the observed [Mg/Fe] trend in ellipticals. In fact, it rather predicts the contrary, as shown by Thomas (1999) and Thomas et al. (2002). In fact, in the hierachical clustering the period of star formation in the most massive ellipticals is predicted to be the longest thus favoring low [Mg/Fe] ratios in massive objects, as shown in Figure 7.

9.7. *Averaged Stellar Metallicities*

The metallicity in elliptical galaxies always refers to the metal content of the stars, in particular of the stellar population dominating in the visual light. For this reason we should define the average stellar metallicity. In particular, the average metallicity of a composite stellar population averaged on the mass is:

$$< X_i >_m = \frac{1}{S_1} \int_0^{S_1} X_i(S) dS \qquad (9.65)$$

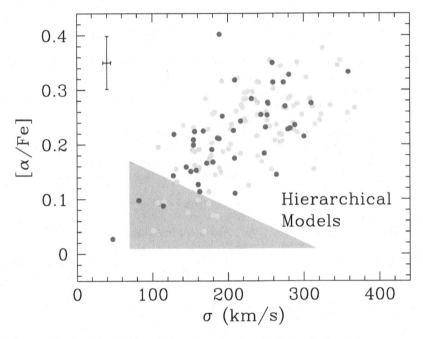

FIGURE 7. Predicted [α/Fe] vs. velocity dispersion for elliptical galaxies. Data are compared with predictions obtained by adopting the star formation history assumed in hierarchical clustering models for galaxy formation. The models and the figure are from Thomas et al. (2002).

where 1 refers to the specific time t_1 and S_1 is the total mass of stars ever born. This is the real average metallicity but in order to compare models and observations it is more appropriate to define the metallicity averaged on the visual light. In particular, the average metallicity of a composite stellar population averaged on the light is:

$$< X_i >_L = \frac{\sum_{ij} n_{ij} X_i L_{Vj}}{\sum_{ij} n_{ij} L_{Vj}} \qquad (9.66)$$

where n_{ij} is the number of stars in the abundance interval X_i and luminosity interval L_{Vj}.

It is worth noting that $< X_i >_m$ is larger than $< X_i >_L$ for galaxies with $M_L < 10^9 M_\odot$, since metal poor giants dominate the visual light whereas they are similar for larger masses (Yoshii & Arimoto 1987).

9.8. *Multi-Zone Models*

There are two multi-zone chemical evolution models for ellipticals available in the literature (Martinelli et al. 1998; Tantalo et al. 1998). Martinelli et al. (1998) assumed that the elliptical galaxy is divided in several concentric shells of thickness ΔR_i. The binding energy of the gas in each shell is computed after assuming a dark matter halo as described before. The model predicts that a galactic wind develops first in the external regions and then gradually in the more internal ones in agreement with the observed Mg_2 vs. v_{esc} relation. As a consequence of this, the star formation lasts for a shorter time in the external regions with the consequence that an abundance gradient is created in agreement with observations. This model also predicts that we should observe higher

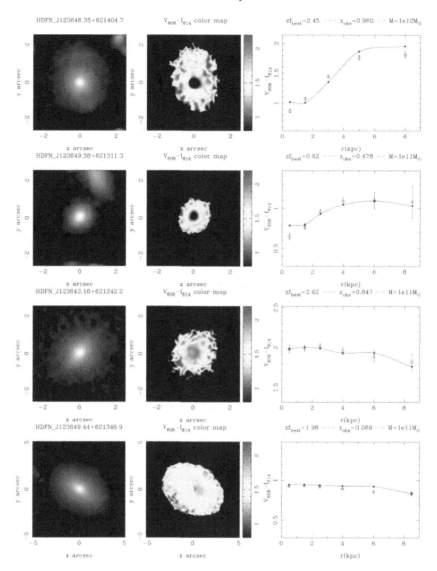

FIGURE 8. Predicted and observed color gradients for some field elliptical galaxies. From Menanteau et al. (2001).Figure shows from left to right: I_{814}-band surface brightness map, V_{606}-I_{814} color pixel map and V_{606}-$I_{814}(r)$ color gradient. Open circles represent observed gradients while solid lines are the model predictions obtained by means of the Martinelli et al. (1998) model.

[Mg/Fe] ratios at larger galactocentric distances, the contrary of what should happens in the disks of spirals. In this case, we can speak of *outside-in* formation. Unfortunately, the available data inside galaxies do not allow yet to observe such an effect. It is worth noting that this model can reproduce very well the observed color gradients in ellipticals, as shown in Figure 8.

10. Conclusions on Ellipticals

The main conclusions that we can draw from comparing theoretical models and observations for ellipticals can be summarized as follows:

• In order to explain the observed $< [Mg/Fe] >_* > 0$ in giant ellipticals the dominant stellar population should have formed on a time scale no longer than $3\text{-}5 \cdot 10^8$ yr, which corresponds to the time at which the SNe Ia rate reaches a maximum in these systems with strong star formation.

• Uncertainties in the stellar yields of Fe and Mg from different authors can change the value of the [Mg/Fe] ratio but do not affect the conclusion above.

• This observational finding argues against a hierarchical clustering formation scenario and favors the fact that ellipticals, especially those in clusters, are mostly old systems (see also Menanteau et al. 2001).

• The increase of the [Mg/Fe] ratio with galactic mass suggests either that more massive ellipticals are older systems or that the IMF is not constant among ellipticals or both.

• Abundance gradients in ellipticals can be produced by biased winds and this would imply that [Mg/Fe] increases with increasing galactocentric distance.

• Better calibrations for metallicity indices are necessary, especially taking into account non-solar ratios in stellar tracks, before drawing firm conclusions.

11. Evolution of Dwarf Galaxies

Dwarf Irregular (DIG) and Blue Compact (BCG) galaxies are very interesting objects for studying galaxy evolution since they are relatively unevolved objects (see Kunth & Östlin, 2000, for a recent exhaustive review). In bottom-up cosmological scenarios they should be the first self- gravitating systems to form, thus they could also be important contributors to the population of systems giving rise to QSO-absorption lines at high redshift (DLAs). In general, they are rather simple objects with low metallicity and large gas content, suggesting that they are either young or have undergone discontinuous star formation activity (bursts). An important characteristic of these systems is that they show a distinctive spread in their physical properties, such as chemical abundances versus fraction of gas. Matteucci & Chiosi (1983) were among the first in studying the chemical evolution of dwarf galaxies. They adopted analytical models as those described in section 3 and showed that closed-box models cannot account for the Z-logμ distribution even if the number of bursts varies from galaxy to galaxy, and suggested possible solutions to explain the observed spread:

• a. different IMFs
• b. different amounts of galactic wind
• c. different amounts of infall

Later on, Matteucci & Tosi (1985) presented a numerical model where galactic winds powered by SNe were taken into account. They concluded that different wind rates from galaxy to galaxy could explain the observed spread in O, N vs. logμ but not the spread in the N/O vs. O/H diagram, suggesting that additional processes could have contributed to that. For example, different amounts of primary N from galaxy to galaxy. Kumai & Tosa (1992) suggested that different fractions of dark matter in different objects could explain the observed spread in the Z-logμ diagram. Pilyugin (1993) forwarded the idea that the spread in the properties of these galaxies (i.e. He/H vs. O/H and N/O vs. O/H) are due to self-pollution of the HII regions coupled with "enriched" or "differential" galactic winds. Differential winds should carry out of the galaxy certain elements more than others, for example the metals ejected by SN II should be favored. In more recent models (Marconi et al. 1994; Bradamante et al. 1998), the novelty was the contribution to the chemical enrichment of SNe of different type (II,Ia and Ib) together with differential winds. In these papers the assumption was made that the products of SN II are lost more easily than the products of stars ending as WDs, such as C and N. Larsen et al. (2001) studied the chemical evolution of gas rich dwarf galaxies and concluded that primary N production from massive stars is not necessary to reproduce the N/O vs O/H and that the spread in this relation is likely to be due to the time-delay effect in the production of N relative to the production of O. In fact, in a starbursting regime when the starburst fades the elements produced by Type II SNe are no more produced whereas the elements produced on long timescales by single stars and Type Ia SNe are still produced. This causes the typical saw-tooth behaviour (see Figure 9). Therefore, the spread can be due to the fact that some objects are observed at different stages of the burst/interburts regime. They also concluded that ordinary winds are better than enriched ones in reproducing the properties of these objects. The existence of a luminosity-metallicity relation (although with spread) can be an indication for galactic winds acting more efficiently in low mass than in high mass high potential well objects.

11.1. *Evidences for Galactic Winds*

Meurer et al (1992), Papaderos et al. (1994), Lequeux et al. (1995) and Marlowe et al. (1995) all suggested the existence of galactic winds in dwarf starbursting galaxies. The evidence is gathered by the indication of outflowing material travelling at a speed larger than the assumed mass of the objects. Papaderos et al. estimated a galactic wind flowing at a velocity of 1320 Km/sec for VIIZw403. The escape velocity estimated for this galaxy being 50 Km/sec. Lequeux et al. (1995) suggested a galactic wind in Haro2=MKn33 flowing at a velocity of $\simeq 200 Km/sec$. Martin (1996, 1998) found supershells in 12 dwarfs including IZw18 and concluded that they imply an outflow which in some cases can become a wind (namely the material is lost from the galaxy).

Recent chemo-dynamical simulations for one instantaneous starburst also suggest the possibiliy of galactic winds and that these winds are metal enriched (MacLow & Ferrara 1999; Recchi et al. 2001).

11.2. *Results for BCG from chemical models*

Purely chemical models (no dynamics) have been computed by several authors by varying the number of bursts, the time of occurrence of bursts, t_{burst}, the star formation efficiency, the type of galactic wind, the IMF and the nucleosynthesis prescriptions (Marconi et al. 1994; Kunth et al.1995; Bradamante et al.1998). The main conclusions of these papers can be summarized as follows:

• The number of bursts should be $N_{bursts} \leq 10$, the star formation efficiency should vary from 0.1 to 0.7 Gyr^{-1} for either Salpeter or Scalo (1986) IMF but Salpeter IMF is favored.

• Enriched winds, carrying out material at a rate proportional to the star formation rate, seem to be preferred (but see Larsen et al. 2001). The observed scatter in the observational properties can be due either to the winds or to the delay in the production of different elements.

• If the burst duration is relatively short (no more than 100 Myr), SNe II dominate

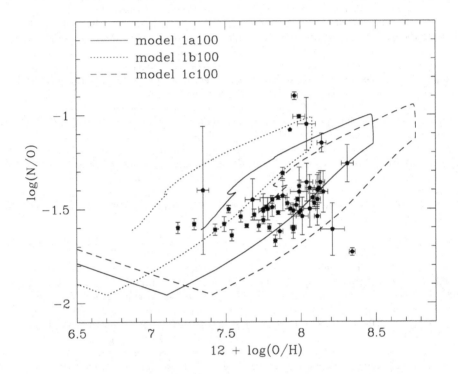

FIGURE 9. The log(N/O) vs. 12 +log(O/H) for a sample of BCG. The data are from Recchi (2002). Overimposed are three models with a single burst of star formation and different star formation efficiency. In particular, the dotted line corresponds to an efficiency $\nu = 1Gyr^{-1}$, the continuous line to $\nu = 2.5Gyr^{-1}$ and the dashed line to $\nu = 5Gyr^{-1}$. The burst duration is 100 Myr. As one can see, the saw-tooth behaviour typical of a bursting mode of star fotmation is evident.

the chemical evolution and energetics of starburst galaxies, while stellar winds seem to be negligible after the onset of SNe II.

• The [O/Fe] ratios tend to be overabundant due to the predominance of Type II SNe during the bursts. Models with a large number of bursts N_{burst}=10 - 15 can give negative [O/Fe].

11.3. *Results from chemo-dynamical models*

Recent chemo-dynamical models (Recchi et al. 2001, 2002) assuming an instantaneous starburst but following in great detail the evolution of several chemical elements (H, He, C, N, O, Mg, Si and Fe) and adopting an efficiency for SN II energy transfer η_{SNII}=0.003 and for SN Ia η_{SNIa}=1.0, suggest that the starburst triggers indeed a galactic wind (see Figure 10). In particular, the metals leave the galaxy more easily than the unprocessed

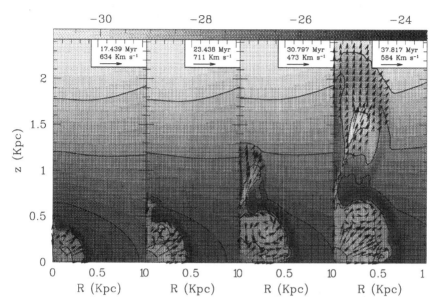

FIGURE 10. Chemo-dynamical simulation of an instantaneous starburst from Recchi et al. (2001). The development of a galactic wind is evident along the z-axis, since the assumed configuration is flattened.The age of the burst is indicated in Myr.

gas and among the metals the SN Ia ejecta leave the galaxy more easily than the SN II ejecta. This is due to the assumed efficiencies for energy transfer for the two types of SNe. This assumption is in turn based on the fact that Type Ia SNe explode in an already heated and rarified medium (thanks to the Type II SNe which explode first) and therefore can transfer all of their energy into the ISM. As a consequence of this type of evolution the following conclusions can be drawn:

• A selective loss of metals seems to occur in dwarf gas-rich galaxies.

• As a consequence of the selective winds, the $[\alpha/Fe]$ ratios inside the galaxy are predicted to be larger than the $[\alpha/Fe]$ ratios outside the galaxy. In fact, the products of SNe Ia are lost more efficiently than those of SN II.

• At variance with previous studies, most of the metals are already in the cold gas phase after 8-10 Myr owing to the fact that the superbubble does not break immediately (the SNe II inject only a fraction of their initial blast wave energy into the ISM) and thermal conduction can act efficiently.

• The model well reproduces the properties of IZw18 (the most metal poor galaxy known locally) if two bursts are assumed and are separated by 300 Myr interval.

11.4. *Dwarf galaxies and DLA Systems*

Finally, before concluding this section, we would like to draw the attention upon the fact that there are similarities between BCG, DIG and DLAs or more in general between DLAs and systems with a low level of star formation.

The nature of DLA systems is under debate and the abundance ratios measured there can be used as a diagnostic to infer their nature and age. Matteucci et al. (1997)

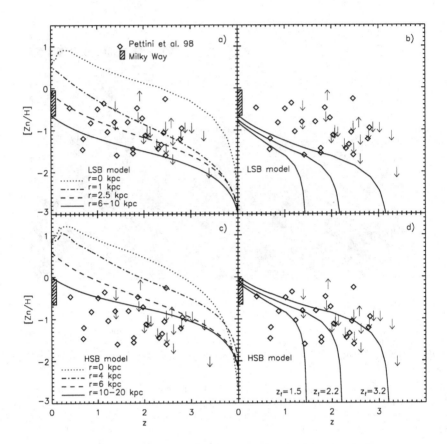

FIGURE 11. Redshift evolution of [Zn/H] for high surface brightness spirals (HSB) (bottom panels) and low surface brightness spirals (LSB) (upper panels) for different galactocentric distances in the disk. a) The evolution in metallicity of LSB formed at z=4 fits well the DLA values measured by Pettini et al. (1999).b) If LSB disks formed late, however, then they do not fit the data. c) Conversely, HSB disks that formed at z=4 become metal rich too quickly to explain the observations; d) only if there is a continuously forming population of HSB disks between z=4 and z=1 then they can account for the metallicities of the DLAs.

suggested that some DLA showing high N/O ratio could be BCG suffering selective winds. In this respect, the similarity between DLAs and IZw18 may suggest that IZw18 is a survivor proto-galaxy which has just started forming stars. If this is true, dwarf irregular galaxies should be born at any time during the age of the universe, unlike elliptical galaxies which appear to have formed a long time ago and in a short time interval (see section 9).

Plots of [α/Fe] vs. [Fe/H] and plots of [α/Fe] vs. redshift should be used to infer the nature and the age of these objects, when compared with chemical evolution predictions.

The observed abundance ratios seem to indicate that DLAs show almost solar [α/Fe] ratios at low [Fe/H] (Pettini et al. 1999; Centurión et al. 2000). This is an indication that they are objects where the star formation proceeded slowly and that they have probably started to form stars long before the redshift at which we observe them. This could indicate that we are looking at the external regions of disks or at dwarf starbursting systems in the interburst phases (see Figure 11). However, more data are necessary to assess this point. One common problem related to the measurement of the abundances in DLAs is that some elements are dust depleted such as Si and Fe and therefore one should try to observe non-refractory elements such as N, O, S and Zn. One should also remember that Si and Ca do not strongly behave as α-elements since they are produced in a non-negligible way in SNe Ia.

12. Chemical Enrichment of the ICM

After having discussed the chemical evolution of galaxies it is important to conclude by studying how the evolution of galaxies can affect the chemical enrichment of the intracluster (ICM) and intergalactic medium. In the past years a great deal of work has been presented on the subject. The first work on chemical enrichment of the ICM was by Gunn & Gott (1972), Larson & Dinerstein (1975), Vigroux (1977), Himmes & Biermann (1988). In the following years, Matteucci & Vettolani (1988) started a more detailed approach to the problem followed by David et al. (1991), Arnaud et al. (1992), Renzini et al. (1993), and many others. The majority of these papers assumed that galactic winds (mainly from ellipticals) are responsible for the ICM chemical enrichment. Alternatively, the abundances in the ICM could be due to ram pressure stripping (Himmes & Biermann 1988) or to pre-galactic Population III stars (White & Rees 1978).

12.1. Models for the ICM

Here we will describe briefly the metodology developed by Matteucci & Vettolani (1988) (hereafter MV88). Starting from SN driven galactic wind models for ellipticals (see section 9) they computed the ejected masses vs. final baryonic total galactic masses for galaxies of different initial luminous masses (from 10^9 to $10^{12} M_\odot$):

$$M_i^{ej} = E_i M_f^{\beta_i}, \tag{12.67}$$

where i refers to either a single chemical element or the total gas. Then, they integrated M_i^{ej} over the cluster mass function obtained from the Schechter (1976) luminosity function (LF) by assuming that only E and S0 galaxies contribute to the chemical enrichment. This assumption was later confirmed by data of Arnaud et al. (1992). The total masses ejected into the ICM in the form of single species i and total gas are:

$$M_{i,clus}^{ej}(> M_f) = E_i f n^* (h^2 k)^{\beta_i} 10^{-0.4\beta_i(M^*-5.48)} \cdot \Gamma[(\alpha+1+\beta_i), (M_f^* h^2/k) 10^{-0.4(M_B^*-5.48)}] \tag{12.68}$$

where α is the slope of the LF, f is fraction of ellipticals plus S0, M^* is the mass at the "break" of the LF and M_B^* is the magnitude at the break.

12.2. MV88 Results

MV88 considered 4 galaxy clusters: Perseus, A2199, Coma and Virgo for which f, n^*, M^*, M_{Fe} and M_{gas} were known. In Table 1 we show the results they obtained for the 4 clusters concerning the total masses ejected by all galaxies in the clusters in the form of total gas and Fe, compared with the observed ones. It is immediate to see from Table 1 that their model could reproduce well the total amount of Fe but they failed in reproducing the total gas mass. However, this is not a failure of the model but the

Table 1. Predicted and observed quantities

Cluster	$(M_{Fe})_{pred}$	$(M_{Fe})_{obs}$	$(M_{gas})_{pred}$	$(M_{gas})_{obs}$
Perseus	$2.6 10^{11}$	$1.9 10^{11}$	$2.0 10^{13}$	$3.0 10^{14}$
A2199	$1.3 10^{11}$	$1.2 10^{11}$	10^{13}	$1.5 10^{14}$
Coma	$2.3 10^{11}$	$3.1 10^{11}$	$1.8 10^{13}$	$1.5 10^{14}$
Virgo	$1.4 10^{10}$	$1.6 10^{10}$	10^{12}	$2.0 10^{13}$

Table 2. Predicted Fe abundances and abundance ratios

Cluster	$X_{Fe}/X_{Fe_\odot}$	[Mg/Fe]	[Si/Fe]
Perseus	0.65	-0.65	-0.080
A2199	0.65	-0.60	-0.096
Coma	0.43	-0.63	-0.110
Virgo	0.53	-0.64	-0.090

indication that most of the gas in clusters has a primordial origin, namely has never been processed inside stars. The same conclusion was reached later by David et al. (1991) and Renzini et al. (1993) among others, but see Chiosi (2000).

Therefore, the $X_{Fe}/X_{Fe_\odot}$ shown in Table 2 was calculated as $(M_{Fe})_{pred}/(M_{gas})_{obs}$. The predicted Fe abundance in the ICM relative to the Sun is in agreement with the observations then and now $(X_{Fe}/(X_{Fe_\odot})_{obs} = 0.3 - 0.5$ (Rothenflug & Arnaud 1985; White 2000). Low values for [Mg/Fe] and [Si/Fe] were predicted due to the assumption that all the Fe, produced by Type Ia SNe, was soon or later ejected into the ICM, as shown in Table 2. With Salpeter IMF, Type Ia SNe contribute $\geq 50\%$ of the total Fe.

12.3. [α/Fe] Ratios in the ICM

Models including Type Ia and II SNe predict an asymmetry in the [α/Fe] ratios (> 0 inside the ellipticals and < 0 in the ICM) due to the different roles of SNe II and Ia in Fe production (Renzini et al. 1993.) ASCA results (Mushotzky et al. 1996) originally suggested $[\alpha/Fe]_{ICM} > 0$ (+0.2 dex). However, Ishimaru & Arimoto (1999) pointed out that $[\alpha /Fe] \sim 0$ in the ICM if the meteoritic Fe abundance is adopted instead of the photospheric value adopted in the other paper. On the basis of the ASCA results on the overabundance of the α-elements, Matteucci & Gibson (1995) discussed how to reproduce the [α/Fe]>0 ratios both in stars and ICM. They adopted the same SN feedback as MV88 and dark matter halos were also included in the galaxy models. They concluded that it is possible to obtain overabundances of the α-elements in the ICM only if not all of the produced Fe is lost from the galaxies (i.e. only early winds). On the other hand, MV88 had assumed that all the Fe is soon or later ejected into the ICM. They concluded that a flat IMF (x=0.95) plus only early winds can reproduce [α/Fe]$_{ICM} > 0$. A similar conclusion was reached by Loewenstein & Mushotzky (1996). However, the situation is not very realistic since the gas in cluster galaxies is likely to be stripped soon or later because of environmental effects, and the [α/Fe] ratios in the ICM are likely to be solar or undersolar. More recently, Martinelli et al. (2000) recomputed the ICM enrichment by adopting a more realistic model for the evolution of ellipticals (multi-zone). The model (already described in section 9) predicts a more extended period of galactic wind and more metals and gas in the ICM than the one-zone model with only early winds. They computed the evolution of abundances vs. redshift for a constant LF and concluded that there is no evolution between z=1 and z=0 and $[\alpha/Fe]_{ICM} \leq 0$ at the present time. Very recently Pipino et al. (2002) computed the chemical enrichment of the ICM as a function of redshift by considering the evolution of the cluster luminosity function and confirmed the conclusions of Martinelli et al. (2000). Very recent data from XMM-Newton (Gastaldello & Molendi, 2002) seem to indicate a low [O/Fe]< 0 ratio for the ICM, thus supporting models where Type Ia SNe are efficient and all the produced Fe is ejected into the ICM. In Figure 12 we show the recent results of Pipino et al. (2002). In the upper panel of Fig. 12 is indicated the predicted evolution as a function of redshift of the thermal energy per particle in the ICM due to galactic winds from E and S0 galaxies. In the lower panel is shown the evolution of the total mass of Fe ejected by the cluster galaxies into the ICM as a function of redshift. It is worth noting that 1 keV per particle is the energy required to explain the observed $L_X - T$ relation in clusters (e.g. Borgani et al. 2001). The model in Figure 12 shows that the energy injected by SNe is not enough (at maximum 0.4 keV) and that other sources of energy are required, such as active galactic nuclei and QSO. Figure 12 also shows how important is the contribution from Type Ia SNe to the chemical enrichment and energetic content of the ICM. In fact, SNe II are important in triggering the galactic wind but after star formation stops the wind is sustained only by Type Ia SNe.

12.4. *[α/Fe] ratios and IMLR*

Abundance ratios and the Iron Mass to Light Ratio (IMLR) are good tests for the evolution of galaxies in clusters since they do not depend on the total cluster gas mass. The IMLR is defined as (Renzini et al. 1993):

$$IMLR = M_{Fe}^{ICM}/L_B \qquad (12.69)$$

and the IMLR $\propto h^{-0.5}$. For $H_0 = 50$ the IMLR=0.02 $(M_\odot/L_\odot)$ (Arnaud et al. 1992) practically constant among rich clusters but it drops for poor clusters and groups. On the one hand, the IMLR can impose constraints on the IMF in cluster galaxies, on the baryonic history of the clusters/groups and on the number of SNe exploded in galaxies. In fact, the most straightforward interpretation of the constancy of the IMLR among rich clusters is that they did not loose Fe at variance with the groups or small clusters having a lower IMLR indicating a loss of baryons (Renzini, 1997). On the other hand, the [α/Fe] ratios can impose constraints on stellar nucleosynthesis, IMF, different roles of SNe and SN feedback.

13. Conclusions on the ICM

The main conclusions on the chemical enrichment of the ICM can be summarized as follows:

• Good models for the chemical enrichment of the ICM should reproduce at the same time the [α/Fe] ratios inside galaxies and in the ICM. They should also reproduce the IMLR.

• Crucial parameters are: SN feedback, SN nucleosynthesis, IMF and whether all the stellar ejecta (produced over a Hubble time) can reach the ICM or remain bound to the parent galaxy.

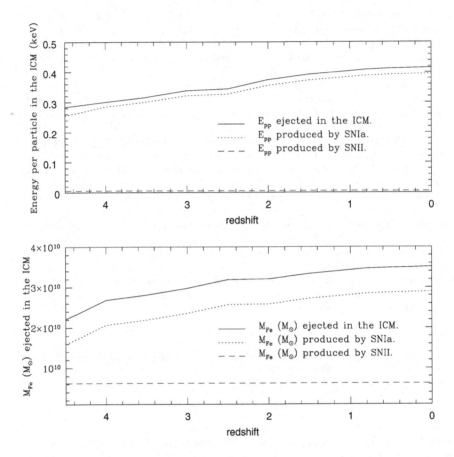

FIGURE 12. The predicted evolution of the thermal content of the ICM (upper panel) as a function of redshift. The different contributions from SNe of different type is indicated. In the lower panel is shown the evolution of the total Fe mass as a function of redshift. The assumed cosmological model is $\Omega_m = 0.3$, $\Omega_\lambda = 0.7$ and h=0.70.

• Type Ia SNe play a fundamental role in the chemical and energy content evolution of the ICM.

• Good models for the ICM enrichment, assuming that the energy transfer from SN II is $\sim 3\%$ and from SNIa is 100%, predict an energy per particle in the ICM of $E_p \sim 0.4$ keV, not enough to break the self- similarity in clusters.

REFERENCES

ALIBÉS, A., LABAY,J. & CANAL, R., 2001, A&A 370, 1103

ARGAST, D., SAMLAND, M., GERHARD, O.E. & THIELEMANN, F.-K., 2000, A&A 356, 873

ARIMOTO, N. & YOSHII, Y. 1987, A&A 173, 23

ARNAUD, M., ROTHENFLUG, R., BOULADE, O.,VIGROUX, L. & VANGIONI-FLAM, E., 1992, A&A, 254, 49

BARBUY, B. 1999 Astrophys. Space Sci. 265, 319

BENDER, R., BURSTEIN, D. & FABER, S. M., 1993, ApJ 411, 153

BENDER, R., BURSTEIN, D. & FABER, S. M., 1992, ApJ 399, 462

BENDER, R., ZIEGLER, B. & BRUZUAL, G., 1996, ApJ 463, L51

BERMAN, B.C. & SUCHOV, A. A., 1991, Astrophys. Space Sci. 184, 169

BERNARDI, M., RENZINI, A., DA COSTA, L. N.., WEGNER, G. & AL., 1998, ApJ 508, L143

BERTIN, G., SAGLIA, R. & STIAVELLI, M., 1992, ApJ 384, 423

BLAIN, A. W., SMAIL, I., IVISON, R. J. & KNEIB, J.-P., 1999, MNRAS 302, 632

BOISSIER, S. & PRANTZOS, N. 1999 MNRAS 307, 857

BORGANI, S., GOVERNATO, F., WADSLEY, J., MENCI, N., TOZZI, P., LAKE, G., QUINN, T. & STADEL, J., 2001, ApJ, 559, 71

BOWER, R. G., LUCEY, J. R. & ELLIS, R. S., 1992, MNRAS 254, 601

BRADAMANTE, F., MATTEUCCI, F. & D'ERCOLE, A., 1998, A&A 337, 338

BRAGAGLIA, A., GREGGIO, L., RENZINI, A. & D'ODORICO, S., 1990, ApJ 365, L13

CARIGI, L. 1996 Rev.MexAA 32, 179

CAROLLO, C. M., DANZIGER, I. J. & BUSON, L., 1993, MNRAS 265, 553

CENTURIÓN, M., BONIFACIO, P., MOLARO, P. & VLADILO, G., 2000, ApJ 536, 540

CHANG, R.X., HOU, J.L., SHU, C.G. & FU, C.Q., 1999, A&A 350, 38

CHIAPPINI, C., MATTEUCCI F. & GRATTON R. 1997, ApJ 477, 765

CHIAPPINI, C., MATTEUCCI, F. & PADOAN, P., 2000, ApJ , 528, 711

CHIAPPINI, C., MATTEUCCI, F., & ROMANO, D., 2001, ApJ 554, 1044

CHIOSI, C., 2000, A&A 364, 423

CLARKE, C. J., 1989 MNRAS 238, 283

COLLESS, M., BURSTEIN, D., DAVIES, R. L., MCMAHAN, R. K., SAGLIA, R. P. & WEGNER, G., 1999, MNRAS 303, 813

COX, D. P., 1972, ApJ 178, 159

D'ANTONA, F. & MATTEUCCI, F., 1991, A&A 248, 62

DAVIES, R. L., SADLER, E. M. & PELETIER, R. F., 1993, MNRAS 262, 650

DAVID, L.P., FORMAN, W., & JONES, C., 1991, ApJ, 376, 380

DONAS, J., DEHARVENG, J. M., LAGET, M., MILLIARD, B. & HUGUENIN, D., 1987 A&A, 180, 12

DOPITA, M. A. & RYDER, S. D., 1994, ApJ, 430, 163

EDMUNDS, M. G. & GREENHOW, R. M., 1995, MNRAS 272, 241

ELBAZ, D., CESARSKY, C. J., FADDA, D., AUSSEL, H. ET AL., 1999, A&A 351, 37

ELMEGREEN, B.G., 1999, ApJ 517, 103

EGGEN, O.J., LYNDEN-BELL, D. & SANDAGE, A. R., 1962, ApJ 136, 748

FERRERAS, I. , CHARLOT, S. & SILK, J., 1999, ApJ 521, 81

FORESTINI, M. & CHARBONNEL, C., 1997, A&ASuppl. 123, 241

FRANX, M. & ILLINGWORTH, G., 1990, ApJ 359, L41

FUHRMANN, K., 1998 A&A, 338, 161

GALLINO, R., ARLANDINI, C., BUSSO, M., LUGARO, M. ET AL., 1998, ApJ 497, 388

GARNETT, D. R., 1998, ASP Conf. Series Vol. 147, p. 78

GASTALDELLO, F. & MOLENDI, S., 2002, ApJ 572, 160

GRATTON, R. G., CARRETTA, E., MATTEUCCI, F. & SNEDEN, C., 2000, A&A 358, 671

GREGGIO, L., 1996 *Interplay between massive star formation, the ISM and galaxy evolution,* IAP Meeting July 1995, Editions Frontieres, p.89

GREGGIO, L. & RENZINI, A., 1983, A&A 118, 217

GONZÁLEZ, J. J., 1993, Ph.D. thesis. Univ. Calif., Santa Cruz

GUNN, J. E. & GOTT, J. R. III, 1972, ApJ 176, 1

HACHISU, I., KATO, M. & NOMOTO, K., 1996 ApJ 470, L97

HACHISU, I., KATO, M. & NOMOTO, K., 1999 ApJ 522, 487

HIMMES, A. & BIERMANN, P., 1988 A&A 86, 11

HOU, J., CHANG, R. & FU, C., 1998, ASP Conf. Series, Vol. 138, 143

IBEN, I. JR. & TUTUKOV, A.V., 1984, ApJSuppl 54, 335

ISHIMARU, Y. & ARIMOTO, N., 1997, PASJ 49, 1

JIMÉNEZ, R., PADOAN, P., MATTEUCCI, F. & HEAVENS, A.F., 1998, MNRAS 299, 123

JORGENSEN, I., 1999, MNRAS 306, 607

JOSEY, S. A. & ARIMOTO, N., 1992, A&A 255, 105

KAUFFMANN, G., CHARLOT, S. & WHITE, S. D. M., 1993, MNRAS 283, L117

KAUFFMANN, G., 1996, MNRAS 281, 475

KENNICUTT, R.C. JR., 1983, ApJ 272, 54

KENNICUTT, R. C. JR., 1998a ARAA 36, 189

KENNICUTT, R. C. JR., 1998b, ApJ 498, 541

KENNICUTT, R. C. JR., 1989, ApJ 344, 685

KOBAYASHI, C. & ARIMOTO, N., 1999, ApJ 527, 573

KODAMA, T., ARIMOTO, N., BARGER, A. J. & ARAGÓN-SALAMANCA, A., 1998, A&A 334, 99

KROUPA, P., TOUT, C.A. & GILMORE, G., 1993, MNRAS 262, 545

KUMAI, Y. & TOSA, M., 1992, A&A 257, 511

KUNTH, D., MATTEUCCI, F. & MARCONI, G., 1995, A&A 297, 634

KUNTH, D. & ÖSTLIN, G., 2000, A&ARv 10, 1

KUNTSCHNER, H., LUCEY, J. R., SMITH, R. J., HUDSON, M. J. & DAVIES, R. L., 2001, MNRAS 323, 625

LARSEN, T.I., SOMMER-LARSEN, J. & PAGEL, B.E.J., 2001, MNRAS 323, 555

LARSON, R. B., 1974, MNRAS 169, 229

LARSON, R. B., 1976, MNRAS 176, 31

LARSON, R. B. & DINERSTEIN, H. L., 1975, PASP 87, 911

LEQUEUX, J., KUNTH, D., MAS-HESSE, J. M. & SARGENT, W. L. W., 1995, A&A 301, 18

LOEWENSTEIN, M., & MUSHOTZKY, F., 1996, ApJ 466, 695

MacLOW, M.-M., FERRARA, A., 1999, ApJ 513, 142

MARCONI, G., MATTEUCCI, F. & TOSI, M., 1994, MNRAS 270, 35

MARIGO, P., BRESSAN, A. & CHIOSI, C., 1996, A&A 313, 545

MARLOWE, A. T., HECKMAN, T. M., WYSE, R. F. G. & SCHOMMER, R., 1995, ApJ 438, 563

MARTIN, C. L., 1996, ApJ 465, 680

MARTIN, C. L., 1998, ApJ 506, 222

MARTINELLI, A., MATTEUCCI, F. & COLAFRANCESCO, S., 1998, MNRAS, 298, 42

MARTINELLI, A., MATTEUCCI, F. & COLAFRANCESCO, S., 2000, A&A 354, 387

MATTEUCCI, F.,1994, A&A 288, 57

MATTEUCCI, F., 2001, *The Chemical Evolution of the Galaxy*, ASSL, Kluwer Academic Publisher

MATTEUCCI, F. & CHIOSI, C., 1983, A&A 123, 121

MATTEUCCI, F. & FRANÇOIS, P., 1989, MNRAS 239, 885

MATTEUCCI, F.& GIBSON, B.K., 1995, A&A 304, 11

MATTEUCCI, F. & GREGGIO, L., 1986, A&A 154, 279

MATTEUCCI, F., MOLARO, P. & VLADILO, G., 1997, A&A 321, 45

MATTEUCCI, F., PONZONE, R. & GIBSON, B.K., 1998, A&A 335, 855

MATTEUCCI, F. & RECCHI, S., 2001, ApJ 558, 351

MATTEUCCI, F., ROMANO, D. & MOLARO, P., 1999, A&A 341, 458

MATTEUCCI, F. & TORNAMBÈ, A., 1987, A&A 185, 51

MATTEUCCI, F. & TOSI, M., 1985, MNRAS 217, 391

MATTEUCCI, F., & VETTOLANI, G., 1988, A& A, 202, 21

McWILLIAM, A. & RICH, R. M., 1994, ApJS 91, 749

MENANTEAU, F., JIMÉNEZ, R. & MATTEUCCI, F., 2001, ApJ 562, L23

MENANTEAU, F., ELLIS, R.S., ABRAHAM, R.G., BARGER, A.J. & COWIE, L.L., 1999, MNRAS 309, 208

MEURER, G. R., FREEMAN, K. C., DOPITA, M. A. & CACCIARI, C., 1992, AJ 103, 60

MO, H. J., MAO, S. & WHITE, S. D. M., 1998, MNRAS 295, 319

MUSHOTZKY, R., LOEWENSTEIN, M., ARNAUD, K. A. ET AL., 1996, ApJ 466, 686

NOMOTO, K., THIELEMANN, F. K. & YOKOI, K. 1984, ApJ 286, 644

NOMOTO, K., HASHIMOTO, M., TSUJIMOTO, T., THIELEMANN, F.-K. ET AL., 1997, Nucl. Phys. A, 616, 79c

OEY, M. S., 2000, ApJ 542, L25

PAGEL, B. E. J., 1997, *Nucleosynthesis and Chemical Evolution of Galaxies*, Cambridge Univ. Press

PAPADEROS, P., FRICKE, K. J., THUAN, T. X. & LOOSE, H.-H., 1994, A&A 291, L13

PARDI, M. C., FERRINI, F. & MATTEUCCI, F., 1995, ApJ 444, 207

PEACOCK, J.A., 1999, *Cosmological Physics*, Cambridge univ. Press

PETTINI, M., ELLISON, S. L., STEIDEL, C. C., & BOWEN, D. V., 1999, ApJ 510, 576

PILYUGIN, I. S., 1993, A&A 277, 42

PIPINO, A., MATTEUCCI, F., BORGANI, S. & BIVIANO, A., 2002, NewAstr. 7, 227

PORTINARI, L. & CHIOSI, C., 2000, A&A, 355, 929

PRANTZOS, N. & BOISSIER, S., 2000, MNRAS 313, 338

RANA, N. C., 1991, ARAA 29, 129

RECCHI, S., 2002, PhD Thesis, University of Trieste

RECCHI, S., MATTEUCCI, F. & D'ERCOLE, A., 2001, MNRAS 322, 800

RECCHI, S., MATTEUCCI, F., D'ERCOLE, A. & TOSI, M., 2002, A&A 384, 799

RENZINI, A., 1994, *Galaxy Formation*, International School of Physics "E.Fermi", Course CXXXII, ed. J. Silk & N. Vittorio, (Amsterdam: North Holland), p. 303

RENZINI, A., 1997, ApJ 488, 35

RENZINI, A. & CIOTTI, L., 1993, 416, L49

RENZINI, A. & VOLI, M., 1981 A&A 94, 175

RENZINI, A., CIOTTI, L., D'ERCOLE, A. & PELLEGRINI, S., 1993, ApJ 416, L49

ROCHA-PINTO, H. J. & MACIEL, W. J., 1996, MNRAS 279, 447

ROCHE, N., RATNATUNGA, K., GRIFFITHS, R. E., IM, M. & NAIM, A., 1998, MNRAS 293, 157

ROMANO, D., MATTEUCCI, F., VENTURA, P. & D'ANTONA, F., 2001 A&A 374, 646

ROSENBERG, A., SAVIANE, I., PIOTTO, G. & APARICIO, A., 1999, AJ 118 2306

ROTHENFLUG, R., & ARNAUD, M., 1985, A&A, 144, 431

SACKMANN, I. J., BOOTHROYD, A.I., 1999 Ap.J. 510, 217

SALPETER, E. E., 1955, ApJ 121, 161

SCALO, J. M., 1986, Fund. Cosmic Phys. 11, 1

SCALO, J. M., 1998, *The Stellar Initial Mass Function*, A.S.P. Conf. Ser. Vol. 142 p.201

SCHECHTER, P., 1976, ApJ 203, 297

SCHMIDT, M., 1959, ApJ 129, 243

SCHMIDT, M., 1963, ApJ 137, 758

SEARLE, L. & ZINN, R., 1978 ApJ 225, 357

SKILLMAN, E. D., KENNICUTT, R. C. JR., SHIELDS, G. A. & ZARITSKY, D., 1996, ApJ 462, 147

STANFORD, S. A., EISENHARDT, P. R. & DICKINSON, M., 1998, ApJ 492, 461

STEIDEL, C. C., GIAVALISCO, M., PETTINI, M., DICKINSON, M. & ADELBERGER, K. L., 1996, ApJ 462, L17

STEIDEL, C. C., ADELBERGER, K. L., DICKINSON, M., GIAVALISCO, M., PETTINI, M., & KELLOGG, M., 1998, ApJ 492, 428

TALBOT, R. J. & ARNETT, D. W. 1973, Ap.J. 186, 69

TANTALO, R., CHIOSI, C., BRESSAN, A., MARIGO, P. & PORTINARI, L., 1998, A&A 335, 823

THIELEMANN, F. K., NOMOTO, K. & HASHIMOTO, M., 1996, ApJ 460, 408

THOMAS, D., 1999, MNRAS 306, 655

THOMAS, D., MARASTON, C. & BENDER, R., 2002, Reviews in Modern Astronomy Vol. 15, Astronomische Gesellschaft, R.E. Schielicke (ed.), p. 219

TIMMES, F. X., WOOSLEY, S. E., & WEAVER, T. A., 1995, ApJS 98, 617

TINSLEY, B. M., 1980, Fund. Cosmic Phys. 5, 287

TINSLEY, B. M. & LARSON, R.B., 1979, MNRAS 186, 503

TOSI, M., 1988, A&A 197, 33

TRAGER, S. C., FABER, S. M., WORTHEY, G. & GONZÀLEZ, J. J., 2000, AJ 120, 165

TRAGER, S. C., WORTHEY, G., FABER, S. M., BURSTEIN, D. & GONZÁLEZ, J. J., 1998, ApJS 116, 1

TSUJIMOTO, T., SHIGEYAMA, T. & YOSHII, Y., 1999, ApJ 519,63

VAN DEN HOEK, L. B. & GROENEWEGEN, M. A. T., 1997 A&AS 123, 305

VAN DOKKUM, P. G. & FRANX, M., 1996, MNRAS 281, 985

VIGROUX, L., 1977, A&A 56, 473

YOSHII, Y. & ARIMOTO, N., 1987, A&A 188, 13

YOSHII, Y. & SOMMER-LARSEN, J., 1989, MNRAS 236, 779

WEISS, A. PELETIER, R. F. & MATTEUCCI, F., 1995, A&A 296, 73

WHELAN, J. & IBEN, I. JR., 1973, ApJ 186, 1007

WHITE, D. A., 2000, MNRAS 312, 663

WHITE, S. D. M., & REES, M. J., 1978, MNRAS 183, 341

WOOSLEY, S. E., AXELROD, T. S. & WEAVER, T. A., 1984 *Stellar Nucleosynthesis*, ed. C. Chiosi and A. Renzini, Reidel (Dordrecht) p.263

WOOSLEY, S. E. & WEAVER, T. A., 1995, ApJSuppl 101, 181

WORTHEY, G. FABER, S. M. & GONZÁLEZ, J. J., 1992, ApJ 398, 69

ZEPF, S. E., 1997, Nature 390, 377

Element Abundances Through The Cosmic Ages

By MAX PETTINI

Institute of Astronomy, University of Cambridge
Madingley Road, Cambridge, UK

The horizon for studies of element abundances has expanded dramatically in the last ten years. Once the domain of astronomers concerned chiefly with stars and nearby galaxies, this field has now become a key component of observational cosmology, as technological advances have made it possible to measure the abundances of several chemical elements in a variety of environments at redshifts up to $z \simeq 4$, when the universe was in its infancy. In this series of lectures I summarise current knowledge on the chemical make-up of distant galaxies observed directly in their starlight, and of interstellar and intergalactic gas seen in absorption against the spectra of bright background sources. The picture which is emerging is one where the universe at $z = 3$ already included many of the constituents of today's galaxies—even at these early times we see evidence for Population I and II stars, while the 'smoking gun' for Population III objects may be hidden in the chemical composition of the lowest density regions of the intergalactic medium, yet to be deciphered.

1. Introduction

One of the exciting developments in observational cosmology over the last few years has been the ability to extend studies of element abundances from the local universe to high redshifts. Thanks largely to the new opportunities offered by the Keck telescopes, the Very Large Telescope facility at the European Southern Observatory, and most recently the Subaru telescope, we find ourselves in the exciting position of being able, for the first time, to detect and measure a wide range of chemical elements directly in stars, H II regions, cool interstellar gas and hot intergalactic medium, all observed when the universe was only $\sim 1/15$ of its present age. Our simple-minded hope is that, by moving back to a time when the universe was young, clues to the nature, location, and epoch of the first generations of stars may be easier to interpret than in the relics left today, some 12 Gyrs later. Furthermore, the metallicities of different structures in the universe and their evolution with redshift are key factors to be considered in our attempts to track the progress of galaxy formation through the cosmic ages.

In the last few years, work on chemical abundances at high redshifts has concentrated on four main components of the young universe: Active Galactic Nuclei (AGN), that is quasars (QSOs) and Seyfert galaxies; two classes of QSO absorption lines, the damped Lyα systems (DLAs) and the Lyα forest; and on galaxies detected directly via their starlight, also referred to as Lyman break galaxies. In these series of lectures I will review results pertaining to the last three; for abundance determinations in the emission line regions of AGN and associated absorbers I refer the interested reader to the excellent recent overview by Hamann & Ferland (1999).

1.1. Some Basic Concepts

For many years our knowledge of the distant universe relied almost exclusively on QSO absorption lines. It is only relatively recently that we have learnt to identify directly 'normal' galaxies; up until 1995 the only objects known at high z were QSOs and powerful

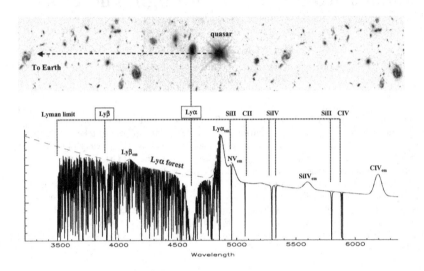

FIGURE 1. The technique of QSO absorption line spectroscopy is illustrated in this montage (courtesy of John Webb). QSOs are among the brightest and most distant objects known. On the long journey from its source to our telescopes on Earth, the light from a background QSO intercepts galaxies and intergalactic matter which happen to lie along the line of sight (and are therefore at lower absorption redshifts, z_{abs}, than the QSO emission redshift, z_{em}. Gas in these structures leaves a clear imprint in the spectrum of the QSO in the form of narrow absorption lines. The task of astronomers working in this field has been to relate the characteristics of the absorption lines to the properties of the intervening galaxies which are normally too faint to be detected directly.

radio galaxies. The technique of QSO absorption line spectroscopy, illustrated in Figure 1, is potentially very powerful. As we shall see, it allows accurate measurements of many physical properties of the interstellar medium (ISM) in galaxies and the intergalactic medium (IGM) between galaxies. The challenge, however, is to relate this wealth of data, which refer to gas along a very narrow sightline, to the global properties of the absorbers. In a sense, all the information we obtain from QSO absorption line spectroscopy is of an *indirect* nature; if we could detect the galaxies themselves, our inferences would be on a stronger empirical basis.

In deriving chemical abundances in QSO absorbers and high redshift galaxies, we shall make use of some of the same techniques which are applied locally to interpret the spectra of stars, cool interstellar gas and H II regions. These methods are discussed extensively in other articles in this volume, particularly those by Don Garnett, David Lambert, and Grazyna Stasińska, and will therefore not be repeated here. The derivation of ion column densities from the profiles and equivalent widths of interstellar absorption lines is discussed in a number of standard textbooks, as well as a recent volume in this series (Bechtold 2002).

When measuring element abundances in different astrophysical environments, we shall often compare them to the composition of the solar system determined either from photospheric lines in the solar spectrum or, preferably, from meteorites. The standard solar abundance scale continues to be refined; here we adopt the compilation by Grevesse &

Sauval (1998) with the recent updates by Holweger (2001). We use the standard notation $[X/Y] = \log(X/Y)_{obs} - \log(X/Y)_{\odot}$ where $(X/Y)_{obs}$ denotes the abundance of element X relative to element Y in the system under observation—be it stars, interstellar gas or the intergalactic medium—and $(X/Y)_{\odot}$ is their relative abundance in the solar system.

Furthermore, it is important to remember that when element abundances are measured in the interstellar gas of the Milky Way, it is usually found that $[X/H] < 0$. This deficiency is not believed to be intrinsic, but rather reflects the proportion of heavy elements that has condensed out of the gas phase to form dust grains (and therefore no longer absorbs starlight via discrete atomic transitions). As discussed in the review by Savage & Sembach (1996), the 'missing' fraction varies from element to element, reflecting the ease with which different constituents of interstellar dust are either incorporated into the grains or released from them. In particular, O, N, S, and Zn show little affinity for dust and are often present in the gas in near-solar proportions; Si, Fe and most Fe-peak elements, on the other hand, can be depleted by large and varying amounts depending on the physical conditions—past and present—of the interstellar clouds under study.

Unless otherwise stated, we shall use today's 'consensus' cosmology (e.g. Turner 2002) with $H_0 = 65 \,\mathrm{km \; s^{-1} \; Mpc^{-1}}$ (and hence $h = 0.65$), $\Omega_{baryons} = 0.022 \, h^{-2}$, $\Omega_M = 0.3$, and $\Omega_\Lambda = 0.7$. Table 1 shows the run of look-back time with redshift for this cosmology. Note that with the above cosmological parameters the age of the universe is 14.5 Gyr, consistent with recent estimates of the ages of globular clusters (Krauss & Chaboyer 2003). When in these lecture notes we refer to 'high' redshifts we usually mean $z = 3 - 4$ which correspond to look-back times of 12–13 Gyr, when the universe had only 15–10% of its present age. Epochs when z was ≤ 1 are generally referred to as 'intermediate' or 'low' redshifts, even though they correspond to look-back times of up to about 60% of the current age of the universe.

Table 1. Lookback time vs. redshift in the adopted cosmology

Redshift	Lookback Time (Gyr)	Lookback time (t/t_∞)
0	0	0
0.5	5.4	0.37
1	8.3	0.57
2	11.0	0.76
3	12.2	0.84
4	12.9	0.89
5	13.3	0.92
6	13.5	0.93
10	14.0	0.97
∞	14.5	1.00

2. Damped Lyα Systems

2.1. *What Are They?*

I am often asked this question, and the truthful answer is: "We do not know". Spectroscopically, DLAs are straightforward to identify (see Figure 2). The large equivalent

Q1331+170 z_{em}=2.084 z_{abs}=1.7764 (WHT)

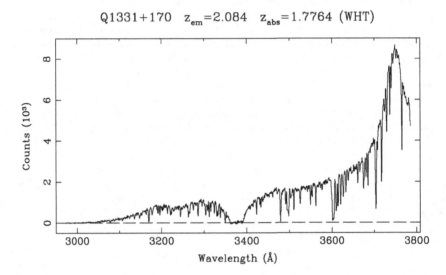

FIGURE 2. The strong absorption feature centred near 3375 Å in the near-ultraviolet (UV) spectrum of the bright QSO Q1331+170 is a good example of a damped Lyα line, in this case produced by a column density of neutral hydrogen atoms $N(\text{H I}) = 1.5 \times 10^{21}\,\text{cm}^{-2}$. This spectrum was recorded in the early 1990s with the Image Photon Counting System on the ISIS spectrograph of the 4.2 m William Herschel telescope on La Palma (Pettini et al. 1994).

widths and characteristic damping wings which signal column densities of absorbing neutral hydrogen in excess of $N(\text{H I}) = 2 \times 10^{20}\,\text{cm}^{-2}$ are easy to recognise even in spectra of moderate resolution and signal-to-noise ratio (significantly worse than those of the spectrum reproduced in Figure 2).

The galaxies producing DLAs, however, have proved difficult to pin down. Wolfe and collaborators, who were the first to recognise DLAs as a class of QSO absorbers of special significance for the study of the high redshift universe (e.g. Wolfe et al. 1986), proposed from the outset that they are the progenitors of present-day spiral galaxies, observed at a time when most of their baryonic mass was still in gaseous form. The evidence supporting this scenario, however, is mostly indirect. For example, Prochaska & Wolfe (1998) showed that the profiles of the metal absorption lines in DLAs are consistent with the kinematics expected from large, rotating, thick disks, but others have claimed that this interpretation is not unique (Haehnelt, Steinmetz, & Rauch 1998; Ledoux et al. 1998).

Imaging studies at high redshift are only now beginning to identify some of the absorbers (Prochaska et al. 2002; Møller et al. 2002). At $z < 1$, where the imaging is easier (an example is reproduced in Figure 3), it appears that DLA galaxies are a very 'mixed bag', which includes a relatively high proportion of low surface brightness and low luminosity galaxies (some so faint that they remain undetected in their stellar populations, e.g. Steidel et al. 1997; Bouché et al. 2001), as well as more 'normal' spirals (Boissier, Péroux, & Pettini 2002).

Much was made in the early 1990s of the apparent correspondence between the neutral gas mass traced by DLAs at high redshift ($\Omega_{\text{DLA}}\,h \simeq 1.2 \times 10^{-3}$, when expressed as a

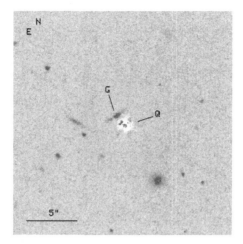

FIGURE 3. (Reproduced from Pettini et al. 2000a). WFPC2 F702W exposure of the field of Q0058+019 (PHL 938). A model point spread function has been subtracted from the QSO image (labelled 'Q'), revealing the presence of a galaxy (labelled 'G') approximately 1.2 arcseconds to the NE of the QSO position. Given its proximity, this is likely to be the damped Ly α absorber at $z = 0.61251$. Residual excess absorption of a diffraction spike cuts across the galaxy image. When this processing artifact is taken into account, the candidate absorber appears to be a low luminosity $(L \simeq 1/6 L^*)$ late-type galaxy seen at high inclination, $i \approx 65°$, at a projected separation of $6 h^{-1}$ kpc from the QSO sightline.

fraction of the critical density) and today's luminous stellar mass, leading to suggestions that the former are the material out of which the latter formed (e.g. Lanzetta 1993). However, the apparent decrease in Ω_{DLA} from $z = 3$ to 0, upon which this picture was based, has not been confirmed with more recent and more extensive samples (Pettini 2001). As can be seen from Figure 4, current data are consistent with an approximately constant value Ω_{DLA} over most of the Hubble time, and this includes the most recent estimates of $\Omega_{H I}$ in the local universe from 21 cm surveys (Rosenberg & Schneider 2002; not shown in Figure 4). Perhaps DLAs pick out a particular stage in the evolution of galaxies, when their dimensions in high surface density of neutral gas are largest, and it may be the case that different populations of galaxies pass through this stage at different cosmic epochs.

2.2. Why Do We Care?

While we would obviously like to know more clearly which population(s) of galaxies DLAs are associated with, this issue does not detract from the importance of this class of QSO absorbers for studies of element abundances, for the following reasons.

First, with neutral hydrogen column densities $N(\text{H I}) \geq 2 \times 10^{20}$ cm^{-2}, DLAs are the 'heavy weights' among QSO absorption systems, at the upper end of the distribution of values of $N(\text{H I})$ which spans 10 orders of magnitude for all absorbers (see Figure 5). Over this entire range, $f(N_{\text{H I}})$—defined as the number of absorbing systems per unit redshift path per unit column density—can be fitted with a single power law of the form

$$f(N_{\text{H I}}) = B \times N_{\text{H I}}^{-\beta} \qquad (2.1)$$

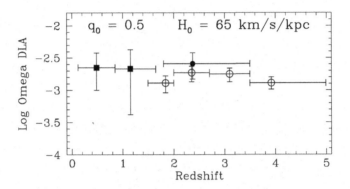

FIGURE 4. Recent estimates of the mass density of neutral gas traced by damped Lyα systems, expressed as a fraction of the critical density. The filled squares are from Rao & Turnshek (2000); the open circles from Storrie-Lombardi & Wolfe (2000) and from Péroux et al. (2002); while the filled circle is from the CORALS survey by Ellison et al. (2001).

with $\beta \simeq 1.5$ (Tytler 1987; Storrie-Lombardi & Wolfe 2000). While the most numerous absorbers are those with the lowest column densities (a turn-over at low values of $N(\text{H I})$ has yet to be found), the high column densities of DLAs more than compensate for their relative paucity. More specifically, so long as $\beta < 2$, the integral of the column density distribution

$$\Omega_{\text{H I}} = \frac{H_0}{c} \frac{\mu\, m_{\text{H}}}{\rho_{\text{crit}}} \int_{N_{\text{min}}}^{N_{\text{max}}} N\, f(N)\, dN = \frac{H_0}{c} \frac{\mu\, m_{\text{H}}}{\rho_{\text{crit}}} \frac{B}{2-\beta} \left(N_{\text{max}}^{2-\beta} - N_{\text{min}}^{2-\beta} \right) \qquad (2.2)$$

is dominated by N_{max}, i.e. by DLAs (Lanzetta 1993). In eq. (2.2), H_0 is the Hubble constant, c is the speed of light, m_{H} is the mass of the hydrogen atom, μ is the mean atomic weight per baryon ($\mu = 1.4$ for solar abundances; Grevesse & Sauval 1998) and ρ_{crit} is the closure density

$$\rho_{\text{crit}} = \frac{3\, H_0^2}{8\, \pi\, G} = 1.96 \times 10^{-29} h^2 \text{ g cm}^{-3} \qquad (2.3)$$

where h is the Hubble constant in units of $100\,\text{km s}^{-1}\,\text{Mpc}^{-1}$.

As a consequence, the mean metallicity of DLAs is the closest measure we have of the global degree of metal enrichment of neutral gas in the universe at a given epoch, *irrespectively of the precise nature of the absorbers*, a point often emphasised by Mike Fall and his collaborators (e.g. Fall 1996). Of course this only applies if there are no biases which exclude any particular type of high redshift object from our H I census.

Second, it is possible to determine the abundances of a wide variety of elements in DLAs with higher precision than in most other astrophysical environments in the distant universe. In particular, echelle spectra obtained with large telescopes can yield abundance measures accurate to 10–20% (e.g. Prochaska & Wolfe 2002), because: (a) the damping wings of the Lyα line are very sensitive to the column density of H I; (b) several atomic transitions are often available for elements of interest; and (c) ionisation corrections are normally small, because the gas is mostly neutral and the major ionisation stages are observed directly (Vladilo et al. 2001). Dust depletions can be a complication, but even these are not as severe in DLAs as in the local interstellar medium (Pettini et al. 1997a) and can be accounted for with careful analyses (e.g. Vladilo 2002a). Thus, abundance

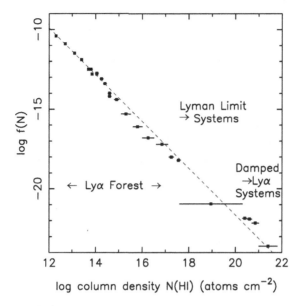

FIGURE 5. (Reproduced from Storrie-Lombardi & Wolfe 2000). The column density distribution function of neutral hydrogen for all QSO absorbers spans ten orders of magnitude, from $\log N(\text{H I}) = 12$ to 22 and can be adequately described by a single power law $f(N_{\text{H I}}) \propto N_{\text{H I}}^{-1.5}$, shown by the dashed line.

studies in DLAs complement in a very effective way the information provided locally by stellar and nebular spectroscopy and, as we shall see, can offer fresh clues to the nucleosynthesis of elements, particularly in metal-poor environments which are difficult to probe in the nearby universe. DLAs are also playing a role in the determination of the primordial abundances of the light elements, as discussed by Gary Steigman in this volume (see also Tytler et al. 2000 and Pettini & Bowen 2001).

2.3. *The Metallicity of DLAs*

Even before the advent of 8-10 m class telescopes, it was realised that the metal and dust content of DLAs could be investigated effectively by targeting a pair of (fortuitously) closely spaced multiplets, Zn II $\lambda\lambda2025, 2062$ and Cr II $\lambda\lambda2056, 2062, 2066$ (Meyer, Welty, & York 1989; Pettini, Boksenberg, & Hunstead 1990). The key points here are that while Zn is essentially undepleted in local diffuse interstellar clouds, Cr is mostly locked up in dust grains (Savage & Sembach 1996). Consequently, the ratio $N(\text{Zn II})/N(\text{H I})$ observed in DLAs, when compared with the solar abundance of Zn, yields a direct measure of the degree of metal enrichment (in H I regions Zn is predominantly singly ionised, and the ratios Zn I/Zn II and Zn III/Zn II are both $\ll 1$). On the other hand, a deficit—if one is found—of $N(\text{Cr II})/N(\text{Zn II})$ compared to the solar relative abundances of these two elements would measure the extent to which refractory elements have condensed into solid form in the interstellar media traced by DLAs and, by inference, be an indication of the presence of dust in these early galaxies (Cr is also singly ionised in H I gas).

From the point of view of chemical evolution, both Zn and Cr closely trace Fe in Galactic stars with metallicities [Fe/H] between 0 (i.e. solar) and -2 (1/100 of solar;

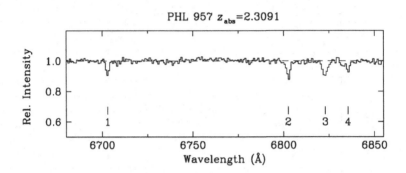

FIGURE 6. Portion of the Palomar spectrum of the bright QSO PHL 957 recorded by Pettini et al. (1990), encompassing the region of the Zn II and Cr II absorption lines in the $z_{abs} = 2.3091$ DLA. The vertical tick marks indicate the positions of the lines as follows. Line 1: Zn II $\lambda 2025.483$; line 2: Cr II $\lambda 2055.596$; line 3: Cr II $\lambda 2061.575$ + Zn II $\lambda 2062.005$ (blended); and line 4: Cr II $\lambda 2065.501$. The spectrum has been normalised to the underlying QSO continuum and is shown on an expanded vertical scale.

Sneden, Gratton, & Crocker 1991, McWilliam et al. 1995). Additional advantages are the convenient rest wavelengths of the Zn II and Cr II transitions, which at $z = 2 - 3$ (where DLAs are most numerous in current samples) are redshifted into a easily observed portion of the optical spectrum, and the inherently weak nature of these lines which ensures that they are nearly always on the linear part of the curve of growth, where column densities can be derived with confidence from the measured equivalent widths (e.g. Bechtold 2002).

All in all, a small portion of the red spectrum of a QSO with a damped Lyα system has the potential of providing some important chemical clues on the nature of these absorbers and the evolutionary status of the population of galaxies they trace. One of the first detections of Zn and Cr in a DLA is reproduced in Figure 6. However, data of sufficiently high signal-to-noise ratio to detect the weak absorption lines of interest (or to place interesting upper limits on their equivalent widths) typically require nearly one night of observation per QSO with a 4 m class telescope. Consequently, it was necessary to wait until the mid-1990s before a sufficiently large sample of Zn and Cr measurements in DLAs could be assembled.

Figure 7 shows the current data set from the survey by our group (e.g. Pettini et al. 1997b; 1999). There are several interesting conclusions which can be drawn from these results.

(*a*) Damped Lyα systems are generally metal poor at all redshifts sampled. Evidently DLAs arise in galaxies at early stages of chemical evolution, since nearly all of the points in Figure 7 lie below the line of solar abundance. The statistic of relevance here is the the column density-weighted mean abundance of Zn

$$[\langle Zn/H_{DLA} \rangle] = \log \langle (Zn/H)_{DLA} \rangle - \log (Zn/H)_{\odot}, \qquad (2.4)$$

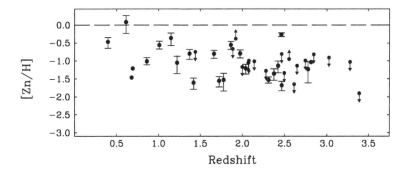

FIGURE 7. Plot of the abundance of Zn against redshift for the full sample of 41 DLAs from the surveys by Pettini and collaborators. Abundances are measured on a log scale relative to the solar value shown by the broken line at [Zn/H] = 0.0; thus a point at [Zn/H] = −1.0 corresponds to a metallicity of 1/10 of solar. Upper limits, corresponding to non-detections of the Zn II lines, are indicated by downward-pointing arrows. Upward-pointing arrows denote lower limits in two cases where the Zn II lines are sufficiently strong that saturation may be important.

where

$$\langle (\mathrm{Zn/H})_{\mathrm{DLA}} \rangle = \frac{\sum\limits_{i=1}^{n} N(\mathrm{Zn\ II})_i}{\sum\limits_{i=1}^{n} N(\mathrm{H\ I})_i}, \tag{2.5}$$

and the summations in eq. (2.5) are over the n DLA systems in a given sample. In this way, by counting all the Zn atoms per unit cross-section (cm^{-2}) and dividing by the total column density of neutral hydrogen we find that DLAs have a typical metallicity of only $\sim 1/13$ of solar ($[\langle \mathrm{Zn/H}_{\mathrm{DLA}} \rangle] = -1.13$).

(*b*) There appears to be a large range in the values of metallicity reached by different galaxies at the same redshift. Values of [Zn/H] in Figure 7 span nearly two orders of magnitude, pointing to a protracted 'epoch of galaxy formation' and to the fact that chemical enrichment probably proceeded at different rates in different DLAs. The wide dispersion in metallicity goes hand in hand with the diverse morphology of the DLA galaxies which have been imaged at $z < 1$, as discussed earlier (§2.1).

When the metallicity distribution of damped Lyα systems is compared with those of different stellar populations of the Milky Way, we find that is broader and peaks at lower metallicities than those of either thin or thick disk stars (Figure 8). At the time when our Galaxy's metal enrichment was at levels typical of DLAs, its kinematics were closer to those of the halo and bulge than a rotationally supported disk. This finding is at odds with the proposal that most DLAs are large disks with rotation velocities in excess of 200 km s^{-1}, put forward by Prochaska & Wolfe (1998).

(*c*) There is little evidence from the data in Figure 7 for any redshift evolution in the metallicity of DLAs. This question has also been addressed using the abundance of Fe which, thanks to its rich absorption spectrum, can be followed to higher redshifts and lower metallicities than Zn (Prochaska, Gawiser, & Wolfe 2001). Once allowance is made for the fraction of Fe in dust grains, Vladilo (2002b) finds a gradient of −0.32 in the linear regression of [Fe/H] vs. z_{abs}. However, while Figure 7 suggests that the chances of finding a DLA with [Zn/H] > −1.0 are possibly greater at $z < 1$, the column density-

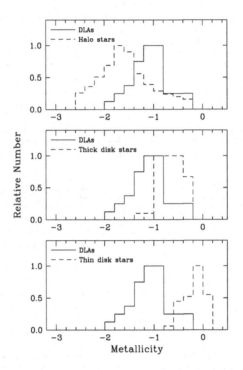

FIGURE 8. Metallicity distributions, normalised to unity, of DLAs at $z \simeq 2 - 3$ and of stars belonging to the disk (Wyse & Gilmore 1995) and halo (Laird et al. 1988) populations in the Milky Way.

weighted metallicities—which measure the density of metals per comoving volume—are consistent with no evolution over the range of redshifts probed so far, irrespectively of whether Zn or Fe are considered (see Figure 9). Evidently, the census of metals at all redshifts is dominated by high column density systems of low metallicity.

The lack of evolution in both the neutral gas and metal content of DLAs was unexpected and calls into question the notion that these absorbers are unbiased tracers of these quantities on a global scale. On the other hand, the paucity of data at redshifts $z < 1$, that is over a time interval of more than half of the age of the universe (Table 1), makes it difficult to draw firm conclusions and it may yet be possible to reconcile existing measurements with models of cosmic chemical evolution (Pei, Fall, & Hauser 1999; Kulkarni & Fall 2002).

2.4. Element Ratios

So far we have considered only the overall metallicity of DLAs as measured by the [Zn/H] ratio. However, the relative abundances of different elements offer additional insights into the chemical evolution of this population of galaxies, as we shall now see. This aspect of the work has really blossomed with the advent of efficient echelle spectrographs on both

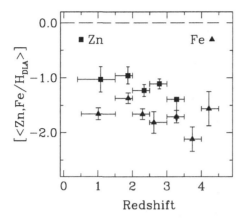

FIGURE 9. Column density-weighted metallicities of DLAs in different redshift intervals, from the surveys by Pettini et al. (1999) for Zn, and Prochaska & Wolfe (2002) for Fe. The lower abundance of Fe relative to Zn probably reflects the presence of moderate amounts of dust in most DLAs (Vladilo 2002b).

the Keck and VLT facilities, which have allowed the absorption lines of a wide variety of elements to be recorded simultaneously, often with exquisite precision.

2.4.1. *Dust in DLAs*

The presence of dust in DLAs can be inferred by comparing the gas phase abundances of two elements which in local interstellar clouds are depleted by differing amounts. The [Cr/Zn] ratio is one of the most suitable of such pairs for the reasons described above. It became apparent from the earliest abundance measurements in DLAs that this ratio is generally sub-solar, as expected if a fraction of the Cr has been incorporated into dust grains. Figure 10 shows this result for a subset of the DLAs in Figure 7; similar plots are now available for larger samples of DLAs and for other pairs of elements, one of which is refractory and the other is not (e.g. Prochaska & Wolfe 1999; 2002).

From this body of data it is now firmly established that the depletions of refractory elements are generally lower in DLAs than in interstellar clouds of similar column density in the disk of the Milky Way. The reasons for this are not entirely clear. The question has not yet been addressed quantitatively; qualitatively the effect is probably related to the lower metallicities of the DLAs and the likely higher temperature of the interstellar medium in these absorbers (Wolfire et al. 1995; Petitjean, Srianand, & Ledoux 2000; Kanekar & Chengalur 2001). Figure 10 does seem to indicate a weak trend of decreasing Cr depletion with decreasing metallicity, also supported by the results of Prochaska & Wolfe (2002).

Typically, it is found that refractory elements are depleted by about a factor of two in DLAs—a straightforward average of the measurements in Figure 10 yields a mean $\langle[\text{Cr/Zn}]\rangle = -0.3^{+0.15}_{-0.2}$ (1σ limits). When we combine this value with the mean metallicity of DLAs, $[\langle\text{Zn/H}_{\text{DLA}}\rangle] = -1.13$, or $\langle Z_{\text{DLA}}\rangle = 1/13\,Z_{\odot}$, we reach the conclusion that in damped systems the "typical" dust-to-gas ratio is only about $\approx 1/30$ of the Milky Way value (although there is likely to be a large dispersion from DLA to DLA, reflecting the range of metallicities evident in Figure 7). In the disk of our Galaxy, there a well determined relationship between the neutral hydrogen column density and the visual

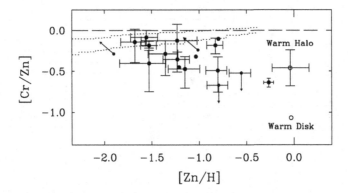

FIGURE 10. (Reproduced from Pettini et al. 1997a). Cr abundance relative to Zn in 18 damped Lyα systems (filled symbols). The region within the dotted lines (reproduced from Ryan, Norris, & Beers 1996) indicates how the [Cr/Fe] ratio varies in Galactic stars in this metallicity regime. The open circles show the typical [Cr/Zn] ratios measured in interstellar clouds in the disk and halo of our Galaxy, where the underabundance of Cr relative to Zn is ascribed to dust depletion (Savage & Senbach 1996).

extinction, $\langle N(\text{H I})\rangle/\langle A_V\rangle = 1.5 \times 10^{21}$ cm^{-2} mag^{-1} (Diplas & Savage 1994), where A_V is the extinction (in magnitudes) in the V band. For the typical damped Lyα system with neutral hydrogen column density $N(\text{H I}) = 1 \times 10^{21}$ cm^{-2} and dust-to-gas ratio 1/30 that of the local ISM, we therefore expect a trifling $A_V \simeq 0.02$ mag in the rest-frame V band. Of more interest is the far-UV extinction, since this is the spectral region observed at optical wavelengths at redshifts $z = 2 - 3$. Adopting the SMC extinction curve (Bouchet et al. 1985)—which may be the appropriate one to use at the low metallicities of most DLAs—we calculate that a damped Lyα system will typically introduce an extinction at 1500 Å of $A_{1500} \simeq 0.1$ mag in the spectrum of a background QSO. Such a small degree of obscuration is consistent with the mild reddening found in the spectra of QSO with DLAs, compared to the average UV continuum slope of QSOs without (Pei, Fall, & Bechtold 1991).

2.4.2. *Alpha-capture elements*

The moderate degree of depletion of refractory elements in DLAs has motivated a number of attempts to correct for the fractions missing from the gas phase (Vladilo 2002a and references therein) and thereby explore *intrinsic* (rather than dust-induced) departures from solar relative abundances. The basic idea, which is discussed extensively in the contribution to this volume by Francesca Matteucci, is that different elements are produced by stars of different masses and therefore different lifetimes. Thus, the relative abundances of two elements can, under the right circumstances, provide clues to the previous star formation history of the galaxy, or stellar population, under consideration (Wheeler, Sneden, & Truran 1989). Such clues are not always easy to decipher, however. For one thing, they rely on our incomplete, and mostly theoretical, knowledge of the stellar yields. Secondly, we must assume a 'standard' initial mass function (IMF) because, if we were free to alter at will the relative proportions of high and low mass stars, then we would obviously be able to reproduce most element ratios, but we could scarcely claim to have learnt anything in the process. Fortunately, all available evidence (including that

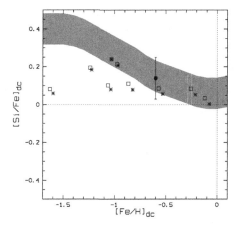

FIGURE 11. (Reproduced from Ledoux et al. 2002). Dust-corrected abundance ratios of Si relative to Fe versus DLA metallicity, as measured from the dust-corrected Fe abundances. Errors are typically ±0.1 dex. Different symbols are used for different dust depletion patterns adopted when correcting the observed abundances. The shaded area shows the region occupied by Galactic stars in the disk and halo over this range of metallicities.

from DLAs, Molaro et al. 2001) points to a universal IMF as a reasonable first order approximation (Kennicutt 1998a).

One of the cornerstones of this kind of approach is the well established overabundance of the alpha-capture elements relative to Iron in metal-poor stars of the Galactic halo. Mg, Si, Ca, and Ti are generally overabundant by factors of between two and three in stars where Fe is below one tenth solar, i.e. $[\alpha/\text{Fe}] = +0.3$ to $+0.5$ when $[\text{Fe}/\text{H}] \lesssim -1.0$ (Ryan et al. 1996). This result can be understood if approximately two thirds of the Fe (and other Fe-peak) elements are produced by Type Ia supernovae (SN) and released into the ISM with a time lag of about 1 Gyr relative the α-capture elements (and one third of the Fe) manufactured by the massive stars which explode as Type II supernovae.

In this picture, 1 Gyr is therefore the time over which the halo of our Galaxy became enriched to a metallicity $[\text{Fe}/\text{H}] = -1$, ultimately reflecting the rate at which star formation proceeded in this stellar component of the Milky Way. Clearly, the situation could be different in other environments (Gilmore & Wyse 1991; Matteucci & Recchi 2001). The thick disk, for example, evidently reached solar abundances of the α-elements in less than 1 Gyr, since the α overabundance—or more correctly the Fe deficiency—seems to persist to this high level of metallicity (Fuhrmann 1998).

Do damped Lyα systems, which as we have seen are generally metal-poor, show an overabundance of the α elements? This question has been addressed by several authors and there seems to be a general consensus that there is not a unique answer. As can be seen from Figure 11, while some DLAs conform to the pattern seen in Galactic stars, many others do not, in that they exhibit near solar values of $[\text{Si}/\text{Fe}]$ even when $[\text{Fe}/\text{H}]$ is $\ll -1$ (Molaro et al. 2000; Pettini et al. 2000a; Ledoux et al. 2002; Prochaska & Wolfe 2002; Vladilo 2002b). Presumably, these are galaxies where star formation has proceeded only slowly, or intermittently, allowing the Fe abundance to 'catch up' with that of the Type II supernova products. The Magellanic Clouds may be local counterparts of these DLAs (Pagel & Tautvaisiene 1998). Thus, the chemical clues provided by the these element ratios are another demonstration, together with the wide range in metallicity at the same epoch (§2.3) and the morphologies of the absorbers (§2.1), that DLAs trace a diverse population of galaxies, with different evolutionary histories. Their common trait is simply a large cross-section on the sky at a high surface density of neutral hydrogen.

2.4.3. *The Nucleosynthesis of Nitrogen*

A case of special interest is Nitrogen, whose nucleosynthetic origin is a subject of considerable interest and discussion. There is general agreement that the main pathway is a six step process in the CN branch of the CNO cycle which takes place in the stellar H burning layer, with the net result that ^{14}N is synthesised from ^{12}C and ^{16}O. The continuing debate, however, centres on which range of stellar masses is responsible for the bulk of the nitrogen production. A comprehensive reappraisal of the problem was presented by Henry, Edmunds, & Köppen (2000) who compiled an extensive set of abundance measurements and computed chemical evolution models using published yields. Briefly, nitrogen has both a primary and a secondary component, depending on whether the seed carbon and oxygen are those manufactured by the star during helium burning, or were already present when the star first condensed out of the interstellar medium.

Observational evidence for this dual nature of nitrogen is provided mainly from measurements of the N and O abundances in H II regions. (For consistency with other published work, we depart here from the notation used throughout the rest of this article, and use parentheses to indicate logarithmic ratios of number densities; adopting the recent reappraisal of solar photospheric abundances by Holweger (2001), we have $(N/H)_\odot = -4.07$; $(O/H)_\odot = -3.26$; and $(N/O)_\odot = -0.81$. In H II regions of nearby galaxies, (N/O) exhibits a strong dependence on (O/H) when the latter is greater than $\sim 2/5$ solar; this is generally interpreted as the regime where secondary N becomes dominant. At low metallicities on the other hand, when $(O/H) \lesssim -4.0$ (that is, $\lesssim 1/5$ solar), N is mostly primary and tracks O; this results in a plateau at $(N/O) \simeq -1.5$ (see Figure 12).

The principal sources of primary N are thought to be intermediate mass stars, with masses $4 \lesssim M/M_\odot \lesssim 7$, during the asymptotic giant branch (AGB) phase. A corollary of this hypothesis is that the release of N into the ISM should lag behind that of O which, as we have seen, is widely believed to be produced by massive stars which explode as Type II supernovae soon after an episode of star formation. Henry et al. (2000) calculated this time delay to be approximately 250 Myr; at low metallicities the (N/O) ratio could then perhaps be used as a clock with which to measure the past rate of star formation, as proposed by Edmunds & Pagel (1978). Specifically, in metal-poor galaxies which have only recently experienced a burst of star formation one may expect to find values of (N/O) *below* the primary plateau at $(N/O) \simeq -1.5$, provided the fresh Oxygen has been mixed with the ISM (Larsen, Sommer-Larsen, & Pagel 2001).

As pointed out by Pettini, Lipman, & Hunstead (1995), clues to the nucleosynthetic origin of nitrogen can also be provided by measurements of N and O in high redshift DLAs. Apart from the obvious interest in taking such abundance measurements to the distant past, when galaxies were young, one of the advantages of DLAs is that, thanks to their generally low metallicities, they probe a regime where local H II region abundance measurements are sparse or non-existent and where the effect of a delayed production of primary nitrogen should be most pronounced.

Figure 12 shows the most recent compilation of data relevant to this question. The fact that all DLA measurements fall within the region in the (N/O) vs. (O/H) plot bounded by the primary and secondary levels of N production provides empirical evidence in support of currently favoured ideas for the nucleosynthesis of primary N by intermediate

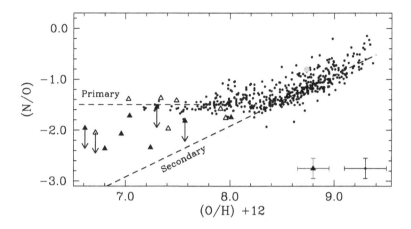

FIGURE 12. Abundances of N and O in extragalactic H II regions (small dots) and damped Lyα systems (large triangles). Sources for the H II region measurements are given in Pettini et al. (2002a). Filled triangles denote DLAs where the abundance of O could be measured directly, while open triangles are cases where S was used as a proxy for O. The error bars in the bottom right-hand corner give an indication of the typical uncertainties; the large dot corresponds to the solar abundances of N and O from the recent reappraisal by Holweger (2001). The dashed lines are approximate representations of the secondary and primary levels of N production (see text).

mass stars. The uniform value $(N/O) \simeq -1.5$ seen in nearby metal-poor star-forming galaxies can be understood in this scenario if these galaxies are not young, but contain older stellar populations, as indicated by a number of imaging studies with *HST*.

It is also somewhat surprisingly to find such a high proportion (40%) of DLAs which have apparently not yet attained the full primary level of N enrichment at $(N/O) \simeq -1.5$. Possibly, the low metallicity regime—where the difference between secondary and primary nitrogen enrichment is most pronounced—preferentially selects young galaxies which have only recently condensed out of the intergalactic medium and begun forming stars. A more speculative alternative, which needs to be explored computationally, is that at low metallicities stars with masses lower than $4M_\odot$ may make a significant contribution to the overall N yield (Lattanzio et al., in preparation; Meynet & Maeder 2002). The release of primary N may, under these circumstances, continue for longer than 250 Myr, perhaps for a substantial fraction of the Hubble time at the median $\langle z \rangle = 2.5$ of our sample.

In concluding this section, it is evident that DLAs are a rich source of information on nucleosynthesis in the early stages of galaxy formation. Element abundances in DLAs are increasingly being taken into consideration, together with stellar and H II region data from local systems, in models of the chemical evolution of galaxies and in the calculation of stellar yields. The chemical clues they provide will be even more valuable once their connection to today's galaxies in the Hubble sequence is clarified.

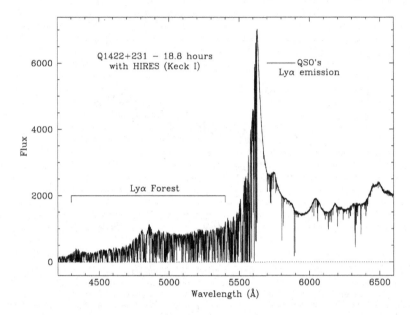

FIGURE 13. (Reproduced from Ellison 2000). This is one of the best QSO spectra ever obtained thanks to the combination of the bright magnitude of the gravitationally lensed QSO Q1422+231 ($V = 16.5$), long exposure time (amounting to several nights of observation), and high spectral resolution offered by the Keck echelle spectrograph (FWHM $\simeq 8\,\mathrm{km\ s^{-1}}$). The signal-to-noise ratio in the continuum longwards of the Lyα emission line is between 200 and 300. At these high redshifts ($z_{\mathrm{em}} = 3.625$) the Lyα forest eats very significantly into the QSO spectrum below the Lyα emission line and, with the present resolution, breaks into hundreds of absorption components (see Figure 14).

3. The Lyman Alpha Forest

The next component to be considered is the all-pervading intergalactic medium which manifests itself as a multitude of individual Lyα absorption lines bluewards of the Lyα emission line of every QSO. As can be appreciated from Figures 13 and 14, the effect is dramatic at high redshift. Observationally, the term Lyα forest is used to indicate absorption lines with column densities in the range $10^{16} \gtrsim N(\mathrm{HI}) \gtrsim 10^{12}$ cm^{-2} (see Figure 5).

Hydrodynamical simulations have shown that the Lyα forest is a natural consequence of the formation of large scale structure in a universe dominated by cold dark matter and bathed in a diffuse ionising background (see Weinberg, Katz, & Hernquist 1998 for an excellent review of the ideas which have led to this interpretation). An example is reproduced in Figure 15. Artificial spectra generated by throwing random sightlines through such model representations of the high redshift universe are a remarkably good match to real spectra of the Lyα forest. In particular, the simulations are very successful at reproducing the column density distribution of H I in Figure 5, the line widths and profiles, and the evolution of the line density with redshift. Consequently, much of what we have learnt about the IGM in the last few years has been the result of a very productive interplay between observations of increasing precision and simulations

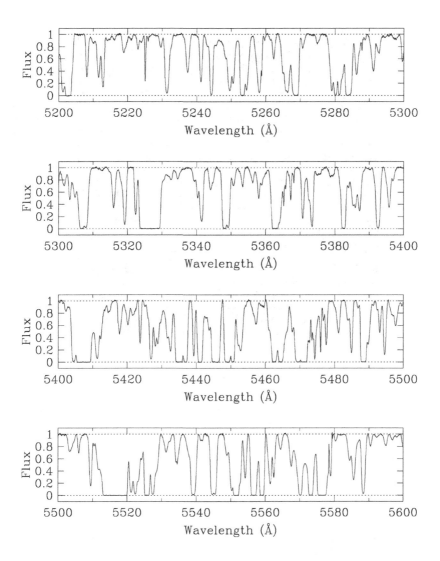

FIGURE 14. (Reproduced from Ellison 2000). Portion of the Lyα forest between $z_{\mathrm{abs}} = 3.277$ and 3.607 in the Keck spectrum of Q1422+231 shown in Figure 13. There are more than 50 individual absorption components in each 100 Å-wide stretch of spectrum.

of increasing sophistication. This modern view of the Lyα forest is often referred to as the 'fluctuating Gunn-Peterson' effect.

There are two important properties of the Lyα forest which we should keep in mind. One is that it is highly ionised, so that the H I we see directly is only a small fraction ($\sim 10^{-3}$ to $\sim 10^{-6}$) of the total amount of hydrogen present. With this large ionisation

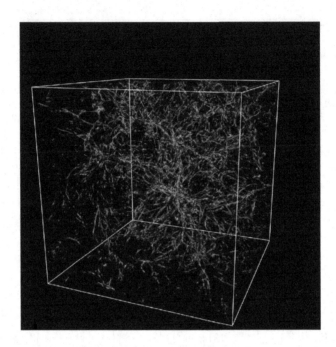

FIGURE 15. (Reproduced from http://astro.princeton.edu/~cen). Distribution of neutral gas at $z = 3$ from hydrodynamic cosmological simulation in a spatially flat, COBE-normalized, cold dark matter model with the cosmological parameters adopted in this article (§1.1). The box size is $25\,\mathrm{Mpc}/h$ (comoving) on the side, and the number of particles used in the simulation is 768^3. The structure seen in this (and other similar simulations) reproduces very well the spectral properties of the Lyα forest when artificial spectra are generated along random sightlines through the box.

correction it appears that the forest can account for most of the baryons at high, as well as low, redshift (Rauch 1998; Penton, Shull, & Stocke 2000); that is $\Omega_{\mathrm{Ly}\alpha} \approx 0.02\ h^{-2}$. Second, the physics of the absorbing gas is relatively simple and the run of optical depth $\tau(\mathrm{Ly}\alpha)$ with redshift can be thought of as a 'map' of the density structure of the IGM along a given line of sight. At low densities, where the temperature of the gas is determined by the balance between photoionisation heating (produced by the intergalactic ionising background) and adiabatic cooling (due to the expansion of the universe), $\tau(\mathrm{Ly}\alpha) \propto (1 + \delta)^{1.5}$, where δ is the overdensity of baryons $\delta \equiv (\rho_b/\langle\rho_b\rangle - 1)$. At $z = 3$, $\tau(\mathrm{Ly}\alpha) = 1$ corresponds to a region of the IGM which is just above the average density of the universe at that time ($\delta \approx 0.6$). The last absorption line in the second panel of Figure 14, near 5395 Å, is an example of a Lyα line with $\tau \simeq 1$. The idea that, unlike galaxies, the forest is an unbiased tracer of mass has prompted, among other things, attempts to recover the initial spectrum of density fluctuations from consideration of the spectrum of line optical depths in the forest (Croft et al. 2002)

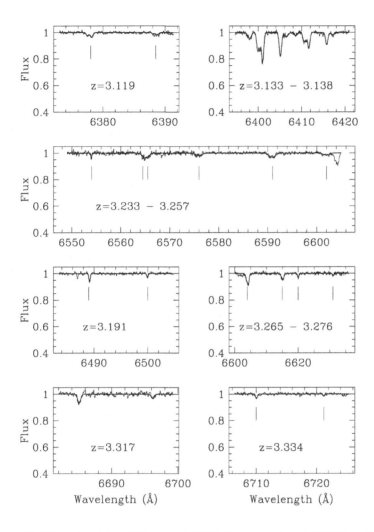

FIGURE 16. (Reproduced from Ellison et al. 2000). Examples of weak C IV lines identified in the spectrum of Q1422+231; most of these would have remained undetected in spectra of lower signal-to-noise ratios. Green (grey) lines show the profile fits used to deduce the column density of C IV. The weakest C IV systems are indicated with tick marks to guide the eye.

3.1. *Metals in the Lyα Forest*

The lack of associated metal lines was originally one of the defining characteristics of the Lyα forest and was interpreted as evidence for a primordial origin of the clouds (Sargent et al. 1980). However, this picture was shown to be an oversimplification by the first observations—using the HIRES spectrograph on the Keck I telescope—with sufficient sensitivity to detect the weak C IV $\lambda\lambda 1548, 1550$ doublet associated with Lyα clouds with column densities $\log N(\text{H I}) \gtrsim 14.5$ (Cowie et al. 1995; Tytler et al. 1995). Typical

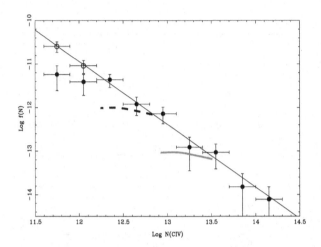

FIGURE 17. (Reproduced from Ellison 2000). C IV column density distribution in Q1422+231 at $\langle z \rangle$ = 3.15; $f(N)$ is the number of C IV systems per column density interval and per unit redshift path. The filled circles are the data; the straight line shows the best fitting power-law slope α = 1.44, assuming the distribution to be of the form $f(N)dN = BN^{-\alpha}dN$. The open circles show the values corrected for incompleteness at the low column density end; with these correction factors there is no indication of a turnover in the column density distribution down to the lowest values of N(C IV) reached up to now. Earlier indications of a turnover shown by the grey (Petitjean & Bergeron 1994) and dashed (Songaila 1997) curves are now seen to be due to the less sensitive detection limits of those studies, rather than to a real paucity of weak Lyα lines.

column density ratios in these clouds are N(C IV)$/N$(H I) $\simeq 10^{-2} - 10^{-3}$, indicative of a carbon abundance of about 1/300 of the solar value, or [C/H] $\simeq -2.5$ in the usual notation, and with a scatter of perhaps a factor of ~ 3 (Davé et al. 1998).

The question of interest is 'Where do these metals come from?'. Obviously from stars (we do not know of any other way to produce carbon!), but are these stars located in the vicinity of the Lyα clouds observed—which after all are still at the high column density end of the distribution of values of N(H I) for intergalactic absorption—or are we seeing a more widespread level of metal enrichment, perhaps associated with the formation of the first stars which re-ionised the universe at $z > 6$ (Songaila & Cowie 2002)?

To answer this question we should like to search for metals in low density regions of the IGM, away from the overdensities where galaxies form. Observationally, this is a very difficult task—the associated absorption lines, if present at all, would be very weak indeed. Ellison et al. (1999, 2000) made some progress towards probing such regions using extremely long exposures with HIRES of two of the brightest known high-z QSOs, both gravitationally lensed: APM 08279+5255 and Q1422+231. The latter set of data in particular (Figures 13 and 14) is of exceptionally high quality, reaching a signal-to-noise ratio S/N $\simeq$ 300 which translates to a limiting rest-frame equivalent width limit $W_0(3\sigma) \lesssim 1$ mÅ; this in turn corresponds to a sensitivity to C IV absorbers with column densities as low as N(C IV) $\simeq 4 \times 10^{11}$ cm^{-2}.

And indeed C IV lines are found at these low levels (see Figure 16), showing that metals are present in the lowest column density Lyα clouds probed, at least down to N(H I) $= 10^{14}$ cm^{-2}. As can be seen from Figure 17, the number of weak C IV lines

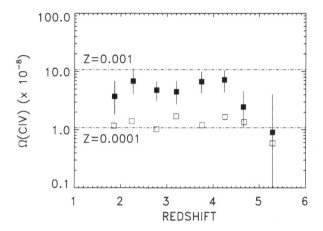

FIGURE 18. (Reproduced from Songaila 2001). Mass density in C IV (expressed as fraction of the closure density—see eqs. 2.2 and 2.3) as a function of redshift. The filled symbols are for C IV absorption systems in the range $12 \leq \log N(\text{C IV}) \leq 15$ (cm^{-2}). The dot-dash lines show values of $\Omega(\text{CIV})$ computed assuming $\Omega_b\, h^2 = 0.022, h = 0.65$, a C IV ionisation fraction of 0.5, and metallicities $Z = 0.0001$ and $0.001\, Z_\odot$, respectively.

continues to rise as the signal-to-noise ratio of the spectra increases and any levelling off in the column density distribution presumably occurs at $N(\text{C IV}) < 5 \times 10^{11}$ cm^{-2}. This limit is one order of magnitude more sensitive than those reached previously. In other words, we have yet to find any evidence in the Lyα forest for regions of the IGM which are truly of primordial composition or have abundances as low as those of the most metal-poor stars in the Milky Way halo. These conclusions are further supported by the recent detection of O VI $\lambda\lambda 1032, 1038$ absorption in the Lyα forest at $z = 2$ by Carswell, Schaye, & Kim (2002). In agreement with the results of Ellison et al. (2000), these authors found that most Lyα forest clouds with $N(\text{H I}) \geq 10^{14}$ cm^{-2} have associated O VI absorption and that [O/H] is in the range -3 to -2. Weak O VI lines from regions of lower Lyα optical depth have not yet been detected directly, but their presence is inferred from statistical considerations (Schaye et al. 2000).

3.2. *C IV at the Highest Redshifts*

A level of metal enrichment of 10^{-3} to 10^{-2} of solar in regions of the IGM with $N(\text{H I}) \geq 10^{14}$ cm^{-2} may still be understood in terms of supernova driven winds from galaxies. The work of Aguirre et al. (2001) shows that such outflows which, as we shall shortly see (§4.5) are observed directly in Lyman break galaxies at $z = 3$, may propagate out to radii of several hundred kpc before they stall. However, if O VI is also present in Lyα forest clouds of lower column density, as claimed by Schaye et al. (2000), an origin in pregalactic stars at much earlier epochs is probably required (Madau, Ferrara, & Rees 2001).

In order to investigate this possibility, Songaila (2001) extended the search for intergalactic C IV to $z = 5.5$, taking advantage of the large number of QSOs with $z_{em} > 5$ discovered by the Sloan Digital Sky Survey. The surprising result, reproduced in Figure 18, is that there seems to be no discernable evolution in the integral of the column density distribution of C IV from $z = 1.5$ to $z = 5.5$. (The reality of a possible drop in $\Omega(\text{CIV})$

beyond $z = 4.5$ is questioned by Songaila because incompleteness effects have not been properly quantified in this difficult region of the optical spectrum, at $\lambda_{obs} > 8500\,\text{Å}$). This finding was unexpected and has not yet been properly assessed. The observed column density of C IV depends not only on the overall abundance of Carbon, but also on the shape and normalisation of the ionising background and on the densities associated with a given $N(\text{H I})$. Thus, we would have predicted large changes in $\Omega(\text{CIV})$ between $z = 5.5$ and 1.5 in response to the evolving density of ionising sources (QSOs) and the development of structure in the universe, even if the metallicity of the IGM had remained constant between these two epochs.

Whatever lies behind the apparent lack of redshift evolution of $\Omega(\text{CIV})$, it is clear that the IGM was enriched with the products of stellar nucleosynthesis from the earliest times we have been able to probe with QSO absorption line spectroscopy, only $\sim 1\,\text{Gyr}$ after the Big Bang. The measurements of $\Omega(\text{CIV})$ in Figure 18 suggest a metallicity $Z_{\text{Ly}\alpha} \gtrsim 10^{-3} Z_\odot$; this is a lower limit because it assumes that the ionisation of the gas is such that the ratio C IV/C_{tot} is near its maximum value of about 0.5. This minimum metallicity can in turn can be used to infer a minimum number of hydrogen ionising photons (with energy $h\nu \geq 13.6\,\text{eV}$, corresponding to $\lambda \leq 912\,\text{Å}$) in the IGM, because the progenitors of the supernovae which produce Oxygen, for example, are the same massive stars that emit most of the (stellar) ionising photons. Assuming a solar relative abundance scale (i.e. [C/O] $= 0$), Madau & Shull (1996) calculated that the energy of Lyman continuum photons emitted is 0.2% of the rest-mass energy of the heavy elements produced.† From this it follows that

$$\frac{N_{\text{photons}}}{N_{\text{baryons}}} \times 13.6\,\text{eV} \simeq 0.002 m_p c^2 Z \simeq 2 \times 10^6\,\text{eV} \times Z \qquad (3.6)$$

(Miralda-Escudé & Rees 1997), where Z is the metallicity (by mass) and m_p the mass of the proton. Since $Z_\odot = 0.02$ (Grevesse & Sauval 1998), if the Lyα forest at $z \simeq 5$ had already been enriched to a metallicity $Z_{\text{Ly}\alpha} \simeq 10^{-3} Z_\odot$, eq. (3.6) implies that by that epoch stars had emitted approximately three Lyman continuum (LyC) photons per baryon in the universe. Whether this photon production is sufficient to have reionised the IGM by these redshifts depends critically on the unknown escape fraction of LyC photons from the sites of star formation.

4. Lyman Break Galaxies

Returning to Figure 1, it must be remembered that everything we have learnt on the distant universe up to this point required 'decoding' the information 'encrypted' in the absorption spectra of QSOs. It is easy to appreciate, therefore, the strong incentive which motivated astronomers in the 1990s to detect high resdshift galaxies directly. After many years of fruitless searches, we have witnessed since 1995 a veritable explosion of data from the *Hubble Deep Fields* and ground-based surveys with large telescopes. The turning point was the realisation of the effectiveness of the Lyman break technique (Figure 19) in preselecting candidate $z \simeq 3$ galaxies (Steidel et al. 1996). Although these galaxies constitute only $\sim 3 - 4\%$ of the thousands of faint objects, at all redshifts, revealed by a moderately deep CCD exposure at the prime focus of a $4\,\text{m}$ telescope (see Figure 20), they can be readily distinguished on the basis of their colours alone, when observed through appropriately selected filters (Figure 21).

The Lyman break technique has been very successful at finding high redshift galaxies

---

† This is a lower limit if [C/O] < 0, as is the case for low metallicity gas in nearby galaxies (e.g. Garnett et al. 1999).

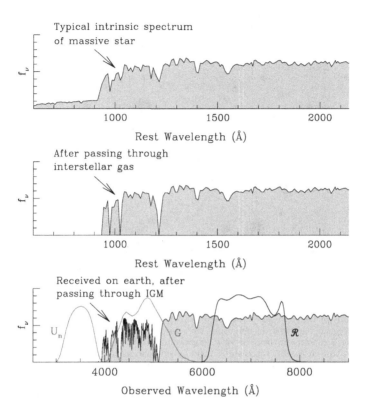

FIGURE 19. (Courtesy of Kurt Adelberger). An illustration of the principles behind the Lyman break technique. Hot stars have flat far-UV continua, but emit fewer photons below 912 Å, the limit of the Lyman series of hydrogen (top panel). These photons are also efficiently absorbed by any H I associated with the sites of star formation (middle panel) and have a short mean free path—typically only ~ 40 Å—in the IGM at $z = 3$. Consequently, when observed from Earth (bottom panel), the spectrum of a star forming galaxy at $z \simeq 3$ exhibits a marked 'break' near 4000 Å. With appropriately chosen broad-band filters, this spectral discontinuity gives rise to characteristic colours; objects at these redshifts appear blue in $(G - \mathcal{R})$ and red in $(U_n - G)$. For this reason, such galaxies are sometimes referred to as U-dropouts. A more quantitative description of the Lyman break technique can be found in Steidel, Pettini, & Hamilton (1995).

thanks to the combination of the increasingly large and UV sensitive CCDs used to identify candidates on the one hand, and the multi-object spectroscopic capabilities of large telescopes required for follow-up and confirmation on the other. Thus, samples of spectroscopically confirmed $z \simeq 3$ galaxies have grown from zero to more than one thousand in the space of only five years. As can be seen from Figure 22, the spectroscopic redshifts generally conform to expectations based on just two colours. Such large samples have made it possible to trace the star formation history of the universe over most of the Hubble time and to measure the large-scale properties of this population of galaxies, most notably their clustering and luminosity functions (see Steidel 2000 for a review).

In parallel with this work on the Lyman break population as a whole, in the last few

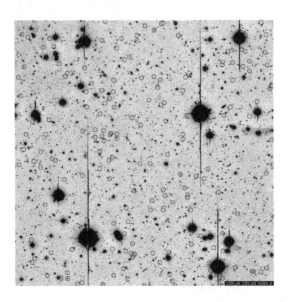

FIGURE 20. A typical deep CCD image recorded at the prime focus of a 4m-class telescope. This particular image (of the field designated DSF2237b) was obtained with the COSMIC camera of the Palomar Hale telescope, by exposing for a total of two hours through a custom made $\mathcal{R}$ filter. In the 9×9 arcmin field of view (corresponding to co-moving linear dimensions of $11.6 \times 11.6 \, h^{-1} \, \text{Mpc}$ at $z = 3$) there are ~ 3300 galaxies brighter than $\mathcal{R} = 25.5$; 140 of these (circled) show Lyman breaks which place them at redshifts between $z = 2.6$ and 3.4.

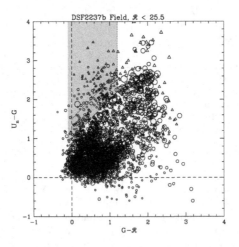

FIGURE 21. All the ~ 3300 galaxies from Figure 20 are included in this colour–colour plot. The shaded region shows how the 140 candidate Lyman break galaxies are selected for subsequent spectroscopic follow-up. The symbol size is proportional to the object magnitude; circles denote objects detected in all three bands, while triangles are lower limits in $(U_n - G)$ for U_n dropouts.

Spectroscopically Confirmed Objects

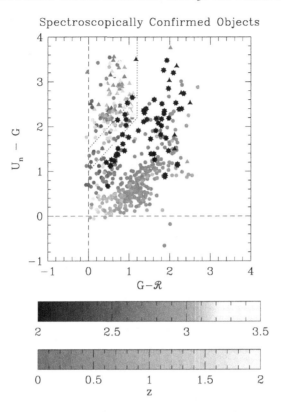

FIGURE 22. The location of spectroscopically confirmed galaxies (from the surveys by Steidel and collaborators) on the $(U_n - G)$ vs. $(G - \mathcal{R})$ plot. Triangles denote objects undetected in the U_n band; stellar symbols are used for Galactic stars.

years we have also begun to study in more detail the physical properties of some of the brighter galaxies in the sample. The questions which we would like to address are:

(*a*) What are the stellar populations of the Lyman break galaxies?

(*b*) What are their ages and masses?

(*c*) What are their levels of metal enrichment?

(*d*) What are the effects of star formation at high z on the galaxies and the surrounding IGM?

Many of these questions link observational cosmology with stellar and interstellar astrophysics, and this will become evident as we now explore them in turn.

4.1. *Stellar Populations and the Initial Mass Function*

Among the thousand or so known Lyman break galaxies (LBGs), one, designated MS 1512−cB58 (or cB58 for short) has provided data of exceptional quality, thanks to its gravitationally lensed nature. Discovered by Yee et al. (1996), cB58 is, as far as we can tell, a typical $\sim L^*$ galaxy at a redshift $z = 2.7276$ magnified by a factor of ~ 30 by the foreground cluster MS 1512+36 at $z = 0.37$ (Seitz et al. 1998). This fortuitous

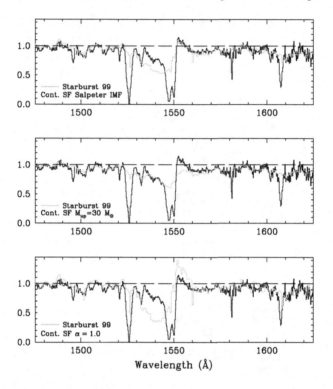

FIGURE 23. Comparisons between *Starburst99* (Leitherer et al. 1999) population synthesis models with different IMFs (grey lines) and the Keck spectrum of MS 1512-cB58 analysed by Pettini et al. (2000b) in the region near the C IV doublet (black histogram). The *y*-axis is residual intensity.

alignment makes it by far the brightest known member of the LBG population and has motivated a number of studies from mm to X-ray wavelengths.

When we record the spectrum of a $z = 3$ galaxy at optical wavelengths, we are in fact observing the redshifted far-UV light produced by a whole population of O and early B stars. Such spectra are most effectively analysed with population synthesis models, the most sophisticated of which is *Starburst99* developed by the Baltimore group (Leitherer et al. 1999). In Figure 23 we compare *Starburst99* model predictions for different IMFs with a portion of the moderate resolution Keck LRIS spectrum of cB58 obtained by Pettini et al. (2000b), encompassing the C IV $\lambda\lambda1548, 1550$ doublet.

It is important to realise that the comparison only refers to *stellar* spectral features and does not include the *interstellar* lines, readily recognisable by their narrower widths (the interstellar lines are much stronger in cB58, where we sample the whole ISM of the galaxy, than in the models which are based on libraries of nearby Galactic O and B stars). With this clarification, it is evident from Figure 23 that the spectral properties of at least this Lyman break galaxy are remarkably similar to those of present-day starbursts—a continuous star formation model with a Salpeter IMF provides a very good fit to the observations. In particular, the P-Cygni profiles of C IV, Si IV and N V are sensitive to the slope and upper mass limit of the IMF; the best fit in cB58 is obtained with a

MS 1512−cB58 C IV Region Keck II + ESI 16,000 s 1.4 Å

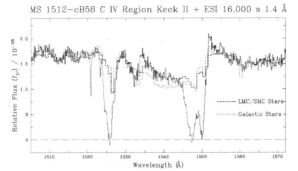

FIGURE 24. A portion of the high resolution spectrum of MS 1512-cB58 obtained by Pettini et al. (2002b), encompassing the C IV λλ1548, 1550 doublet lines, is compared with *Starburst99* synthetic spectra for solar and ∼ 1/3 solar (LMC/SMC) metallicities (Leitherer et al. 2001).

standard Salpeter IMF with slope $\alpha = 2.35$ and $M_{up} = 100 M_\odot$ (top panel of Figure 23). IMFs either lacking in the most massive stars or, conversely, top-heavy seem to be excluded by the data (middle and bottom panels of Figure 23 respectively).

The only significant difference between the observed and synthesised spectra is in the optical depth of the P-Cygni absorption trough which is lower than predicted (top panel of Figure 23). A possible explanation is that this is an abundance effect—the strengths of the wind lines are known to be sensitive to the metallicity of the stars, since it is through absorption and scattering of photons in the metal lines that momentum is transferred to the gas and a wind is generated. The *Starburst99* spectra shown in Figure 23 are for solar metallicity, but a recent update (Leitherer et al. 2001) now includes stellar libraries compiled with *HST* observations of hot stars in the Large and Small Magellanic Clouds, taken to be representative of a metallicity $Z \simeq 1/3 Z_\odot$.

When these are compared with a higher resolution spectrum of cB58, obtained with the new Echelle Spectrograph and Imager (ESI) on the Keck II telescope, the match to the observed spectrum is improved (Figure 24). The emission component of the P-Cygni profile is perhaps underestimated by the model, but we suspect that there may be additional nebular C IV emission from the H II regions in the galaxy, superposed on the broader stellar P-Cygni emission. We take the good agreement in Figure 24 as an indication that the young stellar population of cB58 has reached a metallicity comparable to that of the Magellanic Clouds. This conclusion is reinforced by measurements of element abundances in the interstellar gas, as we shall see in §4.2 below.

Before leaving this topic, it is worth noting that with a modest amount of effort (and luck in the form of gravitational lensing) it is now possible to obtain spectra of high redshift galaxies of sufficient quality to distinguish between the OB stellar populations of the Milky Way and of the Magellanic Clouds. As a matter of fact, it is evident from Figure 24 that the resolution of the ESI spectrum of cB58 is superior to those of Magellanic Cloud stars (and of nearby starburst galaxies) with which we would like to compare it!

4.2. *Element Abundances in the Interstellar Gas*

The ESI spectrum of cB58, which covers the wavelength region from 1075 to 2800 Å at a resolution of $58 \, \text{km s}^{-1}$, is a real treasure trove of information on this galaxy. For example, it includes 48 interstellar absorption lines of elements from H to Zn in a variety of ionisation stages, from neutral (H I, C I, O I, N I) to highly ionised species (Si IV,

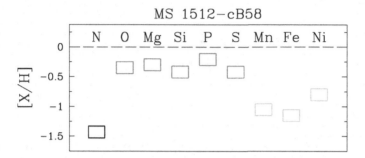

FIGURE 25. Pattern of chemical abundances in the ambient interstellar medium of cB58 deduced by Pettini et al. (2002b). The vertical height of the boxes shows the typical uncertainty in the abundance determinations. O, Mg, Si, P, and S are thought to be synthesised by massive stars which explode as Type II supernovae, while Mn, Fe and Ni are predominantly produced by Type Ia SN. Their release into the ISM, as well as that of N from intermediate mass stars, lags behind that of the Type II SN products by several 100 Myr.

C IV, N V). The lines are fully resolved so that column densities can be derived from the analysis of their profiles. From these data Pettini et al. (2002b) were able to piece together for the first time a comprehensive picture of the chemical composition of the interstellar gas in a Lyman break galaxy (Figure 25) and examine the clues it provides on its evolutionary status and past history of star formation.

As can be seen from Figure 25, the ambient interstellar medium of cB58 is highly enriched in the elements released by Type II supernovae; O, Mg, Si, P, and S all have abundances of $\sim 2/5$ solar. Thus, even at this relatively early epoch ($z = 2.7276$ corresponds to 2.5 Gyr after the Big Bang in our adopted cosmology—see Table 1), this galaxy had already processed more than one third of its gas into stars.

Furthermore, cB58 appears to be chemically young, in that it is relatively deficient in elements produced by stars of intermediate and low mass with longer lifetimes than those of Type II SN progenitors. N and the Fe-peak elements we observe (Mn, Fe, and Ni) are all less abundant than expected by factors of between 0.4 and 0.75 dex. Depletion onto dust, which is known to be present in cB58, probably accounts for some of the Fe-peak element underabundances, but this is not likely to be an important effect for N. On the basis of current ideas of the nucleosynthesis of N, discussed in §2.4.3, it would appear that much of the ISM enrichment in cB58 has taken place within the last 250 Myr, the lifetime of the intermediate mass stars believed to be the main source of N. For comparison, the starburst episode responsible for the UV and optical light we see is estimated to be younger than ~ 35 Myr, on the basis of theoretical models of the spectral energy distribution at these wavelengths (Ellingson et al. 1996).

Taken together, these two findings are highly suggestive of a galaxy caught in the act of converting its interstellar medium into stars on a few dynamical timescales—quite possibly in cB58 we are witnessing the formation of a galactic bulge or an elliptical galaxy. The results of the chemical analysis are consistent with the scenario proposed by Shapley et al. (2001), whereby galaxies whose UV spectra are dominated by strong, blueshifted absorption lines, as is the case here, are the youngest in the range of ages of LBGs. These findings also lend support to models of structure formation which predict that, even at $z \simeq 3$, near-solar metallicities should in fact be common in galaxies with

Q1422+231 D81 z=3.1037 (NIRSPEC R=1550 6300 s)

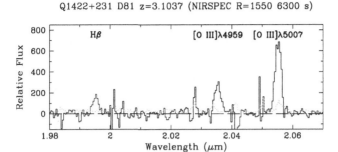

FIGURE 26. Example of a NIRSPEC K-band spectrum of a Lyman break galaxy from the survey by Pettini et al. (2001). The objects targeted typically have $K = 21$ (on the Vega scale) and remain undetected in the continuum. However, the nebular emission lines of [O III] $\lambda\lambda 4859, 5007$, [O II] 3727 (not shown), and Hβ usually show up clearly with exposure times of 2–3 hours. The dotted line is the 1σ error spectrum.

masses greater than $\sim 10^{10} M_{\odot}$ (e.g. Nagamine et al. 2001). The baryonic mass of cB58 is deduced to be $m_{\mathrm{baryons}} \simeq 1 \times 10^{10} M_{\odot}$, from consideration of its star formation history, metallicity, and the velocity dispersion of its ionised gas.

4.3. *The Oxygen Abundance in H II Regions*

How typical are these results of the Lyman break galaxy population as a whole? In the nearby universe, element abundances in star forming regions have traditionally been measured from the ratios of optical emission lines from H II regions. At $z = 3$ these features move to near-infrared (IR) wavelengths and have only become accessible in the last two years with the commissioning of high resolution spectrographs on the VLT (ISAAC) and Keck telescopes (NIRSPEC). Using these facilities, our group has recently completed the first spectroscopic survey of Lyman break galaxies in the near-IR, bringing together data for 19 LBGs; the galaxies are drawn from the bright end of the luminosity function, from $\sim L^*$ to $\sim 4 L^*$ (Pettini et al. 2001). Figure 26 shows an example of the quality of spectra which can be secured with a 2–3 hour integration. In five cases we attempted to deduce values of the abundance of oxygen by applying the familiar $R_{23} = ([\mathrm{O\ II}] + [\mathrm{O\ III}])/\mathrm{H}\beta$ method first proposed by Pagel et al. (1979). We found that generally there remains a significant uncertainty, by up to 1 dex, in the value of (O/H) because of the double-valued nature of the R_{23} calibrator (see Figure 27).

Thus, in the galaxies observed, oxygen could be as abundant as in the interstellar medium near the Sun or as low as $\sim 1/10$ solar. When the R_{23} method is applied to cB58, a similar ambiguity obtains (Teplitz et al. 2000). The results from the analysis of the interstellar absorption lines described above (§4.2) resolve the issue by showing that the upper branch solution is favoured (we have no reason to suspect that the neutral and ionised ISM have widely different abundances). It remains to be established whether this is also the case for other LBGs.

In the near future this work will shift to lower and more easily accessible redshifts near $z = 2.2$ where all the lines of interest, from [O II] $\lambda 3727$ to Hα, fall in near-IR atmospheric transmission windows. Nevertheless, the determination of element abundances from nebular emission lines will remain a time consuming task until multi-object spectrographs

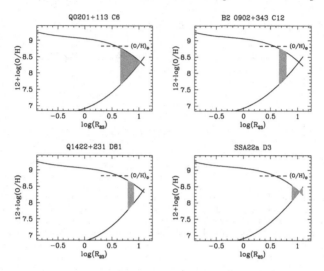

FIGURE 27. Oxygen abundance from the $R_{23} = ([\text{O II}]+[\text{O III}])/\text{H}\beta$ ratio. In each panel the continuous lines are the calibration by McGaugh (1991) for the ionisation index $O_{32} = [\text{O III}]/[\text{O II}]$ appropriate to that object. The shaded area shows the values allowed by the measured R_{23} and its statistical 1σ error. The broken horizontal line gives for reference the solar abundance $12 + \log(\text{O/H}) = 8.83$ from the compilation by Grevesse & Sauval (1998); the recent revision by Holweger (2001) would bring the line down by 0.09 dex.

operating at near-IR wavelengths become available on large telescopes, or until the launch of the *Next Generation Space Telescope* (Kennicutt 1998b).

4.4. *Dating the Star Formation Activity*

Realistically, the detailed spectroscopic analysis described above can only be applied to a subset of LBGs, at the bright end of the luminosity funtion. However, the coarser spectral energy distribution (SED) of Lyman break galaxies still holds important information on the star formation episodes. Broad-band photometry in the optical and near-infrared, spanning the wavelength interval 900–5500 Å in the rest-frame, is now available for more than one hundred galaxies at $z \simeq 3$ (Papovich, Dickinson, & Ferguson 2001; Shapley et al. 2001). The colours over this range (typically four colours are used in the analysis) depend on the degree of dust reddening, $E(B - V)$, and on the age of the stellar population, t_{sf}. The two can be decoupled with some degree of confidence provided that the SED includes the age-sensitive Balmer break near 3650 Å, which at $z = 3$ falls between the H and K bands—hence the need for accurate near-IR photometry. A third parameter, the instantaneous star formation rate, $\Psi(t_{\text{sf}})$, determines the normalisation (rather than the shape) of the SED. The analyses by Papovich et al. (2001) and Shapley et al. (2001) deduced the best-fitting values of $E(B - V)$, t_{sf}, and $\Psi(t_{\text{sf}})$ by χ^2 minimisation of the differences between the observed SEDs and those predicted by the widely used population synthesis code of Bruzual & Charlot (1993 and subsequent updates). The results have turned out to be very interesting—some would say surprising (see Figures 28 and 29).

Evidently, Lyman break galaxies span a wide range of ages. One fifth of the sample considered by Shapley et al. (2001) consists of objects which apparently have just collapsed and are forming stars on a dynamical timescale ($\sim 35\,\text{Myr}$). As we have seen,

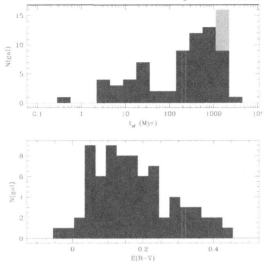

FIGURE 28. Histograms of best-fitting ages and reddening for the sample of 81 $z \simeq 3$ LBGs analysed by Shapley et al. (2001). There is a large spread of ages in the population; the median age is 320 Myr and 20% of the objects are older than 1 Gyr. The light grey bin corresponds to galaxies with inferred ages older than the age of the universe at their redshifts (an indication of the approximate nature of the ages derived by SED fitting). The median $E(B - V) = 0.155$ for the sample corresponds to attenuations by factors of ~ 4.5 and ~ 2 at 1600 Å and 5500 Å respectively.

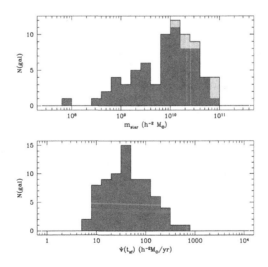

FIGURE 29. Histograms of assembled stellar mass and star formation rates from Shapley et al. (2001). By redshift $z \simeq 3$ a significant fraction of LBGs seem to be approaching the stellar mass of today's L^* galaxies, $m_{\text{star}} \simeq 4 \times 10^{10} M_\odot$.

Velocity Offsets in Lyman Break Galaxies

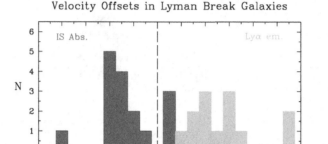

FIGURE 30. Velocity offsets of the interstellar absorption lines (blue or dark grey) and of the Lyα emission line (red or light grey) relative to [O III] and Hβ. Large scale motions of the order of several hundred km s^{-1} are indicated by the systematic tendency for the former to be blueshifted and the latter redshifted relative to the nebular emission lines.

cB58 seems to belong to this class. At the other end of the scale, some 20% of the galaxies at $z = 3$ have been forming stars for more than 1 Gyr, placing the onset of star formation at much higher redshifts ($z > 5 - 10$). Furthermore, there appears to be a correlation between age and star formation rate, with the younger objects typically forming stars at about ten times the rate of the older ones and being more reddened on average. The mean SFR is $\langle \Psi(t_{\rm sf}) \rangle = 210 h^{-2} M_\odot$ yr^{-1} for galaxies with $t_{\rm sf} < 35$ Myr while, for the 20% of the sample with $t_{\rm sf} > 1$ Gyr, $\langle \Psi(t_{\rm sf}) \rangle = 25 h^{-2} M_\odot$ yr^{-1}.

   This range of properties is further reflected in the total formed stellar masses $m_{\rm star}$ obtained by integrating $\Psi(t_{\rm sf})$ over $t_{\rm sf}$. A variety of star formation histories was considered (e.g. star formation which is continuous or decreases with time); in general $m_{\rm star}$ does not depend sensitively on this choice, although an older population of stars which by $z \simeq 3$ have faded at UV and optical wavelengths could remain hidden (Papovich et al. 2001). As can be seen from Figure 29, by redshift $z \simeq 3$ some galaxies had apparently already assembled a stellar mass comparable to that of an L^* galaxy today, $m_{\rm star} \simeq 4 \times 10^{10} M_\odot$, while 20% of the sample have values of $m_{\rm star}$ one order of magnitude smaller. These findings led Shapley et al. (2001) to speculate that we may be beginning to discern an evolutionary sequence in Lyman break galaxies, with the younger, dustier, more actively star-forming objects evolving to the older, less reddened, and more quiescent phase. It remains to be seen how this scenario stands up to the scrutiny of future observations, as we try to link the properties of the stellar populations of *individual* galaxies to other parameters, such as dynamical mass and metallicity.

4.5. *Galactic-Scale Outflows*

The near-IR spectroscopic survey by Pettini et al. (2001) confirmed a trend which had already been suspected on the basis of the optical (rest-frame UV) data alone. When the redshifts of the interstellar absorption lines, of the nebular emission lines, and of the resonantly scattered Lyα emission line are compared within the same galaxy, a systematic pattern of velocity differences emerges in all LBGs observed up to now (see Figure 30). We interpret this effect as indicative of galaxy-wide outflows, presumably driven by the supernova activity associated with the star-formation episodes. Such 'superwinds' appear to be a common characteristic of galaxies with large rates of star formation per unit area at high, as well as low, redshifts (e.g. Heckman 2001). They involve comparable amounts of matter to that being turned into stars (the mass outflow rate is of the same order as the star formation rate) and about 10% of the total kinetic energy delivered by the starburst (Pettini et al. 2000b). These outflows have a number of important astrophysical consequences.

First, they provide self-regulation to the star formation process—this is the 'feedback' required by theorists (e.g. Efstathiou 2000; Binney, Gerhard, & Silk 2001) for realistic galaxy formation models. Galactic winds may well be the key factor at the root of the 'evolutionary sequence' for LBGs just discussed (§4.4).

Second, they can distribute the products of stellar nucleosynthesis over large volumes of the intergalactic medium since the outflow speeds are likely to exceed the escape velocities in many cases. As we have seen, many LBGs are already metal-enriched at $z = 3$ and have by then been forming stars for much of the Hubble time. There is therefore at least the potential for widespread pollution of the IGM with metals, thereby explaining at least in part the results on the metallicity of the Lyα forest described in §3.1 and 3.2).

Third, the outflowing hot gas is likely to 'punch' through the neutral interstellar medium of the galaxies and provide a route through which Lyman continuum photons can leak out of the galaxies, easing the problem of how the universe came to be reionised (Steidel, Pettini, & Adelberger 2001). Indeed it now appears (Adelberger et al. 2002, in preparation) that LBGs have a substantial impact on the surrounding IGM, and that shock-ionisation by their winds leads to a pronounced 'proximity effect'—the Lyα forest is essentially cleared out by these outflows over radii of $\sim 100 h^{-1}$ kpc.

5. Bringing it All Together

5.1. *A Global View of Metal Enrichment in the Universe*
Two Billion Years after the Big Bang

We could briefly summarise everything we have covered so far as follows.

(*a*) The intergalactic medium at $z = 3$ does not consist entirely of pristine material. At least the regions we have been able to probe so far show traces of metals at levels between 1/1000 and 1/100 of solar metallicity. Matter in these regions, however, is still at a higher density than the mean density of the universe at that epoch, and it is unclear at present whether a pre-galactic episode of star formation (often referred to as Population III stars), is required to explain the large scale distribution of elements in the IGM.

(*b*) Damped Lyα systems have abundances similar to those of Population II stars in our Galaxy. Perhaps they represent an early stage in the formation of spiral galaxies, before most of the gas had been converted into stars. It is also clear that a wide variety of galaxy morphologies, including low surface brightness galaxies, share the common characteristic of providing a large cross-section on the sky at high surface densities of H I.

(*c*) Finally, Lyman break galaxies strongly resemble what we call Population I stars in the Milky Way. They are the sites of vigorous star formation which (i) has produced a relatively high level of chemical enrichment at early epochs, (ii) has built up stellar masses of $\gtrsim 10^{10}\, M_\odot$ in a sizeable fraction of the population, and (iii) drives large-scale outflows of gas, metals and dust into the intergalactic medium. All of these characteristics point to LBGs as the progenitors of today's spiral bulges and elliptical galaxies, observed during the most active phase in their lifetimes.

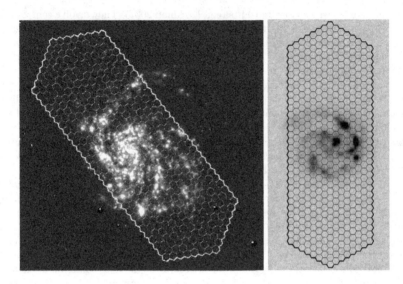

FIGURE 31. (Reproduced from Bunker et al. 2001). *Left:* An Hα image of the local spiral galaxy NGC 4254 as it would appear at $z = 1.44$ with the CIRPASS integral field unit overlaid (using 0.25 arcsec diameter fibres). *Right:* A spiral galaxy at $z \approx 1$ from the Hubble Deep Field B-band. The star-forming H II regions are prominent in the rest-frame UV. CIRPASS will accurately determine the true star formation rates, since (1) the compact knots of star formation are well-matched to the fibre size, reducing the sky background and increasing the sensitivity; (2) the large area surveyed by the integral field unit covers most of a spiral disk and (3) the Hα line is a much more robust measure of the star formation rate than the dust-suppressed UV continuum and resonantly-scattered Lyα.

Table 2. Typical parameters of LBGs and DLAs at $z \simeq 3$

Property	LBGs	DLAs
SFR $(M_\odot \ \mathrm{yr}^{-1})$	~ 50	< 10
$Z \ (Z_\odot)$	$\sim 1/3$	$\sim 1/20$
$\Delta v \ (\mathrm{km \ s}^{-1})$	~ 500	$\lesssim 200$

The connection between LBGs and DLAs is currently the subject of considerable discussion, as astronomers try and piece together these different pieces of the puzzle describing the universe at $z = 3$. Table 2 summarises some of the relevant properties. Lyman break galaxies have systematically higher star formation rates and metallicities, and their interstellar media have been stirred to higher velocities, than is the case in most DLAs. These seemingly contrasting properties can perhaps be reconciled if the two classes of objects are in fact drawn from the same luminosity function of galaxies at $z = 3$. Since they are selected from magnitude limited samples, the LBGs are preferentially bright galaxies—the data in Table 2 refer to galaxies brighter than L^* which corresponds to $\mathcal{R} = 24.5$ (Adelberger & Steidel 2000). If, on the other hand, the H I absorption cross-section decreases only slowly with galaxy luminosity, as is the case at lower redshifts (Steidel,

Abundances at High Redshift (z = 3)

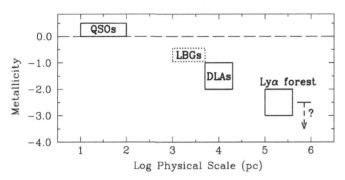

FIGURE 32. Summary of our current knowledge of abundances at high redshift. The 'metallicity' is plotted on the y-axis on a log scale relative to the solar reference; the latter is shown as the broken horizontal line at 0.0 and corresponds to approximately 2% of the baryons being incorporated in elements heavier than helium. The x-axis shows the typical linear dimensions of the strucures to which the abundance measurements refer, from the central regions of active galactic nuclei on scales of 10–100 pc to the intergalactic medium traced by the Ly α forest on Mpc scales. Generally speaking, these typical linear scales are inversely proportional to the overdensities of the structures considered relative to the background.

Dickinson, & Persson 1994), the DLA counts would naturally be dominated by the far more numerous galaxies at the steep ($\alpha = -1.6$) faint end of the luminosity function.

Such a picture finds theoretical support in the results of hydrodynamical simulations and semi-analytic models of galaxy formation (e.g. Nagamine et al. 2001; Mo, Mao, & White 1999). In the coming years it will be tested by deeper and more extensive searches for DLA galaxies, by comparing the clustering of LBGs and DLAs (Adelberger et al. 2002, in preparation), and by more reliable measurements of the star formation activity associated with DLAs. This last project is best tackled in the near-infrared, by targeting the Hα emission line with integral field—rather than slit—spectroscopy, as provided for example by the Cambridge Infrared Panoramic Survey Spectrograph (CIRPASS—see Figure 31).

When we combine the available abundance determinations for Lyman break galaxies, damped Lyα systems and the Lyα forest with those for the inner regions of active galactic nuclei (from analyses of the broad emission lines and outflowing gas in broad absorption line QSOs—see Hamann & Ferland 1999), a 'snapshot' of metal enrichment in the universe at $z \simeq 3$ emerges (Figure 32). The x-axis in the figure gives the typical linear dimensions of the structures to which the abundance measurements refer, from the 10–100 pc broad emission line region of QSOs, to the kpc scales of LBGs revealed by *HST* imaging (Giavalisco, Steidel, & Macchetto 1996), to the 10 kpc typical radii of DLAs deduced from their number density per unit redshift (Steidel 1993), to the 100 kpc dimensions of condensations in the Lyα forest with $N(\text{H I}) \gtrsim 10^{14}\,\text{cm}^{-2}$ indicated by the comparison of the absorption along adjacent sightlines in the real universe (e.g. Bechtold et al. 1994) and in the simulations (e.g. Hernquist et al. 1996).

These different physical scales in turn reflect the depths of the underlying potential wells and therefore the overdensities of matter in these structures relative to the mean density of the universe. Even from such an approximate sketch as Figure 32, it seems clear

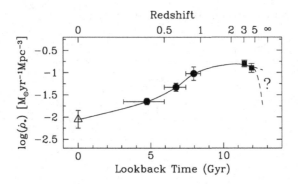

FIGURE 33. The comoving star formation rate density $\dot{\rho}_*$ vs. lookback time compiled from wide angle, ground based surveys (Steidel et al. 1999 and references therein). The data shown here are for a $H_0 = 50$ km s^{-1} Mpc^{-1}, $\Omega_{\mathrm{M}} = 1$, $\Omega_{\Lambda} = 0$ cosmology.

that it is this overdensity parameter which determines the degree of metal enrichment achieved at any particular cosmic epoch. Thus, even at the relatively early times which correspond to $z = 3$, the gas in the deepest potential wells where AGN are found had already undergone considerable processing and reached solar or super-solar abundances. At the other end of the scale, condensations in the Lyα forest which correspond to mild overdensities contained only traces of metals with metallicity $Z \simeq 1/100 - 1/1000\, Z_{\odot}$. The dependence of Z on the environment appears to be stronger than any age-metallicity relation—old does not necessarily mean metal-poor, not only for our own Galaxy but also on a global scale. This empirical conclusion can be understood in a general way within the framework of hierarchical structure formation in cold dark matter models (e.g. Cen & Ostriker 1999; Nagamine et al. 2001).

5.2. *Missing Metals?*

It can also be appreciated from Figure 32 that our knowledge of element abundances at high redshift is still very patchy. This becomes all the more evident when we attempt a simple counting exercise. Figure 33 shows a recent version of a plot first constructed by Madau et al. (1996) which attempts to trace the 'cosmic star formation history' by following the redshift evolution of the comoving luminosity density of star forming galaxies. This kind of plot enjoyed great popularity after it was presented by Madau et al.; more recently astronomers have approached it with greater caution as they have become more aware of the uncertainties involved. In particular, the dust corrections to the data in Figure 33 have been the subject of intense debate over the last five years, as has been the contribution to $\dot{\rho}_*$ from galaxies which may be obscured at visible and ultraviolet wavelengths and only detectable in the sub-mm regime with instruments such as SCUBA. Furthermore, the normalisation of the plot depends on the IMF and on the slope of the faint end of the galaxy luminosity function. Nevertheless, if we assume that we have the story about right, some interesting consequences follow.

The first question one may ask is: *"What is the total mass in stars obtained by integrating under the curve in Figure 33?"* The data points in Figure 33 were derived assuming a Salpeter IMF (with slope -2.35) between $M = 100\, M_{\odot}$ and $0.1\, M_{\odot}$. However, for a more realistic IMF which flattens below 1 $M_\odot$, the values of $\dot{\rho}_*$ in Figure 33 should be reduced by a factor of 2.5.† With this correction:

$$\int_0^{13\ Gyr} \dot{\rho}_*'\ dt \simeq 3.3 \times 10^8\ M_\odot\ \text{Mpc}^{-3} = 0.0043\ \rho_{\text{crit}} \approx \Omega_{\text{stars}}\ \rho_{\text{crit}} \tag{5.7}$$

where $\rho_{\text{crit}} = 3.1 \times 10^{11}\ h^2\ M_\odot\ \text{Mpc}^{-3}$ and $\Omega_{\text{stars}} \simeq (1.5 - 3)\ h^{-1} \times 10^{-3}$ (Cole et al. 2001) is the fraction of the closure density contributed by stars at $z = 0$. Thus, within the rough accuracy with which this accounting can be done, the star formation history depicted in Figure 33 is apparently sufficient to produce the entire stellar content, in disks and spheroids, of the present day universe. Note also that $\approx 1/4$ of today's stars were made before $z = 2.5$ (the uncertain extrapolation of $\dot{\rho}_*$ beyond $z = 4$ makes little difference).

We can also ask: *"What is the total mass of metals produced by $z = 2.5$?"*. Using the conversion factor $\dot{\rho}_{\text{metals}} = 1/42\ \dot{\rho}_*$ to relate the comoving density of synthesised metals to the star formation rate density (Madau et al. 1996) we find

$$\int_{11\ Gyr}^{13\ Gyr} \dot{\rho}_{\text{metals}}\ dt \simeq 4.5 \times 10^6\ M_\odot\ \text{Mpc}^{-3} \tag{5.8}$$

which corresponds to

$$\Omega_Z \simeq 0.035 \times (\Omega_{\text{baryons}} \times 0.0189) \tag{5.9}$$

where $\Omega_{\text{baryons}} = 0.088$ for $h = 0.5$ (§1.1) and 0.0189 is the mass fraction of elements heavier than helium for solar metallicity (Grevesse & Sauval 1998). In other words, the amount of metals produced by the star formation we *see* at high redshift (albeit corrected for dust extinction) is sufficient to enrich the whole baryonic content of the universe at $z = 2.5$ to $\approx 1/30$ of solar metallicity. Note that this conclusion does not depend sensitively on the IMF.

As can be seen from Table 3, this leaves us with a serious 'missing metals' problem which has also been discussed in more detail by Pagel (2002). The metallicity of damped Lyα systems is in the right ballpark, but Ω_{DLA} is only a small fraction of Ω_{baryons}. Conversely, while the Lyα forest may account for a large fraction of the baryons, its metal content is one order of magnitude too low. The contribution of Lyman break galaxies to the cosmic inventory of metals is even more uncertain. The value in Table 3 is a strict (and not very informative) lower limit, calculated from the luminosity function of Steidel et al. (1999), taking into account *only* galaxies brighter than L^* and assigning to each a mass $M^* = 10^{11}\ M_\odot$ (which is likely to be a lower limit, as discussed by Pettini et al. 2001) and metallicity $Z = 1/3\ Z_\odot$. Galaxies fainter than L^* are not included in this census because we still have no idea of their metallicities; potentially they could make a significant contribution to $\Omega_Z(\text{LBG})$ because they are so numerous.

Nevertheless, when we add up all the metals which have been measured with some degree of confidence up to now, we find that they account for no more than $\approx 10 - 15\%$ of what we expect to have been produced by $z = 2.5$ (last column of Table 3). Where are these missing metals? Possibly, $\Omega_Z(\text{DLA})$ has been underestimated, if the dust associated with the most metal-rich DLAs obscures background QSOs sufficiently to make them drop out of current samples. However, preliminary indications based on the CORALS survey by Ellison et al. (2001) suggest that this may be a relatively minor effect (see also Prochaska & Wolfe 2002). The concordance in the values of $\Omega_Z(\text{IGM})$

---

† I have not applied this correction directly to Figure 33 in order to ease the comparison with earlier versions of this plot.

TABLE 3
CENSUS OF METALS AT $z = 2.5^a$

Component	Ω^b	Z^c	Ω_Z^d
Observed:			
DLAs	0.0025	0.07	0.002
Lyα Forest	$0.05 - 0.08$	0.003	$0.002 - 0.003$
Lyman Break Galaxies	?	0.3	> 0.0002
Predicted:			
All Baryons (BBNS)	0.088		
Metals synthesised in			
Lyman Break Galaxies			0.035

^{a}All entries are for $H_0 = 50$ km s^{-1} Mpc^{-1}; $\Omega_M = 1$, $\Omega_\Lambda = 0$.
^{b}In units of the closure density $\rho_{crit} = 3.1 \times 10^{11} h^2 M_\odot$ Mpc^{-3}.
^{c}In units of solar metallicity (0.0189 by mass).
^{d}In units of $\Omega_{Z_\odot} = \Omega_{baryons} \times Z_\odot = 1.7 \times 10^{-3}$.

derived from observations of O VI and C IV absorption in the Lyα forest makes it unlikely that the metallicity of the widespread IGM has been underestimated by a large factor. On the other hand, we do know that Lyman break galaxies commonly drive large scale outflows; it is therefore possible, and indeed likely, that they enrich with metals much larger masses of gas than those seen directly as sites of star formation. This gas and associated metals may be difficult to detect if they are at high temperatures, and yet may make a major contribution to Ω_Z; there are tantalising hints that this could be the case at the present epoch (Tripp, Savage, & Jenkins 2000; Mathur, Weinberg, & Chen 2002).

In concluding this series of lectures, it is clear that while we have made some strides forward towards our goal of charting the chemical history of the universe, our task is far from complete. It is my hope that, stimulated in part by this school, some of the students who have attended it will soon be contributing to this exciting area of observational cosmology as their enter their research careers.

I am very grateful to César Esteban, Artemio Herrero, Ramón García López and Prof. Francisco Sánchez for inviting me to take part in a very enjoyable Winter School, and to the students for their patience and challenging questions. The results described in these lectures were obtained in various collaborative projects primarily with Chuck Steidel, Kurt Adelberger, David Bowen, Mark Dickinson, Sara Ellison, Mauro Giavalisco, Samantha Rix and Alice Shapley; I am fortunate indeed to be working with such productive and generous colleagues. Special thanks to Alec Boksenberg and Bernard Pagel for continuing inspiration and for valuable comments on an early version of the manuscript. As can be appreciated from these lecture notes, the measurement of element abundances at high redshifts is a vigorous area of research. In the spirit of the school, I have not attempted to give a comprehensive set of references to all the numerous papers on the subject which have appeared in recent years, as one would in a review. Rather, I have concentrated on the main issues and only given references as pointers for further reading. I apologise for the many excellent papers which have therefore been omitted from the (already long) list of references. Such omissions do not in any way denote criticism on my part of the work in question.

REFERENCES

ADELBERGER, K.L., & STEIDEL, C.C. 2000, ApJ, 544, 218

AGUIRRE, A., HERNQUIST, L., SCHAYE, J., KATZ, N., WEINBERG, D. H., & GARDNER, J. 2001, ApJ, 561, 521

BECHTOLD, J., CROTTS, A.P.S., DUNCAN, R.C., & FANG, Y. 1994, ApJ, 437, L83

BECHTOLD, J. 2002, in Galaxies at High Redshift, eds. I. Pérez-Fournon, M. Balcells, F. Moreno-Insertis, & F. Sánchez, (Cambridge: Cambridge Univ. Press), p. 131

BINNEY, J., GERHARD, O., & SILK, J. 2001, MNRAS, 321, 471

BOISSIER, S., PÉROUX, C., & PETTINI, M. 2002, MNRAS, submitted

BOUCHÉ, N., LOWENTHAL, J.D., CHARLTON, J.C., BERSHADY, M.A., CHURCHILL, C.W., & STEIDEL, C.C. 2001, ApJ, 550, 585

BOUCHET, P., LEQUEUX, J., MAURICE, E., PREVOT, L., & PREVOT-BURNICHON, M.L. 1985, A&A, 149, 330

BRUZUAL, A.G., & CHARLOT, S. 1993, ApJ, 405, 538

BUNKER, B., FERGUSON, A., JOHNSON, R. ET AL. 2001, in Deep Fields, ESO Astrophysics Symposia, eds. S. Cristiani, A. Renzini, & R.E. Williams, (Berlin:Springer), 330

CARSWELL, R.F., SCHAYE, J., & KIM, T.S. 2002, ApJ 578, 43

CEN, R., & OSTRIKER, J.P. 1999, ApJ, 519, L109

COLE, S., NORBERG, P., BAUGH, C.M., ET AL. 2001, MNRAS, 326, 255

COWIE, L.L., SONGAILA, A., KIM, T.S., & HU, E.M. 1995, AJ, 109, 1522

CROFT, R.A.C., WEINBERG, D.H., BOLTE, M. ET AL. 2002, ApJ, submitted (astro-ph/0012324)

DAVÉ, R., HELLSTEN, U., HERNQUIST, L., KATZ, N. & WEINBERG, D.H. 1998, ApJ, 509, 661

DIPLAS, A., & SAVAGE, B.D. 1994, ApJ, 427, 274

EDMUNDS, M.G., & PAGEL, B.E.J. 1978, MNRAS, 185, 77P

EFSTATHIOU, G. 2000, MNRAS, 317, 697

ELLINGSON, E., YEE, H.K.C., BECHTOLD, J., & ELSTON, R. 1996, ApJ, 466, L71

ELLISON, S.L. 2000, Ph.D. Thesis, University of Cambridge

ELLISON, S.L., LEWIS, G.F., PETTINI, M., CHAFFEE, F.H., & IRWIN, M.J. 1999, ApJ, 520, 456

ELLISON, S.L., SONGAILA, A., SCHAYE, J., COWIE, L.L., & PETTINI, M. 2000, AJ, 120, 1175

ELLISON, S.L., YAN, L., HOOK, I.M., PETTINI, M., WALL, J.V., & SHAVER, P. 2001, A&A, 379, 393

FALL, S.M. 1996, in The Hubble Space Telescope and the High Redshift Universe, eds. N. Tanvir, A. Aragon-Salamanca, & J.V. Wall (Singapore: World Scientific), 303

FUHRMANN, K. 1998, A&A, 338, 161

GARNETT, D.R., SHIELDS, G.A., PEIMBERT, M., ET AL. 1999, ApJ, 513, 168

GIAVALISCO, M., STEIDEL, C.C., & MACCHETTO, F.D. 1996, ApJ, 470, 189

GILMORE, G., & WYSE, R.F.G. 1991, ApJ, 367, L55

GREVESSE, N., & SAUVAL, A.J. 1998, Space Sci Rev, 85, 161

HAMANN, F., & FERLAND, G. 1999, ARA&A, 37, 487

HAEHNELT, M.G., STEINMETZ, M., & RAUCH, M. 1998, ApJ, 495, 647

HECKMAN, T.M. 2001, in ASP Conf. Ser., Gas and Galaxy Evolution, ASP Conference Proceedings, Vol. 240, eds. J.E. Hibbard, M.P. Rupen, & J.H. van Gorkom, (San Francisco:ASP), 345

HENRY, R.B.C., EDMUNDS, M.G., & KÖPPEN, J. 2000, ApJ, 541, 660

HERNQUIST, L., KATZ, N., WEINBERG, D.H., & MIRALDA-ESCUDÉ, J. 1996, ApJ, 457, L51

HOLWEGER, H. 2001, in Solar and Galactic Composition, ed. R.F. Wimmer-Schweingruber, American Institute of Physics Conference proceedings, 598, 23

KANEKAR, N., & CHENGALUR, J.N. 2001, A&A, 369, 42

KENNICUTT, R.C. 1998a, in ASP Conf. Ser. 142, The Stellar Initial Mass Function, ed. G. Gilmore & D. Howell (San Francisco: ASP), 1

KENNICUTT, R.C. 1998b, in The Next Generation Space Telescope: Science Drivers and Technological Challenges, 34th Liège Astrophysics Colloquium, 81

KRAUSS, L.M., & CHABOYER, B. 2003, Science 299, 65

KULKARNI, V., & FALL, S.M., 2002, ApJ 580, 732

LAIRD, J.B., RUPEN, M.P., CARNEY, B.W., & LATHAM, D.W. 1988, AJ, 96, 1908

LANZETTA, K.M. 1993, in The Environment and Evolution of Galaxies, eds. J.M. Shull & H.A. Thronson, (Dordrecht: Kluwer), 237

LARSEN, T.I., SOMMER-LARSEN, J., & PAGEL, B.E.J. 2001, MNRAS, 323, 555

LEDOUX, C., BERGERON, J., & PETITJEAN, P. 2002, A&A, 385, 802

LEDOUX, C., PETITJEAN, P., BERGERON, J., WAMPLER, E.J., & SRIANAND, R. 1998, A&A, 337, 51

LEITHERER, C., SCHAERER, D., GOLDADER, J.D., ET AL. 1999, ApJS, 123, 3

LEITHERER, C., LEÃO, J.R.S., HECKMAN, T.M., LENNON, D.J., PETTINI, M., & ROBERT, C. 2001, ApJ, 550, 724

MADAU, P., FERGUSON, H.C., DICKINSON, M.E., GIAVALISCO, M., STEIDEL, C.C., & FRUCHTER, A. 1996, MNRAS, 283, 1388

MADAU, P., FERRARA, A., & REES, M.J. 2001, ApJ, 555, 92

MADAU, P., & SHULL, J.M. 1996, ApJ, 457, 551

MATHUR, S., WEINBERG, D.H., & CHEN, X. 2002, ApJ 582, 82

MATTEUCCI, F., & RECCHI, S. 2001, ApJ, 558, 351

McGAUGH, S. 1991, ApJ, 380, 140

McWILLIAM, A., PRESTON, G.W., SNEDEN, C., & SEARLE, L. 1995, AJ, 109, 2757

MEYER, D.M., WELTY, D.E., & YORK, D.G. 1989, ApJ, 343, L37

MEYNET, G. & MAEDER, A. 2002, A&A, 381, L25

MIRALDA-ESCUDÉ, J., & REES, M.J. 1997, ApJ, 478, L57

MO, H.J., MAO, S., & WHITE, S.D.M. 1999, MNRAS, 304, 175

MOLARO, P., BONIFACIO, P., CENTURIÓN, M., D'ODORICO, S., VLADILO, G., SANTIN, P., & DI MARCANTONIO, P. 2000, ApJ, 541, 54

MOLARO, P., LEVSHAKOV, S.A., D'ODORICO, S., BONIFACIO, P., & CENTURIÓN, M. 2001, ApJ, 549, 90

MØLLER, P., WARREN, S.J., FALL, S.M., FYNBO, J.U., & JAKOBSEN, P. 2002, ApJ, 574, 51

NAGAMINE, K., FUKUGITA, M., CEN, R., & OSTRIKER, J.P. 2001, ApJ, 558, 497

PAGEL, B.E.J. 2002, in ASP Conf. Series 253, Chemical Enrichment of Intracluster and Intergalactic Medium, eds. R. Fusco-Femiano & F. Matteucci, (San Francisco: ASP), 489

PAGEL, B.E.J., EDMUNDS, M.G., BLACKWELL, D.E., CHUN, M.S., & SMITH, G. 1979, MNRAS, 189, 95

PAGEL, B.E.J., & TAUTVAISIENE, G. 1998, MNRAS, 299, 535

PAPOVICH, C., DICKINSON, M., & FERGUSON, H.C. 2001, ApJ, 559, 620

PEI, Y.C., FALL, S.M., & BECHTOLD, J. 1991, ApJ, 378, 6

PEI, Y.C., FALL, S.M., & HAUSER, M.G. 1999, ApJ, 522, 604

PENTON, S.V., SHULL, J.M., & STOCKE, J.T. 2000, ApJ, 544, 150

PÉROUX, C., MCMAHON, R.G., STORRIE-LOMBARDI, L.J., & IRWIN, M.J. 2002, MNRAS, submitted (astro-ph/0107045)

PETITJEAN, P., & BERGERON, J. 1994, A&A, 283, 759

PETITJEAN, P., SRIANAND, R., & LEDOUX, C. 2000, A&A, 364, L26

PETTINI, M. 2001, in Gaseous Matter in Galaxies and Intergalactic Space, ed. R. Ferlet, M. Lemoine, J.M. Desert, & B. Raban (Frontier Group), 315

PETTINI, M., BOKSENBERG, A., & HUNSTEAD, R.W. 1990, ApJ, 348, 48

PETTINI, M., & BOWEN, D.V. 2001, ApJ, 560, 41

PETTINI, M., ELLISON, S. L., BERGERON, J., & PETITJEAN, P. 2002a, A&A, 391, 21

PETTINI, M., ELLISON, S. L., STEIDEL, C. C., & BOWEN, D. V. 1999, ApJ, 510, 576

PETTINI, M., ELLISON, S. L., STEIDEL, C. C., SHAPLEY, A. E., & BOWEN, D. V. 2000a, ApJ, 532, 65

PETTINI, M., LIPMAN, K., & HUNSTEAD, R.W. 1995, ApJ, 451, 100

PETTINI, M., KING, D.L., SMITH, L.J., & HUNSTEAD, R.W. 1997a, ApJ, 478, 536

PETTINI, M., RIX, S.A., STEIDEL, C.C., ADELBERGER, K.L., HUNT, M.P., & SHAPLEY, A.E. 2002b, ApJ, 569, 742

PETTINI, M., SHAPLEY, A.E., STEIDEL, C.C., CUBY, J.G., DICKINSON, M., MOORWOOD, A.F.M., ADELBERGER, K.L., & GIAVALISCO, M. 2001, ApJ, 554, 981

PETTINI, M., SMITH, L.J., HUNSTEAD, R.W., & KING, D.L. 1994, ApJ, 426, 79

PETTINI, M., SMITH, L.J., KING, D.L. & HUNSTEAD, R.W. 1997b, ApJ, 486, 665

PETTINI, M., STEIDEL, C.C., ADELBERGER, K.L., DICKINSON, M., & GIAVALISCO, M. 2000b, ApJ, 528, 96

PROCHASKA, J.X., GAWISER, E., & WOLFE A. 2001, ApJ, 552, 99

PROCHASKA, J.X., GAWISER, E., & WOLFE A., ET AL. 2002, AJ, 123, 2206

PROCHASKA, J.X., & WOLFE A. 1998, ApJ, 507, 113

PROCHASKA, J.X., & WOLFE A. 1999, ApJS, 121, 369

PROCHASKA, J.X., & WOLFE A. 2002, ApJ, 566, 68

RAO, S.M., & TURNSHEK, D.A. 2000, ApJS, 130, 1

RAUCH, M. 1998, ARA&A, 36, 267

ROSENBERG, J., & SCHNEIDER, S. 2002, ApJ, 567, 247

RYAN, S.G., NORRIS, J.E., & BEERS, T.C. 1996, ApJ, 471, 254

SARGENT, W.L.W., YOUNG, P.J., BOKSENBERG, A., & TYTLER, D. 1980, ApJS, 42, 41

SAVAGE, B.D., & SEMBACH, K.R. 1996, ARA&A, 34, 279

SCHAYE, J., RAUCH, M., SARGENT, W.L.W., & KIM, T.S. 2000, ApJ, 541, L1

SEITZ, S., SAGLIA, R.P., BENDER, R., HOPP, U., BELLONI, P., & ZIEGLER, B. 1998, MNRAS, 298, 945

SHAPLEY, A.E., STEIDEL, C.C., ADELBERGER, K.L., DICKINSON, M., GIAVALISCO, M., & PETTINI, M. 2001, ApJ, 562, 95

SNEDEN, C., GRATTON, R.G., & CROCKER, D.A. 1991, A&A, 246, 354

SONGAILA, A. 1997, ApJ, 490, L1

SONGAILA, A. 2001, ApJ, 561, L153

SONGAILA, A., & COWIE, L.L. 2002, AJ, 123, 2183

STEIDEL, C.C. 1993, in The Environment and Evolution of Galaxies, eds. J.M. Shull & H.A. Thronson, (Dordrecht: Kluwer), 263

STEIDEL, C.C. 2000, in Discoveries and Research Prospects from 8- to 10-Meter-Class Telescopes, ed. J. Bergeron, Proc. SPIE Vol. 4005, 22

STEIDEL, C.C., ADELBERGER, K.L., GIAVALISCO, M., DICKINSON, M., & PETTINI, M. 1999, ApJ, 519, 1

STEIDEL, C.C., DICKINSON, M., MEYER, D.M., ADELBERGER, K. L., & SEMBACH, K.R. 1997, ApJ, 480, 568

STEIDEL, C.C., DICKINSON, M., & PERSSON, S.E. 1994, ApJ, 437, L75

STEIDEL, C.C., GIAVALISCO, M., PETTINI, M., DICKINSON, M., & ADELBERGER, K. L. 1996, ApJ, 462, L17

STEIDEL, C.C., PETTINI, M., & ADELBERGER, K.L. 2001, ApJ, 546, 665

STEIDEL, C.C., PETTINI, M., & HAMILTON, D. 1995, AJ, 110, 2519

STORRIE-LOMBARDI, L.J., & WOLFE, A.M. 2000, ApJ, 543, 552

TEPLITZ, H.I., MCLEAN, I.S., BECKLIN, E.E., ET AL. 2000, ApJ, 533, L65

TRIPP, T.M., SAVAGE, B.D., & JENKINS, E.B. 2000, ApJ, 534, L1

TURNER, M.S. 2002, in Proceedings of the XXth International Symposium on Photon and Lepton Interactions, International Journal of Modern Physics A, 17, 3446

TYTLER, D. 1987, ApJ, 321, 49

TYTLER, D., FAN, X.-M., BURLES, S., COTTRELL, L., DAVIS, C., KIRKMAN, D., & ZUO, L. 1995, QSO Absorption Lines, ed. G. Meylan, (Garching, ESO), 289

TYTLER, D., O'MEARA, J.M., SUZUKI, N., & LUBIN, D. 2000, Physica Scripta T, 85, 12

VLADILO, G. 2002a, ApJ, 569, 295

VLADILO, G. 2002b, A&A, 391, 407

VLADILO, G., CENTURIÓN, M., BONIFACIO, P., & HOWK, J.C. 2001, ApJ, 557, 1007

WEINBERG, D.H., KATZ, N., & HERNQUIST, L. 1998, in ASP Conf. Series 128, Origins, ed. C.E. Woodward, J.M. Shull, & H.A. Thronson, (San Francisco: ASP), 21

WHEELER, J.C., SNEDEN, C., & TRURAN, J.W. 1989, ARA&A, 27, 279

WOLFE, A. M., TURNSHEK, D. A., SMITH, H. E., & COHEN, R. D. 1986, ApJS, 61, 249

WOLFIRE, M.G., HOLLENBACH, D., MCKEE, C.F., TIELENS, A.G.G.M., & BAKES, E.L.O. 1995, ApJ, 443, 152

WYSE, R.F.G., & GILMORE, G. 1995, AJ, 110, 2771

YEE, H.K.C., ELLINGSON, E., BECHTOLD, J., CARLBERG, R.G., & CUILLANDRE, J.-C. 1996, AJ, 111, 1783